America's Secret Eyes in Space
The U.S. Keyhole Spy Satellite Program

Jeffrey T. Richelson

1817

Harper & Row, Publishers, New York
BALLINGER DIVISION

Grand Rapids, Philadelphia, St. Louis, San Francisco
London, Singapore, Sydney, Tokyo, Toronto

"It's nice to see it from the ground."
—former Deputy Director of Central Intelligence, Robert Gates,
on his first trip to Moscow, 1989

International Standard Book Number: 0-88730-285-8

Printed in the United States of America

Library of Congress Cataloging-in-Publication Data

Richelson, Jeffrey T.
   America's Secret Eyes in Space : the U.S. keyhole spy satellite
program / Jeffrey T. Richelson.
      p. cm.
   Includes bibliographical references.
   ISBN 0-88730-285-8
   1. Space surveillance–United States–History. I. Title.
UG1523.R53 1990
358.8'0973–dc20
                                                  89-26698
                                                  CIP

90 91 92 93 HC 9 8 7 6 5 4 3 2 1

CONTENTS

PREFACE

In August 1985 approximately 100 people journeyed to CIA headquarters in Langley, Virginia for an anniversary celebration. The celebration commemorated the first successful CORONA mission 25 years earlier; CORONA was the codename for the CIA's initial program to take satellite photographs of the Soviet Union and its allies.

On August 18, 1960 the Discoverer XIV satellite, the "cover" for CORONA, ejected a capsule containing the pictures taken while it passed over the Soviet Union. The capsule was snagged out of the air by specially designed aircraft and returned to Washington, D.C. for analysis.

The hundred celebrants were instrumental, in a variety of ways, in guiding the United States into the space reconnaissance age. They arrived at Langley from the surrounding local area, Boston, California, New Mexico, and elsewhere. Among those attending were Amrom Katz and Merton Davies, who had been involved in early satellite studies at the RAND Corporation. Other participants included lens designer James Baker, photo-interpreters Arthur Lundahl and Dino Brugioni, camera designer Walter Levinson, and Richard Leghorn, an early proponent of the need for *strategic* (as opposed to tactical/battlefield) reconnaissance. The management side was represented by General Bernard Schriever, who had the initial Air Force responsibility for managing the satellite reconnaissance program, and Richard M. Bissell, Jr., who had directed the CORONA program in his capacity as the CIA's Deputy Director for Plans.

Two notable events occurred that day. The invitees received "Space Pioneer" medals in appreciation of their contribution to the success of CORONA. The Director of Central Intelligence, William J. Casey, was there to present the medals and have his picture taken with each awardee. To illustrate how the U.S. reconnaissance effort had progressed over 25 years the guests were shown pictures of a Soviet airbase taken by each camera system. In addition, Casey authorized the showing of photos from the KH–11, America's most modern reconnaissance satellite as of 1985. The photos of Vladivostok harbor that had been taken at 10:30 that morning were presented late in the afternoon by Edward Aldridge, the Director of the National Reconnaissance Office.

How the United States translated the concept of space photography to an initial operational satellite and then to the sophisticated satellites of today is a story of great importance. For the photo reconnaissance satellite is one of the most important military technological developments of this century, along with radar and the atomic bomb. Without it, the history of this century would be very different. Indeed, without it history might have ceased.

On the one hand, the photo reconnaissance satellite has been a partner to the atomic and nuclear weapons whose use could devastate the civilized world. Both the United States and Soviet Union have relied on their reconnaissance satellites to locate and identify targets to be attacked in the event of war. At the same time, those satellites have played a significant role in preventing the occurrence of war and permitting arms limitation agreements. In the first year of their operation, the CORONA satellites helped dispel America's fear of Soviet strategic superiority that had haunted many Americans since the launch of Sputnik. Since then, they have allowed knowledge to prevail over fear in assessing Soviet capabilities. And the arms limitation agreements of past, present, and future would not be possible without such devices to verify compliance. In the future, they may be significant in helping curb the spread of ballistic missiles and atomic weapons to a variety of Third World countries.

This book covers several aspects of the satellite reconnaissance program: the people, the organizations, the technology and its uses, and the program's impact.

Jeffrey T. Richelson

ACKNOWLEDGMENTS

A variety of sources were crucial to the writing of this book. Forming a necessary foundation were a large number of books, reports, and newspaper, magazine, and trade journal articles that deal with the U.S. imaging reconnaissance satellite program. Information on satellite orbital parameters (apogee, perigee, inclination) are derived, except where noted, from the Royal Aerospace (formerly Aircraft) Establishment's yearly editions of *Table of Earth Satellites*.

Given the secrecy attached to the program, it is not surprising that only a limited number of official documents were available from Presidential libraries or history offices, or through the Freedom of Information Act. Most of those that are available deal with the very early years of the program. For help in identifying and obtaining documents relevant to my research, I gratefully acknowledge the assistance of the NASA History Office and the National Security Archive.

The most important sources of information, in terms of quality if not quantity, were the interviews I conducted with a variety of former government officials, including officials from the National Security Council, Central Intelligence Agency, State Department, Defense Department, and other departments and agencies. I thank all those who took time to answer my questions.

Those interviewees willing to be identified are listed in the Sources section, near the end of the book. Some individuals spoke on a not-for-attribution basis. Material in the book based on such sources is simply referenced as "Interview" in the notes. Although some would argue against the use of anonymous sources, I believe—as both an author and a reader—that reliance on anonymous sources who provide new and correct information is far preferable to recycling previously published inaccuracies. Discussions with several sources indicated that there were significant errors in previous published accounts, including my own, of the program's early years.

viii America's Secret Eyes in Space

I would also like to express my gratitude to several professional colleagues who offered assistance in a variety of ways. William E. Burrows read the manuscript and offered numerous suggestions on how to improve it. Jay Peterzell of *Time*, John Pike of the Federation of American Scientists, Bob Windrem of the NBC Nightly News, and Bob Woodward of *The Washington Post* all had information and/or suggestions to offer.

Special thanks go to Carol Franco, former president of Ballinger Publishing Company, who initiated this project.

CHAPTER 1

Pioneer Reconnaissance

While a teenager, Merton Davies read a book that was to affect the course of his life: *Rockets through Space* by British scientist P. E. Cleator, published in 1936. It helped shift Davies' horizons from earth to space. He proceeded to build a large collection of books on the challenges of outer space, including books by rocket scientists Hermann Oberth and George Goddard.[1]

Rockets through Space was a guide to the solar system and the challenges involved in interplanetary exploration. It also explored the possibilities of an artificial earth satellite and a space station. Ultimately, Davies was to become heavily involved with the NASA space probes that have traveled through the solar system and beyond. However, his first venture into space research concerned another idea discussed by Cleator—the artificial earth satellite.

Davies, who grew up in Palo Alto, received his undergraduate degree in mathematics from Stanford in 1938. After a short stint as a professor at the University of Nevada, he returned to Stanford to pursue further research. Davies, however, had more than an academic career in mind. He had a pilot's license and was intent on joining the Army Air Force. However, at six feet eight inches Davies was four inches above the Army Air Force maximum height for pilots. As an alternative, he traveled to southern California, where he joined the Douglas Aircraft Company during World War II. Working at Douglas's El Segundo facility, Davies designed naval aircraft.[2]

In 1946 Davies heard that another Douglas component was involved in a study concerning the feasibility of artificial earth satellites. That component was Project RAND (Research ANd Development), which had grown out of Douglas's Santa Monica–based research laboratories in March 1946. In April, Major General Curtis LeMay, then Director of Research and Development of the Army Air Force, commissioned RAND to conduct a three-week crash study on the feasibility of a space satellite.[3]

LeMay's request had been stimulated by a basic rule of military life: Don't let a rival service gain control over a new area of operations. A May 1945 report by Wernher von Braun, the former Nazi scientist who had developed the V-1 and V-2 rockets, examined German views of the potential of rocket-launched satellites. Von Braun's report, in part because it was written for the Army, stimulated Navy interest and an October 1945 Navy proposal to develop a satellite. As might be expected, Air Force interest followed soon after. In a November 1945 report General "Hap"Arnold declared that a space ship "is all but practicable today," and a December 1945 report by the Army Air Force Scientific Advisory Group announced that long-range rockets were feasible and satellites were a "definite possibility."[4]

Space satellites thus became a subject of discussion before the Army-Navy Aeronautical Board on April 9. The board decided to reconsider the issue on May 14, after the Army representatives could consult with LeMay. LeMay, possibly in response to the urging of General Carl Spaatz, the Commanding General of the Army Air Force, decided to commission the production of an independent Army Air Force study that would demonstrate an independent competence in space technology and allow the Army Air Force to gain primary responsibility for any military satellite vehicles. RAND was assigned responsibility for the study.[5] Hearing of the RAND study, Davies asked for an interview and successfully requested a transfer from Douglas to Project RAND.[6]

By the time Davies arrived, RAND had published the results of its crash study as the May 2, 1946 *Preliminary Design for an Experimental World Circling Spaceship*. The 324-page study concluded that it was possible to develop a launch vehicle to place a spacecraft in orbit with minor and entirely attainable advances over then-existing technology, although the payload would be limited to less than 2000 pounds until better rockets became available. In addition, the study concluded that such a satellite would be undetectable by then-existing radar systems.[7]

The study examined some components of a satellite system, including propulsion, multi-stage launch vehicles, the dangers of meteors, methods of analyzing trajectories, and problems of recovering space payloads. The study focused on the utility of a satellite for gathering scientific information on cosmic rays, gravitation, geophysics, terrestrial magnetism, meteorology, and properties of the upper atmosphere; but the report also identified potential military missions for a satellite: missile guidance, weapons delivery, weather reconnaissance, communications, attack assessment, and "observation." In a chapter entitled "The Significance of a Satellite Vehicle," Dr. Louis N. Ridenour noted that "the satellite offers an observation aircraft which cannot be brought down by an enemy who has not mastered similar techniques."[8]

Davies was on board by the time the second satellite study was underway. That study, the first to analyze the potential of satellites for reconnaissance, resulted in the February 1, 1947 publication of a series of documents intended to assist contractors in preparing their own preliminary designs and analyses. The studies covered utility for reconnaissance, political and psychological problems, cost estimates, launching sites, communications and observation problems, generating power, structure and weight, stability and control, propellant systems, atmospheric properties at extreme altitudes, dynamics and heat transfer problems, and flight mechanics.[9]

Among the reports was the one edited by the head of the RAND satellite project, James Lipp, who was also head of RAND's missiles division. Lipp's *Reference Papers Relating to a Satellite Study* contained papers by RAND consultants Lyman Spitzer Jr., Luis W. Alvarez, Leonard I. Schiff, Bruno Rossi, and Lipp himself. Two papers dealt with the potential uses of reconnaissance satellites.[10]

The paper by Spitzer, a Yale University astronomy professor, discussed the use of satellites for ocean surveillance, noting that a "ship at sea could, in principle, be detected," and that a "satellite traveling over the poles, with a period of about one and a half hours, would scan the oceans at least once every day."[11]

Lipp, in his paper, "The Time Factor in the Satellite Program," discussed the potential use of satellites to obtain electro-optical images and then transmit them using television-type technology:

> By installing television equipment combined with one or more
> Schmidt-type telescopes in a satellite, an observation and

reconnaissance tool without parallel could be established.
. . . a spaceship can be placed upon an oblique or north-south
orbit so as to cover the entire surface of the earth at frequent
intervals as the earth rotates beneath the orbit.[12]

RAND's work was not ignored by the higher levels of the newly
established Air Force. On September 25, 1947, one week after the
Air Force was established, the Air Staff instructed the Air Materiel
Command (AMC) to evaluate RAND's satellite studies. AMC reported
in December that although a satellite was feasible, its practicality was
questionable, and consequently recommended establishing a project to
prepare Air Force requirements and specifications for satellites.[13]

As a result, in February 1948, the Air Force requested that RAND
establish a satellite project to help stimulate the development of the
components and techniques required for the successful construction and
operation of a reconnaissance satellite. RAND was to "prepare a detailed
specification for the optimum satellite in light of present knowledge,"
to revise continually and alter the specification to stay current with
advancements in relevant areas, and to advise the Air Force regarding
the level of effort and timing for different phases of the project and the
optimum time to begin actual construction of a complete satellite, as
opposed to component development.[14]

RAND relied on a variety of institutions to aid the project, including a defense contractor, an entertainment empire, two universities, a
television station, and several experts. North American Aviation studied
the altitude control system, and RCA researched a television transmission and presentation system. The Ohio University Research Foundation
investigated the impact of altitude and resolution errors on the accuracy of target location, and Boston University was involved in minor
flight tests relative to television equipment. Television station KNBH in
Hollywood conducted television experiments, and experts on trajectory,
atmosphere, attitude control, and solar heating engines were consulted.[15]

The results of RAND's own internal work over the next two years
as well as that of its consultants created a feeling of optimism. Davies
recalls that by 1950,

> The RAND engineers were confident that an operating satel-
> lite could be built and launched into orbit. This led to studies
> of the utility of satellites: Why should they be built? It was
> recognized that a satellite program would be expensive, and

there was no national interest in proving that it could be done. Of course, there were scientific reasons but these could not hope to justify a project of this magnitude. If photographic and television cameras were incorporated into the payload, the satellite would have an observation reconnaissance capability. This mission would be of interest to the Air Force.[16]

The last months of 1950 and first months of 1951 saw RAND and the Air Force pushing for further research into the potential of satellite reconnaissance. In November 1950, RAND recommended an extension of research into specific aspects of the reconnaissance mission for satellites. Responding to RAND's proposal, the Air Force Directorate of Intelligence recommended that a decision as to whether further research and development was warranted be made as quickly as possible.[17]

Following up on the Air Force memo, Colonel Bernard A. Schriever called a February 16, 1951 conference, during which he specified the requirements of an acceptable system. The requirements included: (1) the ability to produce photography of sufficient quality to enable trained interpreters to identify objects such as harbors, airfields, oil storage areas, large residential areas, and industrial areas; (2) the capability to provide continuous daytime observation of the Soviet Union, to cover the Soviet Union in a matter of weeks, and to record the information collected; and (3) the ability to produce a quality photographic product suitable for the revision of aeronautical charts and maps.[18] A satellite that satisfied such requirements could provide information that could be used to prepare assessments of Soviet military and economic capabilities, warn of impending attack, and aid in the selection of targets for U.S. nuclear weapons in the event of war.

A partial test of the ability to satisfy the requirements was conducted at RAND on March 2, 1951 during a visit by Air Force intelligence representatives. A mosaic of Los Angeles was photographed with a standard television camera, relayed to Mt. Wilson, and sent back to the studio, where the image appearing on the monitoring screen was photographed. Air Force photo-interpreters concluded that if pictures of similar quality were produced by the proposed satellite the minimum requirement for photography of sufficient quality would be met.[19]

Two RAND reports issued in April 1951 represent significant milestones in America's progress toward a satellite reconnaissance capability. One, coauthored by William W. Kellogg and Stanley Greenfield, reported on an *Inquiry into the Feasibility of Weather Reconnaissance from a Satellite Vehicle*. Their report examined the basic requirements for a satellite system that would accurately monitor weather and allow for improved weather forecasting, noting that "in the event of armed conflict, aerial weather reconnaissance over enemy territory, similar to that obtained in World War II, will be extremely difficult if not impossible. An alternative method of obtaining this information, however, is thought to be in the use of the proposed satellite vehicle."[20]

The second report, coauthored by James Lipp, Robert Salter, and Reinhart Wehner and containing the contributions of several others, was *Utility of a Satellite Vehicle for Reconnaissance*. The report concluded that "pioneer reconnaissance (general location and determination of appropriate targets) . . . are suitable with the resolving power presently available to a satellite television system."[21]

Target location and determination had become a top priority of the Air Force since the advent of the Cold War. As U.S. nuclear forces expanded, a complementary attempt was made to identify the most appropriate targets in the event of war. Satellite reconnaissance offered a means of far greater potential than the human sources, the captured German documents, and the peripheral reconnaissance missions that the U.S. was relying on in 1950 to provide such information.[22]

The Lipp-Salter-Wehner report examined the basic problems involved in all phases of a satellite reconnaissance project, from development of the basic system to exploitation of the data produced. The conclusions were similar to those of the 1946 report: No radically new developments were required to produce an effective system, only a reconstitution of known theory and art in rocketry, electronics, engines, and nuclear physics.[23]

In studying the problem, the authors and their supporting staff considered a wide array of variations in satellite systems, hoping to identify a combination of subsystems that would optimize system performance. RAND analysts considered and rejected two methods of data storage as alternatives to television transmission. One of the methods, storage of images on film and recovery of the film, was rejected because of the enormous weight of the film that would be required to provide coverage equivalent to that provided by television. The second rejected method involved film-based storage along with a film-scanning system that would allow the imagery to be radioed back to selected ground stations at a later time. Such a method would require either an enormous amount of film per month of operation or reusable film, which didn't exist.[24]

The envisioned system involved a satellite operating at an altitude of 350 miles, transmitting television-like pictures to whatever ground station could receive them. Data from the network of ground stations used to receive the pictures would then be transmitted to a central evaluation station and assembled into an integrated whole.[25]

The authors concluded that a two stage launch vehicle, in contrast to the three- or four-stage vehicle considered in 1946, weighing about 74,000 pounds and carrying a 1000-pound payload, could "satisfactorily" conduct general reconnaissance and detect objects with a maximum dimension of 200 feet. In addition, laboratory experiments indicated that improved television components would allow objects of 100 feet to be detected while still allowing for coverage of targets every other day.[26]

The report did not engender the same enthusiasm among experienced photo-interpreters at the Air Force that it did at RAND, because 200- or even 100-foot resolution promised to provide little interpretable information about what was going on in the Soviet Union. One-hundred-foot resolution was not even close to the resolution level required to provide adequate information. Resolution of 200 to 80 feet was adequate for analyzing cloud formations, but even resolution in the 80- to 40-foot range was considered inadequate for detection and identification of land-based targets.[27]

Such skepticism was evident at a 1951 briefing given by Lipp and Salter at Wright Field. Those present included Amrom Katz of the Air Force Reconnaissance Laboratory and a visiting delegation from the Boston University Optical Research Laboratories (Duncan MacDonald, Director; Walter Levinson, Assistant Director; and Colonel Richard W. Philbrick, the Air Force liaison officer to the lab).[28]

One recipient of the briefing recalls that "Lipp had a nearly zero batting average; no one was [convinced]. A few of us from the Recce Lab gave him a hard time and he was delighted. The fact that we thought that his scheme was not going to deliver usable results was not as important to Lipp as was the fact that we were interested!"[29]

To bury the idea, Katz, Levinson, MacDonald, and Philbrick formed an ad hoc committee. Katz recalls that "We were going to prove that this proposed project was ridiculous. Mind you, we didn't know or care about the incidental problems such as making the launch rocket, achieving stability in orbit and all other important parts of the system. We were fastened on the proposed scale to be delivered to the TV sensor."[30]

In approximately late 1951, Katz arranged for a series of aircraft over-flights of Dayton that were designed to simulate the prospects of satellite reconnaissance. When flown at 30,000 feet, the camera system produced images essentially equivalent to those that would be produced by a satellite system. When he examined the images, Katz could detect the streets and bridges of Dayton, Wright Field, and other key landmarks. From that day on, Katz became a believer in the feasibility of satellite reconnaissance.[31]

Neither RAND's 1951 reports nor Katz's subsequent conversion stopped the Air Force from rejecting RAND's recommendation that the United States begin development of a satellite reconnaissance system. The Air Force believed that such development was premature, and instead suggested a more detailed study of the feasibility of such a system. As a result, RAND's Project FEEDBACK was born.[32]

Over the next two years a clear momentum developed in favor of the satellite project. A 1952 Project FEEDBACK briefing at Wright Field produced a markedly different reaction than the 1951 Lipp briefing had. Robert Perry, who attended that later briefing, recalled ten years later "the general excitement that gripped the audience during much of the presentation. The animated discussion that followed was marked by a complete absence of "it can't be done" sentiment and by free expressions of hope of the success of the RAND program during latter presentations at the decision level in Air Force headquarters."[33]

Both RAND and the Air Research Development Center (ARDC) proceeded to push the satellite project. In June 1952, RAND signed a contract with the Radio Corporation of America (the parent company of the National Broadcasting Company), assigning RCA to study optical systems, television cameras, radiation recording devices, presentation techniques, and reliability aspects of a satellite reconnaissance subsystem. In May 1953, ARDC planners persuaded the Air Staff that ARDC should be responsible for "active direction" of the RAND satellite study by June 1, and by August 1953 ARDC's Lt. Col. Victor L. Genez returned from his initial RAND satellite office visit convinced that an immediate effort should be made to orbit a satellite, even if the reconnaissance subsystem was not yet available.[34]

On September 8, 1953, James Lipp forwarded RAND's preliminary recommendation for development of a satellite to ARDC. RAND recommended that ARDC establish a reconnaissance satellite design contract within one year, thereafter proceeding to full system development, "perhaps immediately following the completion of experimental component tests."[35]

In December 1953, ARDC pulled together the proliferating aspects of satellite work into a single project, tentatively titled Project 409-40, "Satellite Component Study," and unofficially assigned the weapons system designation WS-117L for the ultimate system development. ARDC also ordered Wright Air Development Center to supervise work on demonstrating the feasibility of major satellite components, including the television-optical reconnaissance system, attitude and guidance control equipment, and the auxiliary power plant.[36]

While Lipp, Salter, and others were working on the FEEDBACK study, others at RAND were examining several aspects of military strategy of direct relevance to the utility of satellite reconnaissance for early warning and targeting purposes. Concerns had arisen about the vulnerability to surprise attack of U.S. strategic forces, then consisting solely of B-36, B-47, and B-50 bombers. Such vulnerability was a product of several factors, including the concentration of forces and the extent to which advance warning of attack could be obtained.

It was also a product of the Soviet half of the U.S.–Soviet nuclear arms race. By 1952 that portion of the arms race involved the regular production of atomic weapons, research directed toward development of thermonuclear weapons, vigorous nuclear testing programs, production of the means for the intercontinental delivery of weapons, and, in particular, the production of TU-4 long-range bombers. It appeared that the Soviets had the potential to deliver a devastating initial blow to the U.S. deterrent—an attack that would make Pearl Harbor seem a pleasant memory.

Thus, on June 1, 1952, RAND issued *The Cost of Decreasing Vulnerability of Air Bases by Dispersal—Dispersing a B-36 Wing*. On November 1, 1952, Albert Wohlstetter and Harry Rowen published their research memorandum, *Elements of a Strategic Air Base System*, which recommended restructuring SAC basing systems to reduce vulnerability to surprise attack while performing SAC's deterrence mission.[37]

Another study, prepared by Andrew W. Marshall and James F. Digby, examined the impact of advanced warning of attack on the performance of military forces in wartime. That study, along with a revised version of the then Top Secret *The Military Value of Advanced Warning of Hostilities and its Implications for Intelligence Indicators* (July 1953), discussed the need for specific intelligence collection systems to improve the reliability of pre-hostilities warning.[38] A third study, released in 1953 and authored by RAND's Plans Analysis Section, was the 112-page *Vulnerability of U.S. Strategic Air Power to a Surprise Enemy Attack in 1956*.[39] Still in progress was a study to be published in April 1954 by Albert Wohlstetter and several RAND colleagues. That report, *The Selection of Strategic Air Bases*, drew upon related studies of the vulnerability of U.S. strategic forces as they existed and projected alternative strategies to enhance deterrence in the nuclear bomber and missile age.[40]

The work of Lipp, Salter, and others reached its climax with the March 1, 1954 publication of the *Project FEEDBACK Summary Report*, a culmination of several studies intended to encourage the Air Force to proceed with a major development effort. The report provided an overview of all aspects of any satellite reconnaissance project: the engineering issues, cost projections, launch requirements, subsystem studies, attainable resolution, and data recovery and photo-interpretation. Davies worked on the latter issue, studying the tradeoff between quantity and quality of data.[41]

The authors reported their findings and recommendations in the two-volume summary report and recommended that the Air Force develop an electro-optical reconnaissance satellite. Such a satellite, operating at an altitude of 300 miles, would image an area 375 miles in width and would produce images with a resolution of 144 feet.[42]

But such resolution, it had become clear, was not adequate for intelligence purposes. A satellite with those capabilities could provide cloud cover and weather information but not the high resolution imagery required for detection and analysis of military facilities. The television and videorecorder technology of the 1950s was simply not sufficiently advanced to provide images with enough detail to identify and target strategic forces or to contribute to early warning of impending military hostilities.[43]

Given the feasibility of all other aspects of a space reconnaissance system and a belief that the resolution requirements could, with more work, be attained, RAND recommended that the Air Force undertake "the earliest possible completion and use of an efficient satellite reconnaissance vehicle" as a matter of "vital strategic interest to the United States. " Further, RAND urged that the satellite project be "considered and planned" at a high policy level and conducted under elaborate security precautions to prevent severe international repercussions. Under such conditions RAND believed that the development and initial operation of the reconnaissance satellite could be completed in about seven years.[44]

The FEEDBACK report did not create an immediate rush to develop the system recommended by RAND. Among those favorably impressed by the FEEDBACK report was Lt. Col. Quentin Riepe, the Assistant Librarian at Wright Field who lobbied various ARDC officials concerning the desirability of the RAND concept. When ARDC approved the project, Riepe became its deputy chief.[45]

In July 1954, the Coordinating Committee on Guided Missiles approved the satellite project, then known as Project 1115. By the end of August 1954, the Western Development Division (WDD) of ARDC received authorization to start work on a satellite reconnaissance system. In turn, WDD issued System Requirement Number 5, "System Requirement for an Advance Reconnaissance System," on November 27, 1954, indicating approval of an effort to develop a satellite reconnaissance system.[46]

Several events in early 1955 gave the satellite reconnaissance project more momentum. In January the Strategic Missile Evaluation Group, more commonly known as the Von Neumann committee after its chairman, John Von Neumann, decided that it was feasible and desirable to limit initial work to a satellite vehicle and its contents rather than working on the total reconnaissance system. This strategy would eliminate the possibility of interference with the ballistic missile program.[47]

The committee defined the Air Force objective as a means of providing continuous reconnaissance of "preselected areas of the earth" in order to "determine the status of a potential enemy's warmaking capability." The resolution requirements demanded a significant increase in capability over that projected in the FEEDBACK report. The satellite was to produce imagery of sufficient detail from daylight photography to allow identification of airfield runways and intercontinental missile launch platforms. It was estimated that 20-foot resolution would be required. Two additional capabilities were also specified: the ability to collect electronic intelligence and to provide weather data.[48]

On Valentine's Day 1955, President Eisenhower was presented with the report of the Technological Capabilities Panel. The previous spring, Eisenhower had asked several of his scientific advisers, including James B. Conant and MIT President James R. Killian, Jr., to study the problem of surprise attack and potential solutions. Subsequent to White House consultations in the spring of 1954, Eisenhower invited Killian to become chairman of a Technological Capabilities Panel (TCP), which operated with three committees: one on offensive forces, one on defensive forces, and one on intelligence. Edwin H. "Din" Land, the founder of Polaroid, chaired the Intelligence Committee, known as Project 3. The Land Committee also included James G. Baker, a lens designing Harvard astronomer; Joseph W. Kennedy of Washington University; Allen Latham, Jr. of Arthur D. Little Inc.; Edward M. Purcell of Harvard University; and John W. Tukey of Princeton University. Aside from field trips to the CIA, Pentagon, Strategic Air Command, and other sites, the committee worked behind locked doors manned by Air Force guards.[49]

The TCP's report, *Meeting the Threat from Surprise Attack*, included the finding from Project 3's section of the report that:

> We must find ways to increase the number of hard facts upon which our intelligence estimates are based, to provide better strategic warning, to minimize surprise in the kind of attack, and to reduce the danger of gross overestimation or gross underestimation of the threat. To this end, we recommend the adoption of a vigorous program for the extensive use, in many intelligence procedures, of the most advanced knowledge in science and technology.[50]

Killian and Land together briefed President Eisenhower on the various technological options to increase the U.S. capability to obtain hard intelligence. A recommendation excluded from the report for security reasons, but conveyed orally, was construction of a strategic reconnaissance aircraft. In the actual report they informed the President of other options, including balloon reconnaissance and, eventually, satellite reconnaissance.[51]

A month later on March 16, 1955, the Air Force issued General

Operational Requirement No. 80, officially establishing at high level the requirement for an advanced reconnaissance satellite. According to Robert Perry, the requirement "paralleled the earlier RAND studies. It defined as the Air Force objective a means of providing continuous surveillance of 'preselected areas of the earth' in order 'to determine the status of a potential enemy's warmaking capability.' "[52] It also echoed the requirements laid down by the Von Neumann committee with regard to resolution requirements and electronic and weather intelligence collection capabilities.[53]

By November 1955, 14 basic "in house" technical tasks had been defined and approved. The Air Force opened up a competition, code-named PIED PIPER, for design studies intended to specifically establish the time and technology requirements to complete the project. Competing were the Radio Corporation of America, Glenn L. Martin, and Lockheed Aircraft.[54]

While RCA, Martin, and Lockheed worked on design studies, the Air Force moved slowly ahead. A full-scale development plan for WS-117L received the approval of General Bernard A. Schriever, the commander of the Western Development Division, on April 2, 1956 and the approval of General Thomas Power, the commander of the ARDC, three weeks later. The satellite was to be placed in orbit by an Atlas launch vehicle. The complete system, including the ground facilities for analyzing and disseminating the imagery, was intended to be fully operational by the third quarter of 1963.[55]

While Air Force headquarters approved the plans on July 24, 1956, development was approved with a spending limitation of $3 million for fiscal 1957. The ARDC called this "inadequate initial funding."[56] Such limited funding was, at least in part, a reflection of Department of Defense hostility to space systems. It was believed, although not proclaimed, that satellites were of doubtful value and that until the Navy's Vanguard rocket experiments established feasibility, the WS-117L program should be funded at the "study" level.[57]

An important shift of focus in RAND's satellite work occurred in 1956. Not that RAND was any less committed to the concept of a reconnaissance satellite; indeed, since the completion of the FEEDBACK report it had added some significant staff members to aid in further work. One recruit was Robert W. Bucheim, a guidance and control project engineer from North American Aviation, and the other was former skeptic Amrom Katz of the Air Force Aerial Reconnaissance Laboratory.[58]

Katz had gotten into the reconnaissance business by a fluke. He had studied mathematics and physics at the University of Wisconsin, where he had taken the civil service exam in physics with a specialty in optics. When a job in a particular area became available, the civil service would routinely send out the names of the top three qualifiers, and Katz's name was sent out for a job as an X-ray technician at Wright-Patterson Air Force Base in Ohio.[59]

When he didn't get the job, his name went back onto the list. On his second try Katz landed a job in the photo lab at Wright Field. After initial work on the testing and evaluation of film, he moved into the testing and evaluation of lenses. He made his reputation in shutter testing, developing a method of plotting light through the shutter as it opens and closes.[60]

From there he moved into the field of photo-interpretation. While it was easy for photo-interpreters to determine the dimensions of objects when they were photographed from directly overhead that was not the case with oblique photographs, in which the camera was tilted at, for example, a 45 degree angle at the target. Katz developed slide rules and other computing devices for photo-interpreters that permitted them to make such determinations.[61]

By the end of World War II, Katz had become deeply involved in aerial reconnaissance, and by 1954 he was ready to move on from the Reconnaissance Laboratory. His intended destination was Washington and a job in the intelligence community, which at the time he considered the "highest plateau." But before he could formalize his plans, he received a visit from several RAND representatives to discuss reconnaissance. When informed of his intentions, they suggested RAND as an alternative. Katz journeyed to Santa Monica, looked around, and signed up.[62]

Katz brought with him nearly 15 years of photo-reconnaissance and camera experience. He added a dimension to RAND's satellite project that was clearly needed because he understood the photo-interpretation process and the relation between resolution of imagery and the ability to extract military intelligence information from that imagery.[63]

Katz and Davies were as enthusiastic as ever in 1956 concerning the feasibility and utility of a reconnaissance satellite. In a May 1956 secret

RAND letter to the Western Development Division, Katz pointed out the possible utility of a reconnaissance satellite in detecting and locating intercontinental ballistic missile (ICBM) sites. The letter suggested an active program to use periodic photographs over test sites, missile bases, and especially sites under construction in order to test the conjecture that missile sites might be more readily detected during the construction phase than after they were finished, camouflaged, and integrated into the landscape.[64]

While continuing to push for development of a reconnaissance satellite, RAND had begun to push for development of a *different* type of reconnaissance satellite. The Air Force continued to focus on reconnaissance satellites that returned imagery by television transmission or used on-board processing for relay by radio link, but RAND had shifted its focus to a previously rejected method—a satellite that would actually return its film to earth.

Richard C. Raymond, a physicist who had joined RAND's Electronics Division in 1953, initially proposed the concept of a recoverable satellite. Before joining RAND, Raymond had taught and published papers on information theory at Pennsylvania State University. Using the mathematics of information theory, Raymond calculated the rates of data recovery from electro-optical satellites versus film-stored images returned to earth and concluded that film-stored images provided more information. Raymond proposed using an Atlas booster plus a solid rocket, together with a vertical strip camera.[65]

Raymond's work led to a Top Secret RAND recommendation to the Air Staff for a recoverable reconnaissance satellite system. The 20-page *Photographic Reconnaissance Satellites*, written by Brownlee W. Haydon and RAND President Frank Collbohm, represented the RAND approach as of March 1956.[66]

In short order, however, the recommendation was withdrawn. The reasons for the withdrawal are not known conclusively, since the recommendation and RAND's correspondence with the Air Staff concerning the recommendation were apparently destroyed. Raymond does recall that some Air Force officials were wedded to a near-real-time reconnaissance satellite for targeting and warning missions. The delays involved in a recoverable satellite far exceeded those associated with the FEEDBACK electro-optical satellite concept.[67]

Many Air Force officials considered the recommendation premature because no systematic comparison had been made with FEEDBACK-type systems and because the feasibility of several key aspects of such a system had not been demonstrated. In particular, there was the question of whether a film re-entry capsule could be designed that would not be incinerated as it descended through the atmosphere at high velocity. Many assumed that the problem would be solved in conjunction with the development of the intercontinental ballistic missile.[68]

As of spring 1956, however, no definitive solution had been found. The Defense Department had devoted an entire summer study to the problems of atmospheric re-entry of high-velocity payloads into the atmosphere in 1955, but the study group, chaired by Robert Bacher of the California Institute of Technology, produced no comprehensive solution. Subsequent work by Richard Porter of the General Electric Space Systems Division in Philadelphia had encouraged some optimism, but his solution remained theoretical.[69]

Despite these problems and the obvious desirability of real-time data, the RAND view was that physical recovery was necessary because of the differing picture qualities. The difference was equivalent to the difference between a televised picture of an object and a photograph of that object. The fuzziness associated with television transmission resulted from the process of breaking down the image electronically and then reassembling it after transmission. The image in standard photography is sharper because it is directly imaged.[70]

A June 26, 1956 product of RAND's research in this area was a report titled *Physical Recovery of a Satellite Payload: A Preliminary Investigation*, authored by John H. Huntzicker, Hans A. Lieske, and others. As with much of RAND's work in that area, there were a larger number of "secondary" authors who also made significant contributions.[71]

The report addressed several issues that were unique to a satellite with a recoverable payload: the weight penalty, which either cuts down other payload components or increases thrust requirements, and additional requirements such as retro-rockets, parachute systems and activation devices. Also subject to scrutiny was the recovery of heat-sensitive items such as photographic film.[72]

RAND's concept received an important boost before the summer was over. A second Bacher-chaired summer study of the re-entry vehicle problem for ICBM warheads had apparently produced a solution—ablative nose cones that were supposed to shed layers of plastic while protecting the warheads on their journey through the atmosphere.

Clearly, what could protect a warhead on the way down could also protect a film capsule.[73]

In light of such results Davies and Katz were further encouraged to push for both recoverable and electro-optical systems. They specified the need for both types of systems in an October 12, 1956 memorandum that proposed a RAND project on "pre-hostilities reconnaissance." Their memo noted that they had "for some time been mulling around and considering a generalized study of the pre–D-Day intelligence and reconnaissance which would simultaneously embrace study of national objectives, intelligence requirements, and proposed collection systems of all types as they are likely to become available in the era just ahead."[74]

Two weeks after the October 12 memo the Air Force awarded a contract to the Lockheed Missile Systems Division to develop WS-117L, the Advanced Reconnaissance System. Lockheed was the prime contractor for WS-117L and the upper-stage vehicle, later redesignated Agena.[75]

By the beginning of 1957 considerable work on the development of appropriate camera systems had begun. Davies and Katz encouraged the adaptation of the concept of the panoramic camera to long-focal-length cameras for high-altitude photography. In the nineteenth century the panoramic camera had been widely employed for large group portraits, and at the beginning of the twentieth century George Lawrence had adapted the panoramic camera to balloon-based wide-angle photography over Chicago. In 1949, USAF Colonel Richard W. Philbrick, on assignment to the optical research program at Boston University, demonstrated the wide-angle potential of panoramic photography for horizon-to-horizon coverage by synchronizing the passage of film past a slit, along with the angular rotation of the lens.[76]

Davies and Katz had encouraged the use of the panoramic camera in a balloon reconnaissance system of the time, known as Project 461L.* But it was also, to them, quite applicable for space satellites. Achieving a wide angle of coverage at the altitude of space satellites was not difficult.

---

* In 1956 the United States launched several hundred camera-equipped balloons from Europe over Soviet territory under the GENETRIX program, but the program was canceled within months owing to Soviet protests. In 1958 Eisenhower approved Project 461L based on assurances that the new camera for 461L would allow the balloons to fly high enough to be undetectable. When the Soviets protested the new flights within a month of the initial launches Eisenhower angrily canceled the program.

What was difficult was attaining the ground resolution that would be possible only with long-focal-length lenses with narrow fields of view. Thus, the concept of a panoramic camera had a special application in space-based photographic systems.[77]

During February 1957 Davies and Katz flew to Boston to attend a meeting at the Boston University Physical Research Laboratories on aerial photography, where they discussed their ideas with Duncan Mac-Donald (head of the laboratory), Walter Levinson (responsible for development of photographic systems for 461L), and James G. Baker. Davies recalls that it was "an exciting all-day meeting, exchanging ideas with innovators in aerial reconnaissance."[78]

Levinson described the cameras designed to obtain imagery from high-altitude balloons. The camera would cover a wide angle, about 120 degrees with a lens that was a modification of the Baker spherical lens of World War II. The only moving part was the focal plane shutter, which was to move $2\frac{1}{2}$ inches across the film during exposure.[79]

Several weeks later Davies and Katz went to the annual meeting of the American Society of Photogrammetry. During a social gathering, Fred Wilcox, Vice President of Fairchild Camera and Instrument Corporation, described a new camera, a rotary panoramic design that Fairchild wanted to build and install in fighter aircraft wing pods. The camera had a 45 degree mirror in front of the 12-inch focal length lens, and the entire camera, film and all, rotated about the axis to perform the panoramic scan. During exposure the film was moved past the slit mounted in the focal plane to compensate for the rotation. The slit would thus act as a focal plane shutter.[80]

The meeting with Wilcox had a major impact on Katz and Davies. They became advocates of the spin-stabilized panoramic camera with a long focal length, in which the entire spacecraft spun to obtain the panoramic effect. In developing a photographic system for the 461L balloons Levinson adopted the concept of a panoramic camera with a long focal length, although not that of a spinning camera.[81]

In the summer of 1957, RAND had completed a further satellite design study with the goal of attaining a photographic capability in a short time. The study envisioned placing the satellite in polar orbit with the Thor-Able booster and a small spin-stabilized, solid rocket. The satellite would contain a spinning panoramic camera with 12-inch wide film, which operated by command and clock. The satellite also contained a solid rocket to be fired on command from ground control. Firing the rocket would cause the satellite to deorbit and fall into the Pacific Ocean to await recovery. An automatic radio beacon would aid in the search.[82]

Meanwhile, a study by RCA led to Air Force rejection of the television option. The television option was abandoned in August 1957 after an RCA study had shown that resolution would be extremely poor. Instead, the decision was made to pursue the talk-back approach using a film-scanning technique with a conventional camera to photograph the target. The film would be developed on board and scanned by a fine-light beam, and the resulting signal would be transmitted to a receiving station on the ground, where it would be used to build up a picture. Although the scanning process would degrade the picture quality to some extent, it was expected to be far better than television.[83]

The fall of 1957 produced another major impetus for U.S. satellite efforts. On October 4, 1957, the Soviet Union placed the earth's first artificial satellite, Sputnik ("Traveling Companion"), into orbit. The satellite was relatively small at 184 pounds, but it had a traumatic effect on the public, and even sophisticated scientists were moved to expressions of panic. John Rinehart of the Smithsonian Astrophysical Observatory announced that "no matter what we do now, the Russians will beat us to the moon . . . I would not be surprised if the Russians reached the moon within a week." Edward Teller, a father of the H-bomb, declared on national television that the United States had lost "a battle more important and greater than Pearl Harbor."[84]

Nor was it just the scientific community that was affected. Sputnik had a profound psychological effect on the general populace. George Reedy, an aide to Senator Lyndon Johnson, summed it up in a memo two weeks after the launch: "The American people are bound to become increasingly uneasy. It is unpleasant to feel that there is something floating around in the air *which the Russians can put up and we can't*. . . . It really doesn't matter whether the satellite has any military value. The important thing is that the *Russians have left the earth and the race for the control of the universe has started*.[85]

The implications of the Sputnik launch extended beyond mere propaganda or even scientific accomplishments. Khrushchev had boasted several months earlier that the Soviet Union had developed intercontinental ballistic missiles, and since the basic technology required to boost a satellite into orbit or to deliver a warhead to the other side of the world was similar, Sputnik's launch gave credence to Khrushchev's claim.

Sputnik apparently gave the Russians the means to leap-frog the oceans that had so long protected America as well as the means to bypass the air defense systems that could threaten Soviet bombers.

Sputnik thus increased the desirability of developing a satellite reconnaissance system. It became necessary to closely monitor the Soviet Union for signs of ICBM deployments, because such deployments were thought to be much easier to spot in their initial phases than after camouflage had been deployed. The thought of deployed Soviet ICBMs also raised the fear of war and of the damage that the Soviets could do in a first strike, so the desire to develop a system that could provide the earliest possible advance warning of attack was further enhanced.

One further aspect of Sputnik did work to the U.S. advantage: Not a single government protested the overflight of their territory by a spacecraft of another country. Sputnik had at least given the United States the ammunition to argue that the Soviets had themselves established the precedent of "open skies." At a White House meeting several days later between Eisenhower and Department of Defense officials, the Undersecretary of Defense Donald Quarles noted that "the Russians in fact have done us a good turn, unintentionally, in establishing the concept of the freedom of international space." Appropriately enough, Eisenhower then inquired about the status of the reconnaissance satellite program.[86]

October 1957 was also notable for publication of an article on the satellite reconnaissance program in the trade journal *Aviation Week*, the first in what would be a continuing series of revelations concerning that subject from a journal that would come to be nicknamed "Aviation Leak." The article reported that the Pied Piper program was also designated WS-117L and Advanced Reconnaissance System and that CBS and Eastman-Kodak were probably working with Lockheed. Such a disclosure created concern over the secrecy of the reconnaissance program.[87]

Subsequent months saw several briefings on different reconnaissance satellite concepts, all given a greater urgency by the events of the previous month and by the launch of the much heavier 1121-pound Sputnik II on November 3. On November 5, the Air Force briefed the Armed Forces Policy Council on its reconnaissance satellite program and possible combinations of vehicles that could be used for "cold war and scientific programs."[88]

A week later, on November 12, RAND formally recommended to the Air Staff development of a recoverable reconnaissance satellite. That day, the Committee on Special Capabilities met in RAND's main

conference room to hear a presentation by Colonel Frederick "Fritz" Oder
of the Project 117L Program Office. Oder recommended a recoverable
satellite program using spin stabilization—a presentation that excited
both Davies and Katz because it indicated a willingness of the 117L
office to proceed with both the radio-relay *and* recoverable satellites.
The Committee, chaired by Homer Joe Stewart of Cal Tech, had been
established to decide between boosters to be used in the International
Geophysical Year satellite program, but had subsequently been given a
wider mandate.[89]

RAND's president, Frank Collbohm, provided a cover letter for the
recommendation, *An Earlier Reconnaissance Satellite System*. In the
letter, prepared prior to Oder's revelation that the 117L office would
recommend development of a recoverable satellite, he noted differences
between the 117L radio-relay concept and RAND's recoverable satellite
concept and detailed why RAND believed its concept was superior. He
wrote that:

> In the light of recent events, RAND has reviewed national and
> military intelligence problems, existing and proposed recon-
> naissance systems, and in particular, the current USAF satel-
> lite reconnaissance program (WS-117L). As a result of certain
> technical and conceptual breakthroughs, it is concluded that
> efficient satellite reconnaissance systems of considerable mil-
> itary worth can be obtained earlier and more easily than those
> envisioned in the current 117L program.
>
> The systems proposed in this recommendation differ substan-
> tially from the current 117L system concept.
>
> The proposed systems feature a spin-stabilized payload stage.
> They use a transverse panoramic camera of essentially con-
> ventional design, fixed to spin with the final stage, which
> scans across the line of flight. Either the entire payload or the
> film is recovered.
>
> The first of the proposed systems uses a 12-inch camera,
> carrying 500 feet of 5-inch wide film. . . . It will provide
> sharp photographs of about 60-ft. ground resolution. Each
> exposure, covering some 300 miles across the line of flight,
> will photograph some 18,000 sq. mi. The 500-ft. roll will
> cover some 4,000,000 sq. mi. (almost half the Soviet Union)

and show major targets, airfields, lines of communication, and urban and industrial areas. This satellite could weigh about 300 lbs. and be placed in a polar orbit at 180 miles altitude . . . A one day operation is envisaged, with recovery by command firing of a braking rocket on the 16th pass, so as to impact in a predictable ocean area.

The next, more sophisticated, system would use a 36-inch camera, carry much more film, [and] do more detailed reconnaissance—with a ground resolution of about 20 feet. This system can possibly be Thor-boosted.

A third system—undoubtedly requiring Atlas-type boosting—would use a 120-inch camera and would have very large film capacity. This system will be able to accomplish very high quality photo reconnaissance and, most important, would do it better than any Air Force system now in development or in prospect will be able to do in the 1960s.

The earliest and simplest of the several systems will collect at least as much information in its one-day operation as the "early" 117L vehicle will in its useful life.

Because of our belief that the first system could be available about a year from start of work, the second in less than two years, and the third in about three years, we recommend that the U.S. Air Force begin work immediately to accomplish this program.

Success in this type should result in refocus of the present components of the 117L program to those tasks requiring the communication link and cyclic talk-back facility of 117L—warning, and daily surveillance of selected targets, being the principal of high priority tasks requiring such an operation. Thus this new family of satellites and the type of satellite presently scheduled under 117L program would be mutually complementary and not competitive.[90]

The RAND recommendation was based on the simultaneously published work of Davies and Katz, "A Family of Recoverable Satellites." The 129-page study, classified Secret, concluded that it was possible to orbit a 300-pound photo-reconnaissance satellite at 95 to 190 miles altitude by using a Thor booster and a Vanguard second stage. RAND's engineers calculated that this could be accomplished by slowing the second-stage satellite's forward momentum, turning it around, and ordering it to fire its camera-carrying payload down and out of orbit:

> Descent from orbit is achieved by the command firing of a braking rocket in the satellite. Assume that the satellite is coming over the [North] Pole, that it is picked up by trackers in the north, and that an impact point in the Pacific is desired. The braking rocket is then fired forward and upward, imparting a downward and backward velocity impulse superimposed on the orbital velocity. The resulting velocity vector points downward, so that vehicle is effectively in a ballistic trajectory. . . .

> Tracking of the vehicle immediately after the beginning of descent establishes a predicted vacuum path. This, together with predicted atmospheric effects, makes it possible to predict the approximate impact area. The vehicle is protected against reentry heating by a coating of suitable vaporizing material: 80lb. [cq] of fiberglass-reinforced plastic, such as is used on advanced designs of the ICBM, and on the Jupiter nose cone, is suggested. Impact survival of the casing, payload, batteries, and beacon is made feasible by the proper selection and arrangement of structural components. Search aircraft are used to find and recover the payload. This means that the radio beacon must operate after water impact, and possibly that some type of dye marker should be released upon impact.[91]

The increasingly sophisticated satellites proposed by RAND were intended to allow surveillance at four different "levels" that were defined in the RAND memo. The levels were the product of Katz's thinking about reconnaissance requirements, work he considered to be among the most important he did with the reconnaissance program. The levels were

> A: large area search, measured in millions of square miles;
>
> B: limited area search, measured in hundreds of thousands of square miles;
>
> C: specific-point-objective, measured in hundreds of square miles;

D: technical-intelligence objective photography, providing coverage in blocks of tens of square miles or less in size.[92]

The photographs that would be produced by the earliest system would, according to the RAND memo, "enable us to do a useful reconnaissance job at level A, over areas measuring millions of square miles. The scales and resolution . . . will make it possible to identify major railroads, highways and canals. Urban centers, industrial areas, airfields, naval facilities, seaport areas and the like can be seen. Very likely, defense missile sites of the sort found around Moscow area will also be identifiable. Thus, with repeated surveillance it will be possible to find new major installations, perhaps to learn something about patterns of use of Soviet ICBM systems, and certainly to obtain clues for the direction of other, higher resolution systems that can go back and take another look."[93]

The memo projected that the second system would be available in 18 months and would allow type B reconnaissance, while the third system would allow type C reconnaissance when it became available in three years.[94]

As might be expected, the Army did not let the Air Force and RAND satellite projects go without competition. Andrew Goodpaster recalls the period as being one of "intense interservice argument over satellites" in the aftermath of the successful Sputnik and several failures in the Navy's Vanguard program. The Army was attempting to carve out a place for itself in the space program.[95]

On October 26, 1957, the Army submitted to the Department of Defense a proposal for a military reconnaissance satellite capable of providing complete photographic coverage of the USSR every three days, cloud cover permitting. On November 19, 1957, the Army made its pitch to the Stewart committee. The briefing noted that

Today, when the tremendous destructive effects of thermonuclear weapons and the speed and range of intercontinental ballistic missiles make a surprise attack against the United States entirely feasible, a timely and accurate means of gathering intelligence information from within the Soviet Union is essential to our National security. . . .

> A satellite carrying surveillance equipment can collect
> timely and accurate intelligence information needed. . . . At
> the same time, such a system will be able to provide timely
> and accurate target information and meteorological data
> over the vastly expanded area from which intelligence infor-
> mation must be available in any future war.[96]

However, given that "the most immediate and urgent national requirement is for current intelligence of selected critical areas of the Soviet
Union, such as ICBM launching sites, air defense complexes, ships at
sea, etc.," the Army proposed in the briefing to concentrate initially on
meeting a more limited requirement for current intelligence, which could
be reached on a time scale compatible with the launching capabilities
that existed.[97]

The Army-proposed development program envisioned a progression
from 20- to 100- to 500-pound payloads. Twenty-pound satellites would
be launched as early as January 1958, with the first 100-pound satellite
being launched as early as June 1958 and the first 500-pound satellite in
January 1959.[98]

Initial photographic coverage was to be obtained by the placement of
the 500-pound satellite into a circular orbit of 300 miles. The satellite
would be launched into an 83 degree inclination to ensure that it would
pass within surveillance range of any area of the earth's surface once
every three days, except for a small circle around each pole. Pictorial
data would be transmitted to ground stations in the United States where
it would be recorded and processed. For most areas of the USSR, it
was expected that photo prints could be made available for intelligence
processing approximately 30 minutes after the pictures were taken.[99]

The picture-taking sequence would "be initiated and terminated by an
on-board programmer-timer . . . set for a desired target area by coded
command from the ground during the previous pass of the satellite over
a ground station in the United States." The timer was to incorporate limit
stops to prevent photographs being taken of United States territory.[100]

During the time the pictures were taken the lens system would focus on
an image of a ten by one mile area on the earth's surface with a recording
and storage device employing television recording techniques. Early
operational models of the satellite would incorporate a new miniature
vidicon television tube and a magnetic tape recorder that would allow
pictorial coverage of up to 45,000 square miles per orbit. Later models

were to project the image directly onto an electrostatic storage tape that would act as a combination camera and storage medium.[101]

The first payload with a photo-intelligence capability would be launched in September and October 1958 to demonstrate the feasibility of obtaining pictures from a satellite in orbit. In order to correlate the pictorial data with a known area of the earth's surface and to check the quality of the pictures, the first pictures would be taken over the United States. The first 500-pound satellite carrying a complete photo-intelligence payload with magnetic tape storage was scheduled for launch in March 1959 to provide a system test. The satellite would provide the first pictures from within the Soviet Union, and would be followed in May 1959 with the first operational satellite providing coverage of selected areas of the Soviet Union. A second payload would possibly be launched in July 1959 if required.[102]

In NSC Action No. 1846 of January 22, 1958, the National Security Council assigned highest priority status to development of an operational reconnaissance satellite.[103] The question remained, however, which particular program(s) should be approved for national support.

To resolve this fundamental question Eisenhower met with Good-paster, James Killian and Edwin "Din" Land on February 7, 1958. Killian held several important positions: he was Eisenhower's science adviser, Chairman of the President's Scientific Advisory Board, and a member of the President's Board of Consultants on Foreign Intelligence Activities. Edwin Land was the Chairman of Polaroid. Both had been involved in issues of technology and intelligence since their work on the Technological Capabilities Panel.[104]

A variety of considerations played a part in the decision that was to be made. According to Goodpaster, Eisenhower was "not at all receptive to the Army's attempt to establish a role for themselves in space." The article in *Aviation Week* had raised concerns over the secrecy of the program, or rather the lack of secrecy, and there was concern that the PIED PIPER–type non-recoverable technology would not produce a satellite as quickly as desired.[105]

All those factors led Eisenhower, based on the advice received at the meeting, to assign the CIA responsibility for developing a recoverable satellite system. The agency had experience in the secret development of an overhead collection system, the U-2, and Eisenhower approved Project CORONA with the expectation that it would result in an operational photographic reconnaissance satellite employing a recoverable capsule system by the spring of 1959.[106]

Several other actions were taken in conjunction with the decision. It was decided to use the DISCOVERER satellite program as a cover for the development, testing, and operation of CORONA and to establish a national military research and development agency to handle the public research and development aspects of the project. Thus, on February 7, 1958 the Advanced Research Projects Agency of the Department of Defense was created.[107]

Eisenhower's decision was not at all upsetting to Fritz Oder, the project manager for WS-117L. Oder knew that obtaining sufficient funding for the radio-relay satellite was difficult enough, and obtaining sufficient funding through the Air Force for two satellites would be nearly impossible. In fact, Oder had been making an effort, code-named SECOND STORY, to get the CIA to take over funding of the recoverable satellite.[108]

The CIA official given responsibility for running the CORONA program was Richard Mervin Bissell, Jr., the Deputy Director for Plans. Bissell had entered the cloak and dagger world in 1954 as a special assistant to DCI Allen Dulles and had been involved in both human-source espionage and covert action as well as aerial reconnaissance operations — it was Bissell who had so successfully directed the effort to produce the U-2 spy plane. Prior to taking the job as Dulles' assistant he had no experience in the intelligence world, nor in the military or investigative worlds.

Prior to World War II, Bissell had joined the Yale economics faculty while still a graduate student, and during the war he had served in the Department of Commerce and War Shipping Administration. When the war concluded he moved on to the Economic Cooperation Administration, where, according to former student William P. Bundy, he was "the real mental center and engine room of the Marshall Plan." In his appearances before congressional committees, Bissell inspired enough confidence in his listeners about the utility of the revolutionary foreign-aid program that the committees voted billions of dollars to implement the program.[109]

The impression he made on his colleagues in the intelligence world was similar, partly because of his imposing size and constant motion.

He was six feet three inches tall, and his long legs made him look even taller. He loped rather than walked. In his office he would pace incessantly, often dictating as he moved about his office, and even at his desk he could be found toying with a screw, polishing his glasses, or mangling a paper clip. Of course, such action was not enough to result in the praise often heaped on Bissell. He could absorb and retain enormous amounts of information. The late Robert Amory, Jr., whose tenure as CIA Deputy Director of Intelligence overlapped Bissell's tenure as DDP, remembered him as a "human computer" as well as "very spookish, a glutton for security." To Arthur Lundahl, the CIA's top photo-interpreter from 1953 to 1972, Bissell "could outwit, outspeak, out-think most of the people around him that I was aware of."[110]

In addition to changing the source of funding and management, it was decided for cover purposes to "cancel" the parts of the WS-117L program that corresponded to the CORONA project—in other words, the plan to develop a recoverable satellite. Davies and Katz and many others were told that the Air Force had decided to terminate work on recoverable satellites but were not told about CORONA. As was Bissell's style, people were then let back into the project on a need-to-know basis. Davies and Katz were not let back into the project.[111]

Being shut out of the program was very frustrating. Davies recalls that "We had been extremely active in this field, we had worked and knew of the people who were involved and all of a sudden nobody would talk to us. We were ostracized."[112]

Bissell also made another significant change. Any satellite had to be stabilized to prevent it from tumbling erratically out of control. In the Davies spin-stabilized satellite concept the whole satellite would revolve around its longitudinal axis. A photoelectric cell, roughly the same as a photographic light-meter, was to point out of the side of the capsule in the same direction as the camera. It would register the differing light values as the capsule sped along at about 17,000 miles an hour, noting each time the horizon came into view and disappeared again. The changing light level striking the photocell would cause a pulse of electricity to be sent to a conductor that would engage a clutch and would start the film moving. As the terrain disappeared and was replaced by the darkness of space, the clutch would disengage and stop the movement of the film. The movement of the satellite would create a panoramic scanning effect.[113]

Lockheed began building such a satellite, but it was never completed. Bissell opted instead to switch to a three-axis stabilized satellite. The satellite was kept stationary relative to the horizon by internal flywheels that would spin like gyroscopes inside the Agena while the camera in the satellite moved and clicked away. Bissell recalls that he made the decision in conjunction with the Air Force co-director of the project, Major General O.J. Ritland. After consulting contractors such as Fairchild and Hycon, Bissell became convinced that a three-axis stabilized spacecraft would produce better photography than a spin-stabilized satellite.[114]

In light of the crucial February decisions, the National Security Council (NSC) reaffirmed the priority assigned to the reconnaissance satellite program. On June 20, 1958, the NSC issued NSC 5814, "U.S. Policy on Outer Space." The document noted that "[r]econnaissance satellites would also have a high potential use as a means of implementing the "open skies" proposal or policing a system of international arms control." It suggested that the U.S. should "At the earliest technologically practicable date, use reconnaissance satellites to enhance to the maximum extent the U.S. intelligence effort."[115]

The policy paper also noted that the international politics of reconnaissance was an issue. NSC 5814 noted that "[s]ome political implications of reconnaissance satellites may be adverse . . . studies must be urgently undertaken in order to determine the most favorable political framework in which such satellites would operate" and recommended that "[I]n anticipation of the availability of reconnaissance satellites, seek urgently a political framework which will place the uses of U.S. reconnaissance satellites in a political and psychological context most favorable to the United States."[116]

In November 1958, the Department of Defense announced that its 117L program consisted of three elements: DISCOVERER, SENTRY (the talk-back satellite), and MIDAS (an infrared early warning satellite).[117] The DISCOVERER project objectives were described, in then-classified documents, as:

a) Flight-test of the satellite vehicle airframe; propulsion; guidance and control systems, auxiliary power supply, and telemetry, tracking and command equipment.

b) Attaining satellite stabilization in orbit.

c) Obtaining satellite internal thermal environment data.

d) *Testing of techniques for recovery of a capsule ejected from an orbiting satellite* (emphasis added).

e) Testing of ground-support equipment and development of personnel proficiency.

f) Conducting biomedical experiments with mice and small primates, including injection into orbit, re-entry and recovery.[118]

While the DISCOVERER program did involve biomedical experiments and other non-reconnaissance related activities, its primary function was to serve as a cover for Project CORONA. Its central concerns were the "testing of techniques for recovery of a capsule ejected from an orbiting satellite" and other techniques related to reconnaissance satellite development.[119]

CHAPTER 2

A New Era

January 21, 1959 was to be the day that the U.S. space reconnaissance program began its transition from the offices of the RAND and Lockheed Corporations, the Air Force, and the CIA to the blackness of outer space. Discoverer I sat on the launch pad at Vandenberg Air Force Base, waiting to be shot off into orbit.

There were four basic tasks facing Richard Bissell and those involved in the DISCOVERER/CORONA programs as they prepared for the launching day. The easiest task would be to place the spacecraft into orbit, making it a satellite. Simple physics told the engineers how much thrust they needed to accomplish their objective. But once the satellite was placed in orbit, it had to be stabilized—it had to fly in a manner that would allow its camera to take clear pictures of preselected targets. The slightest vibration or instability in the spacecraft would produce a blurring of the image on film, as if the camera had cataracts.

The third task was to develop a camera system that would produce clear images from a stabilized satellite. Scientists had to eliminate or limit the impact of atmospheric distortion, such as dust and smoke, on image clarity. Finally, a system for recovering the film capsule once it had been ejected from the spacecraft had to be devised.

While all those problems had been solved theoretically by January 21, it still remained to be seen whether those theoretical solutions would work in practice. But no questions were answered that day. It was a day that was to be like many other days in the early DISCOVERER program, a day in which hope gave way to disappointment. In the midst of the countdown, a crewman accidentally fired a subsidiary rocket, leading to the dumping of nitric acid. That acid, in turn, damaged the first stage of the rocket.[1]

The next launch attempt came on February 25, but was stopped a few seconds before blast-off. The abort came after more than four hours of "holds" of the fully fueled rocket. The problems were corrected and on February 28 at 1:49 P.M. Pacific Standard Time, Discoverer I was launched. The satellite was a specially designed 19-foot, 1300-pound upper element of a two-stage rocket. The first stage was a Thor intermediate range (1300 miles) ballistic missile. The satellite included a 40-pound package of instruments that radioed environmental measurements to the ground and an infrared scanner that focused on the horizon to keep the satellite horizontal, but no recovery capsule.[2]

From a press site on a dune two miles away, reporters watched as the mission got off to a successful start: The Thor blasted off without any complications, followed by Agena A separation and then a successful firing to place the spacecraft in orbit. However, the Agena then began tumbling end over end, interfering with the radio transmissions, and for the next several hours no signals were heard from the spacecraft. Tracking stations could not locate the satellite during its first projected circuit of the earth, but then began picking up sporadic signals that lasted from four to six seconds at a time during each six-minute pass. Among the receivers of the signal was a communications ship stationed 950 miles downrange. Finally, after four days of uncertainty, the Air Force announced that Discoverer I was in a 114 by 697 mile orbit.[3]

Some scientists attributed the initial difficulty in locating Discoverer I to excessive rivalry and secrecy. Dr. J. Allen Hynek, associate director of the Smithsonian Astrophysical Observatory, said that if he had been given 24 hours notice, his observatory's optical tracking facilities would have been able to locate the satellite as they had tracked previous U.S. and Soviet satellites.[4]

The launch of Discoverer I did not go unnoticed by the Soviet Union and its allies. Setting the tone for what would be a four-year verbal assault on the U.S. space reconnaissance program, an East German radio broadcast attacked the U.S. for "carrying the cold war into space" and putting a military satellite into orbit "without even talking to other states over whose territories the DISCOVERER is to perform espionage services." The satellite's observations would be used, the East Germans charged, "to make entries on the map and for determining rocket targets." The broadcast also contended that the U.S. was playing a losing game because "the Soviet Union has the experience. . . to adjust its defense industries to the specific features of modern aerial warfare. The objects of most interest to the DISCOVERER would thus remain hidden." On the other hand, U.S. war industries, the broadcast claimed, "are concentrated to an extraordinary degree in various regions, and are absolutely open to view."[5]

Such warnings, although they eventually affected U.S. public relations policy concerning satellite reconnaissance, had no effect on the actual program itself. On April 13, 1959, the United States was ready to proceed with the launch of the 440-pound Discoverer II, the first satellite to carry a recovery capsule. Thirty-three inches long, 27 inches in diameter, and 195 pounds, the capsule was covered by a heat shield and an afterbody that contained a retro-rocket in the nose of the re-entry vehicle. Immediately after the re-entry vehicle separated from the rocket casing, the retro-rocket would be fired to slow the vehicle and to cause it to plunge back toward earth. After re-entry the afterbody would separate from the capsule. Once within the earth's atmosphere, a switch operated by the forces of deceleration would release a parachute that slowed the capsule's descent to tolerable speeds.[6]

To support the launch program a series of ground facilities had been constructed to track the satellites, receive telemetry data, and issue commands. Overall control and orbit prediction were the responsibilities of the Satellite Test Center at Sunnyvale. The launch site at Vandenberg AFB was also responsible for monitoring launch ascent, orbital tracking, telemetry reception, and trajectory measurements. Point Mugu, California was responsible for ascent tracking and telemetry data reception, trajectory measurements (including the time to ignite the second stage), and shutdown of the Agena. Kodiak, Alaska tracked the satellite during its first pass, triggered the ejection of the capsule, and predicted the location of the capsule's impact. Kaena Point, Oahu was responsible for orbital tracking and telemetry data reception while Hickam AFB, Hawaii had overall direction of capsule recovery operations. Actual command executions were the responsibility of the Alaska and Hawaii stations on instructions from the test center. Instructions for initiating recovery would come from the Satellite Test Center after computers in Lockheed's Palo Alto Scientific Research Center calculated the precise time for re-entry.[7]

Two Navy ships which had been modified for use in tracking and recovery operations augmented the ground stations. Although the ships were designated for tracking and recovery operations for several satellite projects, their initial use was for the DISCOVERER program.[8]

On April 13, after a three hour delay because of fog, Discoverer II was launched into a 156 by 243 mile orbit. The satellite placed itself into a tail-first position and held the position for the next 17 orbits. Seven hours after the launching, the Air Force said that all tracking stations were receiving clear signals from the satellite's acquisition radio beacon, its radio telemetering of scientific information, and its radar beacon.[9]

Prior to the launch, William H. Godel, director of planning for the ARPA, told the press that there was only a one-in-a-thousand chance for the capsule's successful recovery. His pessimism proved to be justified. Normally, a timing device could be programmed to begin capsule separation at a specific time. At a ground station at Kodiak, Alaska, a controller would update the timer. But when the controller pushed the button to make the update, a monitor mistakenly indicated that the command was incorrect—a possibility foreseen by the equipment designers, who provided ways to reset the timer. Unfortunately, the controller forgot to press the reset button before transmitting a new command, so the new pulses were added to the earlier ones. This error was not correctable and as a result the capsule impacted near the Arctic circle. A "space watch" was declared, and at the expected time and in the predicted area, observers on the Norwegian island of Spitzbergen saw a star-burst and a descending parachute.[10]

Tracking signals and sightings indicated that the capsule successfully separated and landed on the snow-covered island of Spitzbergen, north of Norway and near the Soviet-held Franz Josef Land. The main enterprises on Spitzbergen were coal mines, separately operated by the Norwegians and the Soviet Union. Col. Charles G. "Moose" Mathison, director of the recovery effort, flew immediately to Norway after calling Major General Tufte Johnson of the Norwegian Air Force North Command. The two men then flew by small chopper to Spitzbergen to supervise a search operation for the capsule. Back in Washington, Richard Bissell assured them that adequate resources would be devoted to the search. As a result skiers, five airplanes, and two helicopters searched the snow-covered hills and surrounding waters. (Bissell didn't send all of it— some equipment was U.S., some Norwegian. It was Bissell's influence that ensured that a search was made using such equipment.) After six days, the hunt was called off; nothing was found except tracks in the snow. The very existence of the tracks indicated that the Soviets had recovered the capsule. Norwegians never left their camp without skis, but many of the Soviets had grown up in the southern part of their country and, not knowing how to ski, made do with snowshoes.[11]*

Despite the loss of the capsule, Discoverer II represented an important milestone in the development of reconnaissance satellites. A satellite had been stabilized in orbit so that instead of tumbling or twisting it maintained a fixed position relative to earth. Stabilization was achieved by the combination of an infra-red horizon scanning device and external jet nozzles. To establish reference points in space, the satellite had an infra-red horizon scanning device that detected the direction of the horizon and an inertial reference system consisting of accelerometers and gyroscopes to keep track of the motion of the vehicle. If the satellite started to tumble or pointed in the wrong direction, the reference devices would relay corrections to a pneumatic jet control system. Compressed gas under high pressure would then be shot out of a series of external jet nozzles, correcting the satellite's position relative to the earth.[12]

Despite the failure to eject the capsule at the proper time and the resulting failure to recover the capsule, much had been learned and there was no reason to despair. The next year, however, was to provide plenty of reason and opportunity for despair. Discoverer III, which carried a sophisticated timer to prevent a recurrence of the Discoverer II problem, was launched on June 3, 1959 after three launches were aborted because of inclement weather and minor technical difficulties, and on June 25, Discoverer IV was launched. Neither reached orbit due to failure of the Agena second stage. As a result, several modifications were planned to increase the probability of attaining orbit, specifically, a change in fuel and a reduction of weight in orbit. The launch of Discoverer V was postponed from July 1 until the review was completed.[13]

On August 13, Discoverer V was successfully placed into a 135 by 454 mile orbit, but after capsule separation on the following day no signals

---

* The incident was undoubtedly the inspiration for Alistair McLean's 1965 novel, *Ice Station Zebra*, in which a British agent journeyed on an American submarine to the Arctic to recover a stray film capsule. The premise—that it was crucial to recover the film because the U.S. satellite had mistakenly taken pictures of U.S. missile silos—was outdated by the time the book appeared. By then the Soviets had developed their own reconnaissance satellites.

were received. Likewise, on August 19, Discoverer VI reached orbit successfully, but capsule separation was again followed by silence.[14]

Five successive failures to recover a capsule dictated that launches be suspended and procedures reviewed. The Air Force and General Electric examined the telemetry data and concluded that the capsules' batteries had become too cold, preventing them from supplying power to the recovery beacon. Painted patterns were then added to the spacecraft, allowing the capsule to absorb heat during the sunlit part of the flight and retain it while the spacecraft was in the earth's shadow.[15]

After an almost three-month suspension of flights, the program resumed on November 7, with the launch of Discoverer VII into a 99 by 519 mile orbit. Initial telemetry indicated that the satellite was operating as planned, but after three orbits, all the ground stations received fluctuations of the beacon signal, indicating that the satellite was in a slow tumble—perhaps three times a minute. According to telemetry signals, the satellite's inverter, which supplied power to the infrared sensor that sent signals to the pneumatic control system stabilizing the satellite, was not functioning. Since the inverter also provided power to operate the re-entry sequence, it was not possible to return the capsule to earth.[16]

Even if separation had been possible, however, Discoverer VII had another problem. Telemetry data indicated that the satellite's small batteries had gotten too cold to operate a radio transmitter that would have guided planes to the falling capsule. The painted geometric patterns that had been designed to allow the capsule to store heat had failed to keep the capsule interior sufficiently warm.[17]

Discoverer VIII followed on November 20, but success was once again beyond reach. The Agena guidance system malfunctioned during launch, putting the satellite into a 120 by 1032 mile orbit. With an apogee twice as high as expected, the recovery plan had to be modified since re-entry on the seventeenth orbit would bring the capsule down too far west of the designated recovery zone. Re-entry was instead to be initiated on the fifteenth orbit so that the capsule would impact only slightly southwest of the original landing area. Despite the last-minute changes, separation and retro-fire were normal, but telemetry indicated that the retro-rocket was separated late. The C-119 recovery aircraft could only pick up signals from the capsule beacon for two minutes rather than the standard 20 or 30 minutes. Apparently the capsule's parachute never opened, resulting in a rapid journey through the atmosphere before the capsule slammed into the ocean at 25,000 miles per hour and sank.[18]

In an effort to further improve the chances of success, modifications to the capsule continued. To make the capsule easier to see at night or in the water, a brilliant strobe light was added. Also added were packages of chaff, which would be released during descent to provide radar targets.[19]

Consideration was given to suspending further flights. On December 15, Bissell met with George Kistiakowsky, Herbert York, the Defense Department's director of research and engineering, and a number of others to discuss the troubles with the program. They decided that it was necessary to continue flight tests of the entire system since no test conditions could be discovered on the ground to simulate the peculiar failures which occurred in orbit.[20]

Flights resumed in February with no more success. On February 4 1960, Discoverer IX saw the Thor shut down early, the Agena separate but end its burn phase prematurely: it never attained orbital velocity. As a result, it impacted 400 miles south of Vandenberg.[21]

On February 19, Discoverer X's launch saw the Thor go off course, and be destroyed 56 seconds after launch. The launch had been troubled from the start. The engine, mounted on gimbals, wobbled after it began to belch yellow fire, and as the missile rose a few hundred feet it trembled and wobbled. Then, instead of nosing over toward the south and its intended polar orbit, it veered northeast toward the seacoast towns of Santa Maria and San Luis Obispo. Fifty-six seconds after launch, at 20,000 feet, it was destroyed by the range safety officer. A huge fireball flared in the sky and was blotted out by a mushroom cloud of black smoke from which sailed chunks of the missile the size of automobiles. The base's loudspeaker ordered personnel to take cover, and some people scrambled under tables, trucks and automobiles. Others just stood tense, however, looking up and ready to dodge if a piece of debris headed their way.[22]

Two months later, during the Discoverer XI mission of April 15, 1960, it appeared that success was at hand. The launch, first stage Thor performance, separation, Agena ignition, and orbital insertion were excellent as was the resulting orbit of 109.5 miles by 380 miles. The satellite was acquired by every tracking station on every pass with the main batteries lasting till the 26th orbit. The capsule was ejected on the seventeenth orbit as planned, but unfortunately went into a high re-entry trajectory which prevented recovery.[23]

The failure of the Discoverer XI mission triggered an intense testing program of the recovery systems, with the timing of the next launch depending upon the study's outcome. Examination of the earlier attempts revealed that when separated from the Agena, the capsule would tumble or wobble due to failure of the spin rockets. To correct this, a system using compressed nitrogen was fitted.[24]

Beginning in June, drop tests of the new system were made at Holloman Air Force Base in New Mexico. After a dummy capsule was carried to 100,000 feet by a Skyhook balloon, it was released and went through retro-fire, spin rocket operation, and finally parachute deployment. An instrument package reported the results. Wind tunnel tests at the Arnold Engineering Center in Tennessee indicated that retro-rocket performance became erratic due to the extreme cold of space.[25]

Based on these test results, several modifications were made in Discoverer XII. The satellite's capsule was fitted with a special telemetry package and the deployment of tracking and recovery forces was altered. A receiving station was set up on Christmas Island to receive signals if the capsule overshot the recovery zone, and equipment was added to the Alaska and Hawaii stations. Five C-54 aircraft and three ships would provide added coverage during the capsule's return.[26]

On June 29, Discoverer XII was launched from Pad Four of Complex 75–3 of Vandenberg Air Force Base after only minor mechanical holds caused by ground support equipment problems and a major hold because of the weather. The flight was to emphasize monitoring of the sequence of events leading to re-entry in order to diagnose the problems which had blocked earlier recovery.[27]

During the Agena's burn, radio-frequency interference from the telemetry caused a horizon scanner to fail, which made the Agena pitch down. The engine burn went as planned, but the incorrect angle caused the satellite to re-enter the atmosphere and burn up.[28]

Like the previous failures, this failure was frustrating because it left the program managers with little to go on to correct the problems. Bissell recalled that

> . . . one after another was a failure. It was a most heartbreak-
> ing business. If an airplane goes on a test flight and something
> malfunctions, and it gets back, the pilot can tell you about

the malfunction, or you can look it over and find out. But in the case of a recce satellite you fire the damn thing off and you've got some telemetry, and you never get it back. There is no pilot, of course, and you have to infer from telemetry what went wrong. Then you make a fix, and if it fails again you know you've inferred wrong. In the case of CORONA, it went on and on.[29]

Finally, on August 10, the launch of Discoverer XIII brought the program a successful recovery. Col. Mathison monitored the launch from the Sunnyvale control center (known as "Moose's Mansion"). According to the Air Force, Discoverer XIII contained no sensor equipment, with all its circuitry being designed to provide a detailed readout of every step in the orbital and re-entry flight sequences. The Air Force was not quite telling the truth, however. There was no camera or film on board, but the satellite did contain a device known as SCOTOP, which registered radar 'hits' so that the CIA could determine if the satellite was being tracked by Soviet radars.[30]

Discoverer XIII was launched at 1:38 p.m. Pacific Daylight Time through a fog bank into a 155 by 429 mile orbit. The Agena A then moved into a tail-first position. Once the satellite's orbit was confirmed, Col. Mathison boarded a C-130 transport for the flight to Hickam Air Force Base, Hawaii, where he monitored the deployment of the recovery fleet for the capsule's expected re-entry the next day.[31]

By noon on August 11, the recovery force had been assembled to cover a zone 60 miles wide and 200 miles long. The force consisted of 20 aircraft, including the C-119 recovery aircraft, EC-121 aircraft, and two U-2s. Also on call were three ships in case sea recovery was necessary. Twenty-four hours and 37 minutes after the launch, on its seventeenth orbit, Discoverer XIII's programmer automatically began the recovery sequence. Atypically, everything went smoothly: the Agena pitched down, the explosive bolts fired, and springs pushed the capsule free. Fifteen minutes later the parachute deployed.[32]

The ground network began receiving signals that set the recovery process in motion. In Alaska, the Kodiak station received signals indicating that the recovery capsule had spun up and the retro-rocket had fired. The capsule then stopped its spin and the retro-rocket was jettisoned. Within a few minutes, the Hawaii tracking station received signals which indicated that the heat shield had been jettisoned and the parachute had opened. Recovery aircraft and ships picked up the capsule's beacon. Although heavy cloud cover prevented aircraft recovery, the capsule was spotted as it splashed down 300 miles northwest of Oahu. Col. Mathison flew out to the capsule on a seaplane, but high waves made it impossible to land safely.[33]

While two C-119s circled the floating capsule, a call went out to the USNS Haiti Victory to send out a helicopter. From the time it received the call, the Haiti Victory steamed for an hour and ten minutes until it finally came within helicopter range. The helicopter arrived about three hours after splashdown. Lieutenant Albert C. Pospieil flew over the strobe light and yellow-green dye marker released by the capsule when it hit. Bosun's Mate, Third Class Robert W. Carroll, 22 years old and clad in bathing trunks, swim fins and a lifevest, jumped into the 12-foot high swells, swam to the capsule, put the parachute into a bag, and attached a line to the capsule. The capsule was then winched aboard the helicopter and Carroll followed; then the helicopter flew back to the Haiti Victory.[34]

Colonel Mathison had switched from seaplane to helicopter and flew out to the ship to pick up the capsule. With Col. Mathison as its escort, the capsule, with its recovery beacon still operating, was flown by helicopter to Hawaii and loaded aboard a C-130 for the flight back to the West Coast, but not without an interservice struggle. The ship's commander suggested that it might be risky to take the capsule on a chopper and that it would be safer to let the Haiti Victory transport it to Hawaii. Mathison, however, insisted on taking the capsule with him and did so. While legend has it that the disagreement between Mathison and the captain became quite serious, with Mathison unhooking the holster of his .45 to make his point, Mathison recalls the disagreement as being far less heated.[35]

Once back in the continental United States, the capsule was flown to Andrews Air Force Base, where the reception committee was headed by Air Force Chief of Staff Gen. Thomas D. White and Lt. Gen. Bernard Schriever, director of the Air Force Ballistic Missile Division. On Monday, it went to the White House, where General White presented President Eisenhower with a flag the capsule had carried. After several more stops, it was turned over to the Smithsonian Institution.[36]

Such publicity was not what Richard Bissell had had in mind, and Mathison's seizure of the capsule and its delivery to the Air Force brought more attention to the project than he considered desirable. In addition, Mathison did not know about SCOTOP. It had to be removed from the capsule while Mathison was otherwise occupied.[37]

While the Discoverer XIII capsule was touring Washington, the Air Force and CIA were readying Discoverer XIV for launch. Having proven that a spacecraft could be placed in orbit and stabilized, and that a capsule ejected from a spacecraft could be recovered, there was only one more element that had to be added to have an operational photographic reconnaissance satellite: they needed a camera that could take pictures from space. Discoverer XIV would carry a modified version of the lightweight HYAC camera used in the WS-461L balloon reconnaissance program. Its major target was the suspected ICBM complex at Plesetsk—the same main target as that of Francis Gary Powers on the last U-2 flight across the Soviet Union.[38]

Discoverer XIV was scheduled for launch on August 18, and if all went well it would be the first operational CORONA mission. Just before 1 P.M., with only a 15-minute delay caused by the passing of the empty Agena stage from Discoverer XIII through the projected flight path, the first fully equipped CORONA satellite climbed into the California sky. Its early afternoon launch time meant that photography of tall objects would show shadows—making detection and measurement easier.[39]

No sooner was Discoverer XIV in orbit than trouble signs appeared. While still in its first orbit, data received by the Kodiak station indicated that the satellite was at an incorrect altitude and using control fuel at a high rate. Fortunately, the satellite soon stabilized and was able to begin taking photographs from its 113 by 502 mile, 80 degree inclined orbit.* As it began to pass over Soviet territory, its camera, as programmed, began to operate. Its first photos were of Mys-Schmidta air base** in the Soviet Far East—about 400 miles from Nome, Alaska.[40]

A day after its launching, on its seventeenth orbit, the programmer began the recovery sequence and the 84-pound instrument capsule was ejected from Discoverer XIV as it sped over Alaska. A timing device

---

* The inclination of a satellite is the angle its path makes with the equator. Given the rotation of the earth under the satellite, an inclination of 80 degrees means that the satellite will eventually pass over all territory between 80 degrees north latitude and 80 degrees south latitude. The northernmost Soviet territory is found at approximately 75 degrees north latitude.

** Mys-Schmidta has become a traditional first stop for all new types of reconnaissance satellites. Today it is one of five Arctic Control Group staging bases.

triggered the operation of gas jets to pitch the vehicle 60 degrees down from its horizontal path. As with the previous Discoverers, a series of explosive bolts and springs were fired to separate the capsule from the main satellite body, followed by the firing of a retro-rocket within the capsule to slow the vehicle to re-entry velocity. This allowed it to assume the proper trajectory for re-entry. Before it had descended to 60,000 feet, a switch was activated by deceleration forces that released a parachute.[41]

As usual, ground stations, aircraft, and ships were all involved in the recovery effort. Also included were a radar-equipped Lockheed RC-121 for airborne control of the recovery aircraft, two Douglas JC-54s, and two Liberty ships from the Military Sea Transport Service to gather telemetry from the descending capsule. Four ground stations tracked the capsule on its passage through the atmosphere and provided initial guidance to the recovery aircraft.[42]

Within the 200 by 60 mile rectangular recovery area, six C-119s and a C-130 patrolled. Three other C-119s, including the actual recovery aircraft, patrolled an "outfield" area embracing an additional 400 miles. The aircraft, operating at altitudes of up to 16,000 feet, flew a maximum rescue search pattern, making 90 degree turns at approximately 10-minute intervals while maintaining a specified separation from each other.[43]

It was not long before the first stable signal was received from the capsule's telemetry beacon. The pilot of Pelican 9, Captain Harold E. Mitchell, 35, of Bloomington, Indiana, began his first 360 degree orientation turn. Within three minutes he began a second 360 degree turn to obtain positive identification, and ten minutes later he sighted the capsule and parachute at approximately 16,000 feet. With the capsule falling at about 1500 feet per minute, Mitchell had about 10 minutes to snatch the capsule before it fell below the aircraft's minimum operational altitude. As Mitchell lined up on the parachute, the winch operator, Technical Sergeant Louis F. Bannick, told him, "For God's sake, Captain, don't hurt it." He missed on the first pass by only six inches, and his second attempt at about 10,000 feet missed by two to three feet. With two decks of scattered clouds below at 7000 feet and at 2000 feet, Mitchell banked steeply at 8500 feet. On the third pass the trapeze-like hook dangling from the plane's belly hooked the capsule's parachute—they were about 200 miles south of the predicted impact zone.[44]

The parachute and capsule were reeled aboard and placed into a container. In the aircraft there was elation: two years of practice had paid off. When the crew landed at Hickam Air Force Base, they were met by General Emmett O'Donnell, Jr., 200 Air Force personnel, and their wives. Captain Mitchell was awarded a Distinguished Flying Cross and the other nine crew members received air medals.[45]

The capsule was sent immediately to the West Coast. In December 1960, the Air Force Museum received what they were told was the Discoverer XIV capsule although it appears the capsule had actually been destroyed by Air Force officers soon after its arrival on the West Coast. High officials, concerned by the publicity generated by the Discoverer XIII capsule, ordered that the capsule be kept under lock and key. By the time the oral orders made their way down the chain of command, however, it was believed that the instructions demanded that the capsule be destroyed. The officers spent several hours pounding the capsule into oblivion with hammers, then loaded it into a helicopter and dropped it into the Santa Barbara Channel. In November 1963, the real C-119 joined the "Discoverer XIV" capsule in the collection.[46]

In addition to photographing Mys-Schmidta, Discoverer XIV's cameras had taken pictures of the suspected ICBM base at Plesetsk. Naturally, the photos were quickly dispatched to the CIA's Photographic Interpretation Center (PIC). The briefing of the photo-interpreters at PIC prior to their beginning examination of the photos represents one of the most memorable moments of the KEYHOLE program. The routine that had been followed after U-2 missions involved a nighttime meeting in the PIC auditorium once the film had arrived. The assembled photo-interpreters were briefed by CIA representatives about the targets of importance and were shown a map of the Soviet Union with a squiggle on it indicating the route of the U-2. Blow-ups of particular target areas would also be shown.[47]

There was a dramatic flair to the CORONA/Discoverer XIV briefing. After the photo-interpreters filled the auditorium, PIC Director Arthur Lundahl announced that it was "something new and great we've got here." Jack Gardner, who was Lundahl's deputy, opened a curtain that showed a map of the Soviet Union. Instead of a single squiggly line across the map, there were six or seven vertical stripes emanating from the poles and moving diagonally across the Soviet Union. The interpreters knew that those stripes represented the portions of the Soviet Union that had passed under Discoverer XIV's camera and their immediate reaction was to cheer. After being briefed on what to look for, especially for missile sites at Plesetsk, they began work on OAK-8001, the first photo-interpretation report based on satellite photography.[48] The photos were dark and of poor quality, with a resolution in the area of 50 to 100 feet, but to Andrew Goodpaster they were "like the dog that walks on its hind legs, remarkable that it happens at all."[49]

The success of the Discoverer XIII and XIV missions was followed by significant organizational changes in the reconnaissance program. The impetus for such changes stemmed from the problems that had plagued the CORONA program, the difficulties with the Air Force's radio-relay program, which had been assigned the new codename SAMOS,* and high-level concern about management arrangements for the entire reconnaissance program.

On February 5, 1960, the day after the Discoverer IX failure, George Kistiakowsky had met for a half hour with Gordon Gray, Eisenhower's national security advisor, to discuss the state of the SAMOS program. Kistiakowsky told Gray that the Air Force plan to develop a system with instantaneous transmission of information by means of film readout was one which the technical people didn't feel would be effective for many years. As a result Kistiakowsky argued that the emphasis should be placed on the film-return program.[50]

Over three months later, on May 26, Kistiakowsky met with Gordon Gray, General Goodpaster, and the President, and began what he intended to be a detailed account of the problems plaguing both the Discoverer and SAMOS programs. The problems had taken on additional urgency due to the shooting down of Francis Gary Powers and his U-2 on May 1. The loss of the plane and capture of its pilot by the Soviets meant that the United States had to cease overflights of the Soviet Union, depriving the CIA of its means of obtaining photographic evidence of developments in the Soviet interior. Kistiakowsky was cut short, however, for Eisenhower needed only a fraction of the information Kistiakowsky had prepared to firmly decide that corrective action was required. He instructed Goodpaster to prepare a directive for a special ad hoc group that would study the issues involved and advise him on possible remedies.[51]

---

* It has generally been reported that SAMOS was an acronym for Satellite and Missile Observation System. In fact, the name SAMOS was chosen by 117L Project Director Col. Fritz Oder in the belief that no one could produce an acronym from it. Noting that MIDAS had already been selected as the name for the early warning satellite he noted that MIDAS lived on the isle of SAMOS. It was not long before the press started reporting SAMOS as an acronym.[52]

Kistiakowsky suggested a group under Defense Secretary Thomas Gates while Eisenhower suggested Kistiakowsky for the job. Although he did not resolve the issue immediately, Eisenhower did tell Goodpaster to "tell the Defense Department that I won't approve of any money for the projects until I have the information [on the status of the programs]."[53]

Two weeks later, on June 10, Goodpaster sent Gordon Gray copies of the President's directives to Gates and Kistiakowsky for the study of reconnaissance satellite projects.* The directives specified no target date for their completion, and Goodpaster suggested that Gray work out a target date with those involved in the study as soon as they had an initial evaluation of the scope of the studies.[54]

Gates appointed a team of three individuals: Under Secretary of the Air Force Joseph Charyk, Deputy Director of Defense Research and Engineering John H. Rubel, and George Kistiakowsky.[55] The study group made many in the Pentagon nervous. On August 1, 1960, Kistiakowsky received phone calls from Air Force Under Secretary Joseph Charyk, Aerospace Corporation President Ivan Getting, and others as the result of a rumor spreading through the Pentagon. The rumor reported that the reconnaissance study group was going to recommend that responsibility of the management of SAMOS should be transferred to the CIA. Kistiakowsky assured everyone that the rumor was false, but urged Charyk "that the organization should have a clear line of authority and that on the top level the direction be of a national character, including OSD [Office of the Secretary of Defense] and CIA and not the Air Force alone." Kistiakowsky confided to his diary that "obviously, the Air Force is trying to freeze the organization so as to make a change more difficult by the time the NSC is briefed."[56]

On August 25, between 8:15 and 8:30 A.M., Kistiakowsky met in the President's office with Dulles, Killian, Land, and Gray, to show Eisenhower certain intelligence information and to remind him of certain specifics related to SAMOS. Those specifics probably related to planned

---

* The state of the SAMOS program was also then the subject of public concern. On June 12, the Senate Appropriations Committee issued what the *New York Herald Tribune* termed an "unusual report." The Committee said it was "a matter of national emergency to move forward as rapidly as possible with the reconnaissance satellite program." The committee proposed to increase the fiscal year budget for SAMOS by $83.8 million more than requested by Eisenhower. During a television appearance on the same day, Senator Lyndon Johnson proposed a "crash program" to develop a reconnaissance satellite system.[57]

operations, which they were not planning to discuss at the following briefing, but which did influence their recommendations. The meeting began with Edwin Land rolling a duplicate spool of Discoverer XIV across the floor of the Oval Office and announcing "Here's your pictures, Mr. President."[58]

The special NSC meeting took place from 8:30 to 9:30 A.M. to discuss capabilities, organization, and processing. Some humor was injected by Land when Bureau of the Budget Director Maurice Stans was temporarily called out and Land remarked that Mr. Stans was leaving while it was still cheap. When the group recommended that the line of command be directed from the Secretary of the Air Force to the officers in charge of the project, Eisenhower remarked that that was the correct way to do it. At the end Eisenhower said that he approved all the recommendations and that the only unfortunate thing was that they didn't make the recommendations two years ago, so that he could see "the lovely pictures they would be making." Instead, Eisenhower continued, it would probably be Nixon, and he wouldn't even have the clearances necessary to see them. After the meeting, Nixon complimented the group on the presentation and recommendations.[59]

The result of the briefing, according to an official Air Force history, was "a key decision by the NSC and the President which, eliminating previous uncertainties, signaled the start of a highest priority project reminiscent of the wartime Manhattan effort."[60]

That key decision was the creation of the National Reconnaissance Office (NRO), a national-level organization to be responsive to supradepartmental authority. As noted above, the national-level character of the organization was a major point of importance to those involved in its formation. This was a reflection not only of Kistiakowsky's concern but of the President's, who "wanted to make damn sure" that the arrangement did not result in Air Force control. One reason why such a framework was desired was to be certain that the utilization of the photographic "take" not be left solely in the hands of the Air Force. Such fears were not without foundation. In response to OSD requests, the Air Staff prepared two plans for the operation of SAMOS and both called for the Strategic Air Command to command and operate SAMOS.[61]

Although NRO was created as a national level organization with Air Force, CIA, and Navy participation, the Air Force role was substantial, because it provided the organization with a director (Joseph Charyk) and support staff. Charyk's appointment, while he maintained his position as Under Secretary of the Air Force, resulted in part from his experience in the development of high-altitude reconnaissance planes. Charyk had suggested to Richard Bissell that Bissell himself would be the appropriate choice to head NRO, and Bissell, who saw the organization more as a central repository and coordinating unit than as an operational agency, was interested in becoming NRO director while remaining as the CIA's Deputy Director of Plans. However, Director of Central Intelligence Allen Dulles told Bissell that he could not have a CIA official take "line control" over DOD personnel. The CIA official would be a potential scapegoat for any foul-up or disaster such as the U-2 incident, and offering the President or National Security Council the head of a CIA official in response to some failure might just be too tempting for a Defense Secretary. Hence, Charyk became the first in a series of Air Force officials who wore the "black hat."[62]

To provide cover for NRO, an Office of Missile and Satellite Systems within the Office of the Secretary of the Air Force was established on August 31,1960. Its primary responsibility was "assisting the Secretary in discharging his responsibility for the direction, supervision and control of the SAMOS Project." Brigadier General Robert E. Greer was named Director of the SAMOS Project, with offices at the Air Force Ballistic Missile Division, El Segundo, California. The SAMOS Project Office was designated a field extension of the Office of the Secretary of the Air Force, and its director was made directly responsible to the Secretary.[63]

Air Force Secretary Dudley Sharp also established two advisory bodies. The first, the Satellite Reconnaissance Advisory Group, consisted of a standing committee of four leaders in the fields of electronics, photography, and data handling. The committee would be augmented as occasion demanded by an assembly of technical experts who would consider specific matters and make recommendations. The second group, the Satellite Reconnaissance Advisory Council, would include Sharp's top civilian aides; the newly created Director of the Office of Missile and Satellite Systems; and three Air Staff members: the Vice Chief of Staff, the Deputy Chief of Staff for Development, and the Assistant Chief of Staff for Intelligence. The role of the council was "to provide assistance, advice, and recommendations as required."[64]

On September 13, Secretary Sharp informed the Chief of Staff that "no intermediate review or approval channels would exist between the

SAMOS field office and USAF." This was re-emphasized in discussion among Andrew Goodpaster, Air Force General R. D. Curtin, and the Secretary of Defense, who stated that the command and control line for SAMOS was not to involve the Strategic Air Command but was to be direct from the Secretary of the Air Force to field headquarters. In addition, briefings would be given on a strict need-to-know basis to Air Staff and other USAF representatives as required for SAMOS support purposes or in the coordination of related matters.[65]

The remainder of 1960 saw continued launchings of Discoverer/-CORONA satellites, the first attempted SAMOS launch, and another propaganda blast from the Soviet Union. Discoverers XV–XIX apparently added little intelligence to that returned by Discoverer XIV. The Discoverer XV launch on September 13 began without incident—both launch and injection into orbit took place without problems—but, after the capsule was separated on the seventeenth orbit, it began to use up its control fuel at too high a rate. The result was that although the capsule made a successful reentry and splashdown, it landed about 900 miles south of the recovery zone. Search planes and a recovery ship picked up its radio signal as they headed towards it, and two recovery aircraft spotted the capsule's flashing light. Unfortunately, the impact area was swept by heavy rain squalls and high seas, preventing recovery.[66]

Discoverer XVI represented the first launch employing the Agena B upper stage, which at 26.6 feet was seven feet longer than Agena A. The additional fuel carried by the Agena B increased the engine's burn time and payload. The October 26 launch ended as a failure, however, when the new upper stage failed to separate from the Thorad booster and both fell into the Pacific.[67]

Discoverer XVII was successfully launched into a 118 by 615 mile orbit on November 12, 1960. Reports suggested that it carried test equipment for the MIDAS and Navy navigation satellite programs. After it made 13 orbits, the Air Force announced the flight would be extended a day and on November 14, while it passed over Alaska on its thirty-first orbit, a radio command from the Kodiak station began the separation process. Ten minutes later, the capsule was spotted by the crew of the Pelican 2 as it descended at 30,000 feet in its gold and orange parachute. On his first pass, the pilot of Pelican 2, Captain Gene W. Jones of Walla Walla, just missed the capsule at 11,000 feet and caught it on his second try at 9500 feet.[68]

The Discoverer XVII launch also represented the first successful use of the Agena B second stage. The Agena B had a "dual burn" capability, a rocket that could be shut off and then reignited. Though announcing that there were no plans to put that capability to immediate use, Maj. General Ritland, deputy CORONA program manager, noted that in future launches "with more sophisticated systems, it will permit us to control a satellite's course to a much greater degree—and even to change its orbit."[69]

The Discoverer XVIII launch of December 7 represented the last CORONA launch of the year. The launch was notable on several counts. An improved Thor booster with 10 percent more thrust was used and the satellite remained in orbit nearly three days, with separation on the forty-eighth orbit. Once the capsule was spotted at 25,000 feet, Pelican 3 turned toward it. The pilot, again Captain Gene Jones, lined up on the parachute and caught it on the first try, the first time only one try was necessary.[70]

While the Discoverer program continued, so did the Soviet propaganda campaign against the satellites. Dr. G. P. Zhukov, academic secretary of the Space Law Scientific Research Committee of the Soviet Academy of Sciences, published a paper in the Soviet journal *International Affairs*, which discussed the SAMOS, Discoverer, and MIDAS programs. In the paper, "Space Espionage Plans and International Law," he argued that "the main purpose of space espionage is to increase the efficiency of surprise attack, making it possible to knock out enemy missile bases at the very start and thereby avoid a retaliatory blow" and that "Claims by Pentagon leaders and other U.S. officials that space espionage is needed to prevent a so-called surprise attack by establishing the location of Soviet missile bases are absolutely untenable and are designed to justify the long discredited brinkmanship policy." Zhukov went on to note that the "American plans of space espionage directed against the security of the USSR and the other socialist countries are incompatible with the generally recognized principles and rules of international law, designed to protect the security of states against encroachments from outside, including outer space" and that "In case of need, the Soviet Union will be able to protect its security against any encroachments from outer space as successfully as it has done with respect to airspace."[71]

Zhukov also quoted Khrushchev as saying that "information about the location of such bases can be of importance not for a country concerned with its defense requirements, but solely for a state which contemplates aggression and intends to strike the first blow and destroy the missile bases so as to avoid retaliation after attack." He ingenuously ignored the fact that a nation concerned about being attacked needed to determine the nature of a potential enemy's offensive arsenal. The author also noted Khrushchev's statement that "the Soviet Union has everything necessary to paralyze U.S. military espionage, both in the air and outer space. If other espionage methods [besides the U-2s] are used, they will also be paralyzed and rebuffed."[72]

Such Soviet statements were of great concern to U.S. authorities. A more immediate concern, however, was the state of the SAMOS program, for the creation of NRO had not produced any miraculous solutions to the problems plaguing that program.

On September 29, Kistiakowsky spoke to William Baker, the President of Bell Labs and an influential scientific and intelligence adviser to the President, about General Greer's proposed plan for developing SAMOS satellites. In September Greer and his staff had briefed Kistiakowsky on their plan, a plan which shocked Kistiakowsky because it involved development of at least nine different satellite payloads. According to Kistiakowsky, it "looked like an appalling proliferation and is certainly different from the straightforward plan recommended by our panel." Greer told Kistiakowsky that he had briefed Charyk on his plan the day before and Charyk had approved in general. But Charyk had left for the Far East and couldn't confirm Greer's claim.[73]

The previous weekend Kistiakowsky had traveled to Cambridge and had a long session with Edwin Land in which Land also expressed misgivings about the plan. Baker's reaction was even stronger, so Kistiakowsky took Baker to talk to Goodpaster. Goodpaster decided to telephone Air Force Secretary Joseph Douglas and warn him about Greer's plan. In addition to the number of payloads, the plan also involved some programs "in the black"—programs whose existence was secret. Eisenhower considered such programs improper for the military, which he considered incapable of keeping developments secret because of an irresistible urge for publicity.[74]

On October 3, 1960, Kistiakowsky went to see Secretary Douglas and John Rubel about General Greer's plans for SAMOS. Douglas was not very explicit, but conceded that Greer may have been spreading out too much and agreed to hold off endorsement of the program until the return of Charyk, as Kistiakowsky requested.[75]

Kistiakowsky won his battle to keep SAMOS simple, or as simple as a film-scanning satellite could be. But before that issue was fully resolved, SAMOS 1, which had originally been scheduled for an April 1960 launch, lifted off from Port Arguello, California on October 11, 1960. However, the umbilical cord that attached the Agena upper stage of the Atlas Agena booster to the launch tower failed to separate until after the missile became airborne, pulling off part of the Agena second stage and causing it to malfunction. As a result the spacecraft did not make it into orbit.[76]

In 1961 a new administration came into power. The leader of that administration, John F. Kennedy, had made his charge of Republican weakness in defense matters, and specifically a missile gap, a prime aspect of his campaign. The information produced by space reconnaissance was to have a great impact on the administration's perception of the missile gap, and the administration was to have a profound effect on space reconnaissance activities.

Prior to the first SAMOS launch, the DOD issued a fact sheet that acknowledged that the instrument package included test photographic and photo-related equipment. The State and Defense Departments agreed, as they explained in a memorandum to the Assistant Secretary of Defense (International Security Affairs), that the "information should be supplied in response to the inevitable queries by newsmen, many of whom have stated without official authority, but on the basis of information acquired unofficially, that SAMOS will contain photographic equipment."[77]

SAMOS 2, launched on January 31, 1961, orbited at an inclination of 97.4 degrees and with a period (the time necessary to circle the earth) of 95 minutes.[78] Launched from the Naval Missile Facility at Point Arguello, California, the length of SAMOS 2 was 22 feet, with a diameter of five feet and a weight at launch time of 11,000 pounds. Approximately seven thousand pounds of this constituted the fuel supply. Its perigee was 248 miles, and its apogee 296 miles.[79]

The SAMOS 2 launch marked the beginning of SAMOS' "going black." The Assistant Secretary of Defense for Public Affairs, Arthur Sylvester, had noted in a memorandum for President Kennedy shortly before the SAMOS 2 launch that the information released "represents a severe reduction from what had previously been issued. Eliminated entirely from former procedures are four pages comprising 22 questions and answers. Press briefings before and after launching have been eliminated. We have reduced the advance notice to correspondents from five to one day. New official photographs will not be volunteered as has been done heretofore."[80] Thus, the system that President Eisenhower had declared should contain no black components was under President Kennedy going black in its entirety.

NRO's role in the limitation of information was evident in the memo's statement that "Dr. Charyk has reviewed these changes and is satisfied that they meet all his security requirements and those of his SAMOS Project Director, Brigadier General Greer. Dr. Charyk agreed to make public this amount of data in view of the great volume of previous news stories which have already been written about SAMOS."[81]

The memo had noted that the reduction of information was also Kennedy's preference. Fears expressed by Killian and other Eisenhower assistants to the incoming administration as well as the findings of the *Report to the President-Elect of the Ad Hoc Committee on Space*, written under the supervision of MIT's Jerome Weisner, led to a rapid change in policy. Protection of U.S. reconnaissance satellites by terminating official publicity became a priority concern, particularly in light of Soviet statements on the "illegality" of such activities and increasingly credible threats to shoot such satellites down.[82]

Such threats were taken seriously given the Soviet downing of the U-2 less than a year before. Many of those responsible for the CORONA and SAMOS programs had been involved with the U-2 program and were therefore inclined to take such threats as more than mere propaganda. Further, the implications of the loss of the U-2 were clear to Kennedy. A major means of collecting intelligence about the Soviet Union had been lost and the international standing of the United States was damaged, as was Eisenhower's personal prestige and image. The domestic political costs were also significant, and so in January 1961, McGeorge Bundy, the Assistant to the President for National Security Affairs, and Defense Secretary Robert McNamara initiated a review of public information policies concerning SAMOS launches.[83] The results were immediate and eventually drastic.

Restrictions barred military officers, particularly Air Force officers, from mentioning the SAMOS program by name or mission. They were further prevented from making public statements dealing with the subject without obtaining prior approval. General Bernard Schriever, who had previously discussed the SAMOS program in congressional testimony and public forums, removed all references to SAMOS in public statements. Nor would Major General Ritland announce, as he did before SAMOS 1, that a camera-carrying satellite launched from Vandenberg AFB into polar orbit would provide maximum coverage of the Soviet Union or that "Through SAMOS . . . we can hurdle the Iron Curtain and peer down on our planet from the watchtower heights of space."[84]

The new secrecy meant oblivion for the Air Force film *To Catch a Falling Star*. The film, which had been made on an unclassified basis for public release, was a documentary concerning the effort to recover Discoverer capsules from space. Four hundred ninety-seven prints of the film were made and sent to the Air Force Media Depository to await public release, but in light of the new secrecy restrictions the film was withdrawn overnight. In addition, public information officers were instructed by their superiors that if asked for information about the film "you know of no such motion picture" and to avoid further discussion. Although a subsequent film was made on the recovery efforts, it was a classified effort for training purposes.[85]

As the SAMOS program progressed, the secrecy option was exercised with increasing vigor. SAMOS 3 and SAMOS 4, launched on September 9, 1961 and November 22, 1961, failed to orbit. The curtain of secrecy descended further when SAMOS 4 was launched without any press release or publicity. After the launch of SAMOS 5 exactly one month later, officials would no longer admit the existence of the SAMOS project.[86]

While the policy of complete secrecy was strongly supported by the highest officials in the White House (the President and McGeorge Bundy), the Department of Defense (McNamara), the NRO (Charyk), the CIA (Bissell), and subsequently the new DCI (John McCone), it was not a policy without critics both in and out of government. The trade journal *Missiles and Rockets* noted that the Soviet Union could track reconnaissance satellites and determine their orbital parameters. One governmental critic, Solis Horowitz, then the Director of Organizational and Management Planning for the State Department, noted that the launches "could be observed by our press and the Russians. We knew the Russians were observing them and could observe them. . . . We were not fooling anybody except our own people."[87]

The main source of opposition within the Executive Branch resided in the State Department, where Dean Rusk, Adlai Stevenson, and others argued that the blackout could produce the very outcome it was intended to avoid. In the State Department's view, the United States was in a politically vulnerable position legally with regard to satellite reconnaissance activities—as it had been with the U-2—and needed to establish the legitimacy of overhead satellite reconnaissance. This could best be done, they argued, by a policy of openness and by smoothing the distinction between military and civilian space programs. As a result, the State Department did not base its argument on Soviet tracking capabilities, a point conceded by the advocates of secrecy but considered irrelevant. Rather, it shared with the advocates of secrecy a diplomatic concern— an interpretation of the cessation of the U-2 program as being, in large part, the result of public and diplomatic pressure. However, the proposed solution was completely different.

The State Department view was not persuasive to Kennedy or the intelligence and defense establishments. The Soviets would be less inclined to interfere with U.S. reconnaissance satellites, Kennedy believed, if the U.S. ability to penetrate their Iron Curtain was not flaunted before the world. Given the Soviet belief in secrecy as a prime requirement of national security and the role it played in the justification for the way in which the Soviet Union was governed—depriving the common citizen of even the most rudimentary knowledge of Soviet military forces— parading the U.S. capability to photograph any military facility it chose could motivate the Soviet leadership to take extreme measures to eliminate that capability. Furthermore, U.S. statements concerning satellite reconnaissance activities might well be taken as a political challenge by the Soviets and further motivate a response on their part.[88]

Others, including the three armed services and assorted sympathetic congressmen and senators, believed that there must be a military component to the protection of U.S. space assets, especially reconnaissance satellites. Just as the U.S. required a nuclear retaliatory force that could devastate the Soviet landmass in response to a Soviet attack on the United States, so, its advocates argued, did the United States require an anti-satellite (ASAT) capability that could devastate Soviet space platforms in retaliation for any Soviet attack on U.S. space assets.[89]

That the Soviets would attempt such actions was made all the more plausible to the advocates of ASAT because they considered it reasonable for the United States to attack any Soviet satellites passing over the United States. According to General James Gavin in his 1959 book, *War and Peace in the Space Age*: "It is inconceivable to me that we would indefinitely tolerate Soviet reconnaissance of the United States without protection, for clearly such reconnaissance has an association with an ICBM program. . . . It is necessary, therefore, and I believe urgently necessary, that we acquire at least a capability of denying Soviet overflight—that we develop a satellite interceptor."[90]

Air Force General Thomas White, testifying earlier that year before a Congressional appropriations committee, was even more explicit. "The Soviets could be expected to respond to our reconnaissance satellites as we would respond to theirs, and if the enemy develops a reconnaissance space vehicle we probably would want to take him out of space if we could," he explained more than a year before the first successful Discoverer launch. With the development of functioning reconnaissance satellites, General White added, "there immediately becomes a requirement to intercept it and to nullify it."[91]

By 1961, work on development of an ASAT had begun. On August 25, 1960, the same day that he helped brief President Eisenhower on the need to establish a National Reconnaissance Office, Deputy Director of Defense Research and Engineering John Rubel gave his final approval for Project SAINT to proceed. SAINT, for Satellite Inspector, was to be a satellite payload containing at least a TV camera and radar sensors, although later models might also employ infra-red, X-ray, and radiation sensors.[92]

The SAINT vehicle was to be placed into orbit slightly ahead of a target satellite and then employ its own propulsion unit to maneuver close to the target for inspection purposes. Planned demonstrations called for SAINT vehicles to move to within 50 feet of their targets, reported to be inflatable spheres. Many within the Air Force hoped that later models would be equipped with a kill mechanism.[93]

As SAMOS began to publicly vanish, CORONA continued to operate as a classified program, although it was obvious to anyone interested that many Discoverer launches carried photographic equipment. The Discoverer launches of early 1961 did little to add to the quantity of intelligence held by the U.S. intelligence community. Discoverer XX, the largest and heaviest satellite in the Discoverer series yet, was successfully launched on February 17, 1961, but failed to stabilize in orbit. Discoverer XXI carried test equipment for MIDAS. Discoverer XXII, launched on March 30 into a gray and overcast sky, did not reach orbit because of the failure of the Agena. Discoverer XXIII of April 8 reached orbit but then began to tumble. Recovery was attempted after two days, but the capsule went into higher orbit. Discoverer XXIV of June 8 did not achieve orbit: the Agena malfunctioned and impacted at sea 1000 miles downrange.[94]

From mid-June 1961 (with the launch of Discoverer XXV on the sixteenth) through the end of the year, good photographs were recovered from several satellites. Discoverer XXV was placed into a 139 by 251 mile orbit on June 18 and after 33 orbits, the capsule was separated from the satellite. The capsule re-entered successfully but landed north of the recovery zone, where it was spotted by a C-119 who then radioed for a team of divers. Three frogmen parachuted into the ocean and secured the capsule on a twenty-man life raft. An amphibious plane dispatched to the capsule turned back when darkness overtook it. The frogmen waited out the night until they could be picked up by a destroyer.[95]

Discoverer XXVI, launched July 7, completed 32 orbits before the capsule was commanded by the Kodiak ground station to re-enter. It was caught 29 minutes later on the first try by a C-119 flown by Captain Jack R. Wilson, one of eight C-119s sent out from Hickam AFB in Hawaii to attempt a catch.[96]

Discoverer XXIX was launched into a 140 by 345 mile orbit on August 30, with a mission to photograph Plesetsk among other targets. After 33 orbits, the capsule was returned but it re-entered outside the recovery zone. It took three divers 40 minutes before they had the capsule secured in a raft.[97]

The pictures from Discoverer XXIX provided enough detail of the Plesetsk site to confirm it as the first Soviet ICBM site and to establish what a Soviet ICBM site looked like. The photographs showed the SS-6 in a launching site identical to the configuration for the missile that existed at Tyuratam.[98]

Discoverer XXX was successfully orbited on September 12 and its capsule brought down after 33 trips around the earth. Eighteen orbits after its launch on October 13, Discoverer XXXII's capsule was ejected and caught by a JC-130 aircraft. JC-130s had by that time replaced the the C-119s. The Air Force would not discuss the payload, beyond stating that it was planned "to assist engineers in design of advanced space vehicles to perform sophisticated tasks in space." Finally, two more capsules were recovered in 1961—from Discoverer XXXV of November 15 and Discoverer XXXVI of December 12.[99]

The impact of the CORONA photography was devastating. For several years the National Intelligence Estimates (NIE) had predicted quite sizable Soviet ICBM deployments by the early 1960s, deployments that threatened the security of the United States. By the time NIE 11-4-57 was published in November 1957, the Soviets had already tested an ICBM and might have tested about 10 prototype ICBMs available for operational use in 1959 or possibly earlier, depending upon Soviet requirements for accuracy and reliability.[100]

NIE 11-4-57 estimated that the Soviets could have 500 operational ICBMs before the end of 1962, or on a crash basis, by the end of 1961. It was also surmised that the Soviets could probably produce about 20 nuclear powered submarines by mid-1962 and have a total of 50 submarines equipped with guided missiles.[101] A little over a year later, in NIE 11-4-58 of December 23, 1958, the intelligence community believed that the Soviets intended to acquire a sizable ICBM operational capability at the earliest practicable date. They also pointed out the absence of sufficient evidence to judge conclusively the magnitude and pace of the Soviet program to produce and deploy ICBMs. However, based on indirect evidence including production capacity and capacity to construct launch facilities, to establish logistics lines, and to train operational units, the intelligence community believed the Soviets "could achieve an operational capability with 500 ICBMs about three years after the first operational date [1959]."[102]

Over the next year the intelligence community revised its judgment downward in the absence of intelligence to sustain the earlier high estimates of the pace of Soviet ICBM deployments. NIE 11-4-59 estimated that the Soviets might have 140 to 200 ICBMs on launchers by mid-1961 and, speculatively, 250 to 350 by mid-1962 and 350 to 450 by mid-1963. Though smaller than earlier estimates, they left open the possibility of an effective Soviet missile attack destroying the vulnerable SAC bases, particularly since it was believed that improvements in the accuracy and reliability of Soviet ICBMs had sharply reduced the number required to effectively attack the U.S. target system.[103]

The 1960 NIE, NIE 11-4-60, reflected different estimates by various intelligence authorities. As might be expected, the Air Force took the most pessimistic view, predicting 200 ICBMs by mid-1961, 450 by mid-1962 and 700 by mid-1963; the CIA predicted 150, 270, and 400 for the same periods. At the low end were the Air Force's military competitors. The Army and Navy predicted Soviet ICBM deployments of 50, 125, and 200.[104]

Between the issuing of NIE 11-4-60 and NIE 11-4 of 1961, a new administration had taken office, one whose defense policy was based on eliminating the missile gap it had loudly and persistently proclaimed existed during the campaign. Among the first things that Robert McNamara did upon being sworn in as Secretary of Defense on January 20 was to go with his Deputy Secretary, Roswell Gilpatric, a former Under Secretary of the Air Force and a true believer in the missile gap, to the Air Force intelligence office on the fourth floor of the Pentagon.[105]

McNamara wanted to see if the Soviets had truly attained strategic superiority and if they had the extent of that superiority so the U.S. could take measures to close the gap. He and Gilpatric spent hours at a time, for several days over a period of three weeks, examining the CORONA photos. The images failed to support the estimates of Air Force intelligence. The Soviet ICBM, the SS-6, was monstrously huge, heavy, and cumbersome. It required an extensive support and security apparatus, and would have to be moved about on railroad tracks or extremely heavy roads. CORONA was photographing all along and around the railroad tracks and major highways of the Soviet Union, and finding no missiles.[106]

McNamara's conclusions were made public at a February news conference, which McNamara had been prodded into by Defense Department spokesman Arthur Sylvester. McNamara resisted Sylvester's initial suggestion, feeling he didn't yet know his way around and was not ready to meet the press. Yielding to Sylvester's persistence, he sat down in his office for what he thought would be an off-the-record conference. The first question he was hit with was whether there was a missile gap. His response—that "there were no signs of a Soviet crash effort to build ICBMs" and that no missile gap existed—set off a stampede of reporters to the phones. McNamara remembers that "all hell broke loose" and "you couldn't hold the door locked." That night the *Washington Evening Star* headline was "McNamara Says No Missile Gap." The next day Senator Everett Dirksen, possibly tongue-in-cheek, called on President Kennedy to resign because he had won his office on the basis of a fraudulent claim of a Republican-caused missile gap.[107]

Of course, the President did not resign and his administration continued to accumulate new evidence of just how small an ICBM force the Soviets had deployed. NIE 11-8-61 of June 7, 1961 argued that the Soviets might already have 50 to 100 operational ICBM launchers and, therefore, the ability to bring all SAC operational air bases under attack with their missile force. In any case, the estimate concluded that they would have 100 to 200 operational launches within the next year and would almost certainly be able to do so then.[108]

By September of 1961, in NIE 11-8/1-61 of September 21, 1961, the critical intelligence brought back by CORONA led to an official estimate of only 10 to 25 Soviet ICBMs deployed and no significant increase likely in the forthcoming months. The expected number of Soviet ICBMs by mid-1963 was 75 to 125. The Soviets had apparently chosen to deploy only a small number of first generation ICBMs and to concentrate their efforts on developing a smaller, second generation system for deployment, probably in 1962.[109]

Although CORONA dispelled the myth of the missile gap in 1961 and brought great relief to the administration, there were still reasons to be concerned over potential Soviet actions. In April the new President suffered through the humiliation of the Bay of Pigs fiasco, and a June meeting in Vienna between Kennedy and Khrushchev did not go particularly well. And even though Khrushchev backed off from threats that could have led to a major confrontation over Berlin, a future confrontation was not precluded.

But the September NIE indicated a way out; U.S. resolve to maintain its presence in Berlin could be backed up by its strategic forces. It was thus important to top U.S. officials to get the point across that the United States was strategically superior, knew it was superior, and could not be intimidated by empty bluffs. The vehicle for the warning was what would have otherwise been an obscure event—a speech by Deputy Secretary of Defense Gilpatric to the Business Council of Hot Springs, Va. The speech that Gilpatric gave actually began as a speech written by Daniel Ellsberg for President John Kennedy. The speech was approved by McNamara but rejected by Kennedy as too hawkish. When Ellsberg discovered that a speech was being written for Gilpatric with a similar theme he offered his speech to Defense Department official Tim Stanley, who was writing the speech for Gilpatric. The speech was passed to Gilpatric, who reviewed it in separate meetings with Secretary of State Dean Rusk, McGeorge Bundy, and Robert McNamara, with Rusk and Bundy making some changes. In his remarks Gilpatric noted that the United States "has a nuclear retaliatory force of such lethal power that an enemy move which brought it into play would be an act of self-destruction on his part." To make it clear that his statement was not an idle boast he also noted that although the Soviets used rigid secrecy as a military weapon, "their Iron Curtain is not so impenetrable as to force us to accept at face value the Kremlin boasts." Gilpatric's speech was promptly denounced by Khrushchev and his Defense Minister, Rodion Malinovsky, but they never again claimed superiority. And they stopped talking of concluding a treaty with East Germany before the end of the year.[110]

Several elements of the reconnaissance program fell into place by the end of 1961. While the SAMOS program had yet to produce results of value, the CORONA satellites were producing regular photography and two new programs had been established. The Army's continued efforts to find a place in the space reconnaissance program resulted in the development of the KH-5 mapping camera* for use in a mapping program known as ARGON.[111]

Another program, LANYARD, involved the redesignation of the basic E-5 SAMOS camera as the KH-6. The major objective of the LAN-YARD program was to obtain close-look photography of the site near the Estonian city of Tallin, where 1961 CORONA photos showed possible ABM deployments. NPIC's interpreters concluded that the photos showed construction for the deployment of the SA-5 GAMMON interceptor missile, with three batteries of six launchers arranged around a single engagement radar. The exact purpose of the Tallin line, as it was called, was to be hotly debated within the intelligence community. The LANYARD program was designed to help determine whether those deployments belonged to an ABM system or simply an advanced air defense system.[112]

In addition to establishing new systems, a policy of tight secrecy had been established to protect the program. Organizationally, a central organization for overhead reconnaissance, NRO, completed its first full year of operation. Finally, a national organization for the interpretation of the photography had been formalized.

As noted above, an early concern of Baker and Kistiakowsky was that

----

* The KH-5 camera was also used by NASA for moon mapping activities. A DOD-NASA agreement on the "Manned Lunar Mapping and Survey Program" specified that "DOD security classifications and procedures, as prescribed by the Air Force for application to mapping and survey equipments furnished under this agreement, will be observed by both agencies." The signator for the Air Force was NRO Director Brockway McMillan.

interpretation not be left in the hands of the Air Force as the chief of Air Force intelligence at the time was advocating.[113]

The National Photographic Interpretation Center, NPIC, was formally established by National Security Intelligence Directive (NSCID, pronounced N-Skid) No. 8 of January 18, 1961. NPIC was to be run by the CIA as a service of common concern for the entire intelligence community—interpreting both aerial and satellite imagery.

NPIC was put under the command of Arthur Lundahl. Lundahl had a long career in the photo interpretation business. After graduating from Tilden Technical High School in Chicago he went on to the University of Chicago, where he majored in geology. His involvement in aerial photography began during his years as a research assistant, performing geological research in Ontario, Canada. Such research required the production of maps based on aerial photography.[114]

Based on his background Lundahl was requested, as war approached, to help train some of the photogrammetrists that would be needed to interpret aerial reconnaissance photos. After conducting training courses and having gotten his Master of Science degree, Lundahl joined the Navy, and was eventually assigned to Adak, Alaska, where he interpreted the photographs produced by aerial photographic missions over enemy targets in the Aleutian Islands, Japan, and the Kuriles.[115]

Lundahl was one of over 800 World War II Navy photo-interpreters. According to Lundahl, the Navy did the best work in photo-interpretation for two reasons. First, the Navy was committed to photo-interpretation as a full-fledged discipline. In addition, the Navy looked for young officers with backgrounds in forestry, engineering, geology, and other natural sciences. The Army Air Force, in contrast, sought higher-ranking, and thus older officers, with industrial experience. The Army believed that those who had run steel mills would be best able to identify them. But the older officers tired easily and were less flexible than the Navy photo-interpreters.[116]

At the end of the war Lundahl found himself back in Washington, and in 1946 he was requested to help write the charter for the Naval Photographic Interpretation Center, a combined military-civilian interpretation organization. But the photo-interpretation center was soon transferred from the Office of Naval Intelligence to the Bureau of Aeronautics, which had little appreciation for the intelligence field. After seven years at the Center, he became convinced that the Navy had "dropped the ball" and was "going to go nowhere in photo-interpretation as it was then structured." He was also approached by the CIA. His initial reaction was one of skepticism. Years later he recalled telling the CIA, "I don't know anything about you guys. If you're going to parachute me into Salerno or somewhere forget it. I'm a scientist." CIA official Otto Guthe, he remembered, reassured him that the CIA "was going into the photo-interpretation business and they wanted someone to come over and run it who had experience and the right credentials for doing the job. They wanted me to do it."[117]

Lundahl joined the CIA in early 1953 as head of the CIA's Photographic Intelligence Division with 13 people and a few hundred square feet of floor space in the Steuart Building at 5th and K Street N.W. The operation had the CIA codename HT/AUTOMAT: HT from the initials of security officer Henry Thomas and AUTOMAT because Lundahl conceived of the division as a place where intelligence consumers could come and pick up whatever interpreted photography they needed. Lundahl recalls that "there was a very narrow front. It was very inconspicuous. It had . . . some funny little sign on the door, and there was some kind of turnstile about 100 feet in from the door and a glass cage where a security guard sat and a dirty little elevator which ran slowly at best, and of course, there was no place to park and no place to eat."[118] Expansion came a few years later in the wake of the U-2: by 1956, 150 interpreters were occupying 50,000 square feet in the Steuart Building. In 1958 the PID became the Photographic Interpretation Center.[119]

With the initial expansion to 150 personnel came military people detailed from the various armed services. Lundahl sought to make them feel like coproprietors of a national facility, since he believed that if he "had Army, Navy, Air Force, State, NSA, CIA people in there, I would have all the ingredients for nationalization, although this was nowhere in the cards in 1956."[120]

Several years later it was in the cards. In the fall of 1959, Maurice Stans, director of the Bureau of the Budget, had gone to the President with a list of 18 studies the Bureau felt needed to be conducted on various aspects of government. Two were on intelligence. Several months later, in March of 1960, Thomas Gates suggested to Eisenhower the need for a study of the defense intelligence establishment, which he described as a huge conglomerate spending $1.5 to $2 billion annually. Although Eisenhower reacted favorably to such proposals, it was the shooting down of the U-2 which provided the necessary impetus.[121]

A May 6 meeting between DCI Allen Dulles, Gates, Stans, Gordon Gray and the President's Board of Consultants on Foreign Intelligence Activities resulted in a decision to establish an ad hoc Joint Study Group to review various aspects of the U.S. foreign intelligence effort. Beginning on July 10, 1960, the group met 90 times for periods ranging from two to nine hours each and spoke to 320 individuals, including representatives of 51 organizations involved in the production or consumption of intelligence.[122]

Recommendations contained in the group's December 15, 1960 report covered all aspects of U.S. intelligence operations: collection, analysis, and the roles of the DCI and USIB. Its 43 recommendations included a centralized approach to defense intelligence, establishment of a central requirements facility to coordinate human and signals intelligence requirements, action by the DCI to achieve more effective coordination, and better CIA and DOD coordination concerning intelligence R&D activities.[123]

Several recommendations were based on the many hours the group spent discussing the problem of processing and interpreting overhead photography. The report noted agreement "in most of the community that a central photographic intelligence center should be established," although it noted that opinions varied as to how much interpretation and analysis should take place at such a center and who should run it. The report recommended that the DCI and Secretary of Defense should determine the details concerning the center's executive direction and that a National Security Council Intelligence Directive should be drafted establishing a National Photographic Interpretation Center.[124]

The question of executive direction of the center was debated at three United States Intelligence Board meetings. The predictable Air Force attempt to seize control of the national center was resisted by the CIA, the State Department, the NSA, the Navy, and the Army. However, given the Air Force claims, the matter was sent to the National Security Council for consideration.[125]

At that meeting Eisenhower asked for the opinion of several participants, none more important to him than that of George Kistiakowsky. His science adviser told Eisenhower that the agency had done a fine job; that he had been to the center and been briefed there many times; that the center had young men with bright ideas, and that the people there were careerists in a growing field. The last point was both important and critical in making a choice between a civilian-run national center and a military-run center. The field demanded careerists, people with scientific backgrounds, Kistiakowsky explained—people who would ride with it right to the ultimate end of whatever it was going to become; whereas the military, bright as they might be and good as they might be, for military career purposes they'd be rotated all the time. As a result excellent officers could be gone in a day.[126]

Eisenhower's response was simple: "Well that settles it Allen . . . Allen Dulles, you're going to run this thing, so carry on." Lundahl was Dulles' choice to head the center. Dulles then offered the military the option of providing the deputy director. Naturally, the military accepted Dulles' offer. Just as naturally, they began squabbling among themselves as to which service should supply the deputy. Eventually, in order to settle the dispute, they appointed Marine Corps General Graves Erskine to head a panel that would make the decision. Based on the financial and human resource support the Army had given to Lundahl's PIC, the Army was awarded the deputy's slot.[127]

CHAPTER 3

Problems in Space, Confrontations on Earth

By the end of its first year in office, the Kennedy administration had largely achieved two major objectives concerning satellite reconnaissance. Most importantly, it was getting a regular supply of photographs from the CORONA satellites, allowing the CIA, the newly created Defense Intelligence Agency, and the military intelligence services to stay up to date with the latest in Soviet military developments. In addition, it had transformed the reconnaissance effort from a public effort to one conducted under the highest degree of official secrecy.

In 1962 even the cover for CORONA, Discoverer, faded to black: officially, the program ceased after the launch of Discoverer 38 on February 27, 1962. In reality, however, as part of the new security procedures it was assigned a program number along with all other programs—Program 162* in its case—and its launches went unidentified.[1]

The security procedures included those specified in a March 23, 1962 memo, a memo written by NRO Director Joseph Charyk and signed by Deputy Secretary of Defense Gilpatric that directed that *all* DOD space activities be classified Secret. No individual projects were to be identified, nor was any individual—civilian or military—to be identified with any specific project. All military launches were to be classified and identified only by registry letter and dates. It was hoped that a total blackout of all military space activities would make it that much harder to pick out the reconnaissance satellites.[2]

Another part of the new security arrangements was a system for the designation of the reconnaissance systems and their product. The codename KEYHOLE was to refer to all photographic reconnaissance

---

* Program 162 continued until April 27, 1964, ending with a total of 78 launches.

65

satellites. Access to the photographs produced by the U-2 or KEYHOLE (KH) satellites would require a TALENT–KEYHOLE (TK) clearance. Satellite photographs or the information derived from them would be classified TOP SECRET RUFF. In addition, the camera systems to be used on the satellites would be distinguished by KH designations. Since the camera system used on the 1962 CORONAs was the fourth since the program began, it was designated the KH-4, with its predecessors being retroactively labeled the KH-1, KH-2, and KH-3. Finally, a system of codenames for specific photographic and electronic reconnaissance satellite programs was established under the general codename BYEMAN. Individuals might be granted access to the product of a reconnaissance satellite but still would not be given access to information about the satellite itself—its codename, orbital parameters, or capabilities—unless they had the appropriate BYEMAN clearance.

For many involved with the satellite reconnaissance program since its inception, 1962 was as bad a year as they had had since the beginning of the program. Eighteen launches of the SAMOS, KH-4 CORONA, KH-5 ARGON and KH-6 LANYARD were conducted throughout 1962, using Thor-Agena B or Thor-Agena D rockets. However, each program was plagued with high failure rates, causing despair among top defense and intelligence officials.[3]

From other sources, such as signals intelligence, it was apparent that significant military developments were occurring in the Soviet Bloc. The Soviets were deploying the SS-7 SADDLER ICBM, a large, two-stage rocket that used storable liquid fuel, had a range of 6000 miles, and carried a 3 megaton warhead. To facilitate defense against bomber attacks and make reloading easier, the initial SS-7 deployments involved complexes with two above-ground pads, arranged in clusters with eight or more missiles.[4]

Likewise, significant developments in the Chinese missile and nuclear weapons program led McNamara and McCone to bemoan the problems with the satellites. In the late 1950s, China, with Soviet assistance, began concrete work on the various aspects of bomb and missile production. In April 1958, China began to build the Chenaxian Uranium Mine in southeast China. Construction on another seven major mines also began in the late 1950s. Steps were taken to build the necessary facilities to process uranium and produce weapons grade material, and August 1958 saw the beginning of construction of the Hengyang Uranium Hydrometallurgy Plant (Plant 414). October marked the beginning of construction on Plant 202 in Baotou, Inner Mongolia for the manufacture of uranium tetrafluoride. In February 1960, ground was broken for the first plutonium production reactor—the Jiquauan Atomic Energy Complex—and in the spring construction began on a gaseous diffusion plant at Lanzhou, deep within the Chinese interior. By that time China had already settled on a nuclear test site. The search that began in August of 1958 concluded in October 1959 with the formal establishment of the 100,000 square kilometer Lop Nur Nuclear Weapons Test Base at Lop Nur in Xinjiang province.[5]

In December 1960, in National Intelligence Estimate 13-2-60, "The Chinese Communist Atomic Energy Program," the CIA pointed to 1962 as an important year:

> Chinese development of uranium resources and their probable construction of ore concentration and uranium metal plants certainly would imply an intended use for the uranium in plutonium production. . . . we estimate that a first Chinese production reactor could attain criticality in late 1961, and the first plutonium might become available late in 1962.[6]

The estimate also concluded that:

> On the basis of all available evidence, we now believe that the most probable date at which the Chinese communists could detonate a first nuclear device is sometime in 1963, though it might be as late as 1965, or as early as 1962, depending on the actual degree of Soviet assistance.[7]

And 1962 did turn out to be a particularly interesting year in the Chinese missile and nuclear weapons program, although no devices were detonated until 1964. Several delays in construction meant that some facilities intended to be operating earlier did not achieve initial operating capability until 1962. The eight major mines under construction in the late 1950s did not reach full operational status until 1962 or 1963, and likewise, the Hengyang Uranium Hydrometallurgy Plant encountered obstacles in 1960 and did not begin operating properly until the spring of 1962. Plant 202 did not begin operation until December 1962.[8]

However desperately U.S. officials desired photographs of these facilities, the CORONAs produced little in 1962. A variety of problems plagued the previously productive satellites. For example, the film brought back by one of the final three CORONA missions of 1962—most likely the July 18 or November 11 mission—was affected by a series of high-altitude nuclear tests conducted by the United States. Known as the FISHBOWL series, the tests involved the detonation of a 1.4 megaton and four submegaton devices at high altitudes in the vicinity of Johnston Island, 700 miles southwest of Hawaii. The first explosion in the series was the detonation of the 1.4 megaton STARFISH PRIME device on July 9, 1962, at an altitude of approximately 248 miles. The four submegaton device,—CHECKMATE, BLUE GILL, KINGFISH, and TIGHTROPE—were detonated at altitudes of tens of miles between October 20 and November 4, 1962.[9] One satellite passed through a high concentration of radioactive material over the south Atlantic, irradiating the film and degrading the quality of the pictures. Precautions were taken to avoid similar problems in the future.[10]

Nor did SAMOS fill any of the gaps. The program was terminated in 1962, a horrendous failure. Newspaper and other accounts linking SAMOS to the discovery of ballistic missile defense emplacements around Leningrad, as well as possible discovery of Chinese construction of a gaseous diffusion plant for the production of nuclear weapons material, were apparently based on a confusion of SAMOS and CORONA. Richard Bissell, recalls that as of 1963 SAMOS "had not produced a single useful photograph." The late Robert Amory, the CIA's Deputy Director of Intelligence at the time, recalled that "hundreds of millions and billions of dollars were spent on [SAMOS], but the bloody thing never was workable." SAMOS returned only one set of photographs, with a resolution of about 150 feet. The photographs were of Asian territory, but analysts were unable to determine whether the photos were of Soviet or Chinese territory.[11]

Nor did the other programs fill the gap. ARGON, the program designed for mapping purposes, was terminated in 1962 after 14 spacecraft had been built and 11 orbited. Only three of those eleven actually worked. The LANYARD program was also terminated with a less than inspiring record: five spacecraft were constructed, three orbited, one satellite returned imagery.[12]

Although those operations took place without any comments by U.S. officials, Soviet representatives were not so reticent. In 1962, the Soviets began a new international offensive in the United Nations and in publications, denouncing U.S. reconnaissance satellites as illegal and defending their right to take action against them. Considering the increasing Soviet capabilities to conduct space operations, it was feared that the Soviets were trying to build international acceptance of any attempt they might make to shoot down or intercept U.S. reconnaissance satellites.[13]

In March 1962, the U.N. Committee on the Peaceful Uses of Outer Space formed subcommittees to examine scientific, technical, and legal issues. The committees would be potential arenas of combat over the issue of U.S. reconnaissance satellites.[14] On May 26, 1962, in preparation for such diplomatic combat, President Kennedy signed National Security Action Memorandum 156, "Negotiation on Disarmament and Peaceful Uses of Outer Space." The memorandum instructed Secretary of State Rusk to form an interagency committee to study the political aspects of U.S. policy concerning reconnaissance satellites.* The "NSAM 156 Committee," as it came to be known, was formally the Interdepartmental Committee on Space. Its existence, as well as its functions and work, were Top Secret.[15]

In order to be properly prepared to blunt the Soviet diplomatic offensive, U.S. officials needed a consistent position on the subject of space reconnaissance. One task of the committee was to help establish such a position. The extent of official secrecy concerning CORONA and SAMOS had begun to cause problems. U.S. representatives at the U.N. and in the arms control area were not briefed and were therefore talking in ignorance. The resulting proposals from the Arms Control and Disarmament Agency, NASA, and U.S. representatives at the United Nations concerning the peaceful uses of space, according to former NSAM 156 committee chairman and Deputy Under Secretary of State U. Alexis Johnson, "scared the devil out of our people in the NRO."[16]

One example where lack of a consistent position and coordination had resulted in great aggravation for the managers of America's spy satellites had occurred in late 1961. A State Department proposal that

---

* The publicly available version of the memorandum is still heavily deleted more than 27 years after its issuance. All that remains after the censor's editing are the two opening sentences—"We are now engaged in several international negotiations on disarmament and peaceful uses of outer space. These negotiations are likely to continue for a long time, and may well grow in scope."—and the two concluding sentences—"Accordingly, I request that the Department of State organize a committee for this purpose, with representatives of all addressees with sufficient standing to permit them to be fully cognizant of all our programs in this area. I wish to receive your report and recommendations by 1 July."

those nations launching objects into orbit should provide information on orbital characteristics to the United Nations went through State and Defense departmental reviews and was introduced to United Nations' First Committee. It was adopted unanimously by the General Assembly on December 20.[17]

The U.N. registration agreement was a very unpleasant surprise to Joseph Charyk and the NRO. The State Department's consultation with Defense had not included the NRO, which had become an "entirely separate structure" with the result that "the normal staff processes did not disclose certain potential difficulties with the registry proposal."[18]

The task before Johnson's committee was to develop a policy that would allow the U.N. and ACDA people to participate in both internal policy formulation and international forums without damaging the vital reconnaissance programs. In addition to protecting the functioning of the programs, NRO was interested in shielding the technology employed and safeguarding information about U.S. achievement in resolution. Those who ran the reconnaissance program believed that U.S. technology was superior to anything the Soviets had at that time and was beyond what NASA thought possible.[19]

The members of the committee consisted of senior representatives from all the major entities of the national security and space communities. Committee chairman Johnson had returned in April 1961 from serving as an ambassador in Bangkok to become Deputy Under Secretary, historically the senior political career job. In his new position Johnson was responsible for relations with the Pentagon, the CIA, and the intelligence community and space council. Representatives of other agencies were Paul Nitze, Assistant Secretary of Defense for International Security Affairs (usually represented by his deputy John McNaughton); Herbert Scoville, the CIA's Deputy Director for Research; NRO Director and Air Force Under Secretary Joseph Charyk; NASA Deputy Director Robert Seamans attending for Director James Webb; ACDA Deputy Director Adrian Fisher; and White House representatives Carl Kaysen (McGeorge Bundy's Deputy at the NSC) and Dr. Jerome Weisner, the Science Advisor to the President. Raymond Garthoff served as Executive Secretary. Other attendees included Dr. Edward Welsh, Executive Secretary of the National Aeronautics and Space Committee; Deputy Director Robert Akers of USIA; State Department intelligence chief Roger Hilsman; and State Deputy Legal Adviser Leonard Meeker.[20]

In order to develop a consistent U.S. policy on satellite reconnaissance and the use of outer space, the committee had to investigate several issues, including what should be made public, what should be made available to allies, and what should be shown to the Soviets in certain situations.

The differences that had existed in 1961 about the best ways to protect the reconnaissance program were still present in May and June 1962. Abraham Chayes, Legal Adviser to the State Department, recalls that on the whole, the State Department still believed that openness was the superior strategy: "We thought that you gained something politically by being somewhat more open about our operations and developing a climate of legitimacy about them instead of trying to keep them completely secret."[21]

In addition to hearing the views of participants advocating more openness, the committee also studied a memo written in April of 1959 by Richard S. Leghorn. Leghorn, formerly an Air Force Colonel and President of the Itek Corporation when he wrote the memo, was one of the little known fathers of strategic reconnaissance. The memo was based on his view that given the size of the superpowers' nuclear arsenals, the only reasonable form of deterrence was "retaliatory deterrence" as opposed to "warfighting deterrence." Leghorn believed that warfighting deterrence—deterrence based on an ability to prevail in a nuclear war—required secrecy and superiority. On the other hand, retaliatory deterrence—deterrence based on an ability to inflict severe damage in retaliation for nuclear attack—required openness about weapons and information systems. Thus, his memo, "Political Action and Satellite Reconnaissance," noted that East Germany had already attacked Discoverer I as an espionage activity and that "we can anticipate powerful Soviet political countermeasures to the Discoverer/Sentry series." Further, attempting to maintain such programs as Top Secret activities was futile, Leghorn wrote, and created an espionage context that was damaging politically while failing to prevent Soviet military countermeasures. The solution, the Itek president believed, was to put reconnaissance satellites "in the white" through a political action campaign to "blunt in advance Soviet political countermeasures" and "gain world acceptance for the notion that surveillance satellites are powerful servants of world peace and security."[22]

According to Chayes, the opposing view was "let's keep them very, very secret." Three familiar justifications were given for secrecy. First was that "nobody knows about them. What they don't know won't hurt them." Second was "nobody knows how good they are." Third, "to the extent that this gets out in public it forces the Russians to make a challenge of some kind because they can't accept the fact they are being observed."[23]

The completed study, *Political and Informational Aspects of Satellite Reconnaissance Programs*, was delivered to Dean Rusk on July 1, 1962, and contained 17 unanimous recommendations and two others that in some ways reconciled those competing views. The committee recommended continuation of tight security around the program, to avoid presenting the Soviet Union or anyone else with a direct challenge and to avoid disclosing exactly what the United States could and could not detect from above.[24] There were to be no direct public statements concerning organization, capabilities, or operations of the reconnaissance program.

At the same time, the committee concluded that there was a need to brief allies on U.S. activities so that their statements on the use of outer space would not undercut the U.S. position. The result was a series of briefings of the leaders of the closest U.S. allies, especially Konrad Adenauer, Charles DeGaulle, and Harold MacMillan. The leaders were extensively briefed, sometimes by the president on state visits, but more often by special teams who talked only to the head of state, foreign minister, or defense minister. They tried to give them a sense of the scope of the U.S. program, its successes, and its close connection to the overall strategic situation. The leaders were also briefed on Soviet space activities.[25]

The committee also recommended the release of NASA photography of foreign territory to demonstrate the benefits of satellite observation and to accustom foreign leaders to such observation of their territory. The photography would demonstrate that satellite observation was of great value for weather forecasting, mapping, natural resources, surveys, and crop prediction.[26]

At the same time the committee suggested a limit on the resolution for the various NASA civilian programs. The resolution for the planned NASA mapping/economic resources program (98 feet) was acceptable but at the limit.[27]

Although it was agreed that the American public should be kept in the dark about the program, consideration was given to showing some samples of photography to Soviet leaders. Such a display would demonstrate that the United States had hard evidence that no missile gap existed and could not be bluffed and would indicate that the United States was keeping a close watch on Soviet military developments. One option was to show the Soviets photos at a summit. A more highly regarded possibility was to present the photos at a special meeting between Khrushchev and Kennedy, and a third option called for the U.S. ambassador to show Khrushchev such photos, as had been considered in the fall of 1961 in response to the Berlin crisis. Ultimately, the committee decided it was an issue that needed to be considered at a higher level and turned it over to Dean Rusk, Assistant Secretary for European Affairs Foy D. Kohler, and Ambassador to the Soviet Union Llewellyn E. Thompson for consideration.[28]

As well as recommending courses of action for information activities, the committee recommended that the United States adopt the position that outer space be considered free like the high seas, for all peaceful purposes, and that any observation of Earth be considered peaceful.[29]

Recommendation 18 of the report advised against the United States making or endorsing proposals for advance notification of missile and space vehicle launchings. The committee feared that notification of space vehicle launchings would facilitate passive and active countermeasures against reconnaissance satellites. Advance notification of missile launchings would also be ruled out because space vehicle launchings might become a collateral issue.[30]

William C. Foster, the ACDA Director, questioned Recommendation 18 in a July 6, 1962 memorandum to the president. Foster disputed the committee's recommendation on three grounds. First, he argued that past publicity concerning reconnaissance satellites was sufficient to arouse Soviet interest in passive countermeasures such as camouflage and concealment. Advance notification of launchings was unlikely to significantly add to the Soviet propensity to undertake such countermeasures. Further, advance notification need not be so precise as to aid Soviet ASAT targeting.[31]

Secondly, Foster argued, advance notification was consistent with the U.S. position that the objects it was placing in orbit represented legitimate uses of outer space. The committee's recommendation could damage the U.S. position: "By carrying our concern regarding our space vehicles to the point where we are unwilling to discuss advance notification of missile launchings, we may encourage the conclusion that we are attempting to shield activities which we ourselves regard as suspect. It is difficult to see how such an approach could contribute to the political and legal defense of satellite reconnaissance."[32]

Third, advance notification would represent an important step in arms control in that it would provide greater assurance that a surprise missile attack could not take place in a way that would catch the United States off guard. It would also help decrease tensions as missile inventories increased.[33]

Recommendation 19 questioned whether the United States should continue to seek a prohibition on the deployment of nuclear bombs in outer space if that prohibition was separate from a more general arms control package. Representatives of the Defense Department, Joint Chiefs of Staff, NRO, and the CIA opposed a separate ban. The JCS believed that the United States should keep the option of such deployment. In most, if not all other cases, opposition was based on the verification measures that would probably be required and the general risk of negotiations somehow interfering with other aspects of military space operations, especially satellite reconnaissance.[34]

Here again Foster challenged the committee's recommendation. In his memo to the President, he stated his belief that preventing extension of the arms race to outer space should be an important objective of arms control. The two objections raised to a separate ban were not convincing to Foster. Claims that the Soviets would gain a better knowledge of U.S. satellite reconnaissance capabilities through verification activities did not, to Foster, "seem like a compelling argument against our acceptance of inspection procedures, particularly in view of present Soviet ability to estimate the capabilities with a substantial degree of accuracy, a fact that is noted in the Satellite Reconnaissance Report."[35]

Nor did the ACDA chief find convincing the argument that discussion of a separate prohibition might involve the United States in a disadvantageous public discussion of satellite reconnaissance. He warned that the Soviets had raised the issue in the U.N. and were likely to raise it again. By shifting the focus of outer space discussions to the Geneva arms control conference, the United States could plausibly ignore U.N. activities concerning similar issues or even veto condemnatory U.N. principles of space law.[36]

The committee report and Foster's memo were considered by the NSC on July 10, 1962. Kennedy then issued NSC Action 2454, approving the first 18 recommendations of the committee but sending recommendation 19 back to the committee for further consideration.[37]

By the time the report had been completed the Soviets had already moved to outlaw space reconnaissance. In the spring of 1962, the U.N. had established a Committee on the Peaceful Uses of Outer Space (COP-UOS) to develop international principles to govern activity in space. On June 7, the Soviets presented a nine-point proposal to the committee's first meeting in Geneva. Among the nine points was a ban on the use of satellites for reconnaissance. The committee's first meeting adjourned with the United States unwilling to accept the Soviet proposed ban.[38]

When the committee met again in September, the Soviets tabled a Draft Space Treaty which sought to ban intelligence collection from space.[39] In December, at the United Nations, the Soviets resumed their offensive. In a December 3 statement to the First Committee of the General Assembly the Soviet representative to the outer space committee, Planton D. Morozov, announced that

> According to our proposals, foreign space ships, satellites and capsules found by contracting states on their territories or salvaged on the high seas, would be returned without delay to the launching States if they bear markings indicating their national origin and if the launching States have officially announced the launching of the devices involved. The only exception would be made in the case of a vehicle aboard which devices have been found for the collection of intelligence data from the territory of another State. This exception we believe is fully consonant with the general policy that should be followed in opposing the use of outer space for objectives that are incompatible with the United Nations Charter, especially those involving the violation of the sovereignty of another State. It is indubitable that espionage is such a violation, even if is effected from space.[40]

The U.S. Delegate, Albert Gore, offered a different view:

> . . . any nation may use space satellites for such purposes as observation and information-gathering. Observation from space is consistent with international law, just as observation from the high seas. . . . Observation satellites obviously have military as well as scientific and commercial applications. But this can provide no basis for objection to observation satellites.

With malice toward none, science has decreed that we are to live in an increasingly open world, like it or not, and openness, in the view of my Government, can only serve the cause of peace. The United States, like every other nation represented here in the Committee, is determined to pursue every non-aggressive step which it considers necessary to protect its national security and the security of its friends and allies, until that day arrives when such precautions are no longer necessary.[41]

As might be expected Gore's logic did not change the view of the Soviet delegate, who responded that

We cannot agree with the claim that all observation from space, including observation for the purpose of collecting intelligence data, is in conformity with international law. . . . Such observation is just as wrong as when intelligence data are obtained by other means, such as photographs made from the air. The object to which such illegal surveillance is directed constitutes a secret guarded by a sovereign State, and regardless of the means by which such an operation is carried out, it is in all cases an intrusion into something guarded by a sovereign State in conformity with its sovereign prerogative.[42]

Although U.S. and Soviet representatives continued to debate the legitimacy of space reconnaissance throughout most of 1963, the CIA and NRO were proceeding with their programs. On January 1, 1963, the new headquarters of the National Photographic Interpretation Center were dedicated. After the shabbiness of the Steuart Building, NPIC personnel were delighted to move to Building 213 of the Washington Navy Yard at 1st and M Streets in the southwest corner of Washington. NPIC's move had been the result of 1962 visits by McNamara and the President's Foreign Intelligence Advisory Board (PFIAB) to the Steuart Building. Shocked by the environment at 5th and K, they advised the president that a new building was needed.[43]

Kennedy promptly informed DCI John McCone, "I want you to get them out of that structure" and wanted to know how soon a move could be accomplished. In response, McCone informed Kennedy that the Naval Gun Factory appeared to be a reasonable choice but that it would require a year to refurbish it. Kennedy's reply was "All right, you do it."[44]

Moving into their new building was described by McCone as a "rags to riches" situation. The 200,000 square feet of floor space meant that hundreds more people could be added. The building had big elevators, air conditioning, and good security. Most of all, it was the national center that Lundahl had envisioned almost ten years earlier. The lion's share of people in the building worked for the CIA. The people who typed the letters, drove courier trucks, ran the computers and library searches, and produced the graphics all worked for the CIA.[45] But the photo-interpreters came from the CIA, DIA, Army, Navy, Air Force, and other organizations. An Air Force interpreter who studied photos of Soviet silos might ride the elevator with a CIA interpreter who pored over photos of Chinese nuclear facilities and a Navy representative whose safe was filled with the latest photography of Soviet submarines.

CORONA operations continued. After two unsuccessful attempts to place spacecraft into orbit using a new first stage, the Thrust-Augmented Thor, eight CORONAs were shot off between May 18, 1963 and December 21, 1963. The satellites operated with mean perigees and apogees of 115 by 212 miles respectively and a mean inclination of 76.4 degrees. Use of the Thrust-Augmented Thor was tied to the CORONA program's new camera system, the KH-4A, which had a resolution of about 10 feet.[46]

Additionally, a newcomer joined the CORONA program in the U.S. reconnaissance arsenal—the Air Force's KH-7 GAMBIT. The first launch of the KH-7 came on July 12, 1963, employing an Atlas-Agena D booster.[47] Three more launches followed on September 6, October 25, and December 18. The four spacecraft had mean lifetimes of 4.25 days and mean inclinations of 94 degrees.

The mean perigee for the four GAMBIT missions was 92 degrees, with two of the spacecraft dipping as low as 76 and 87 miles. The difference in the mean perigees of the 1963 GAMBIT and CORONA launches provided a clue to their different missions. GAMBIT was the first "close-look" satellite. Whereas the KH-4 camera system and the CORONA orbit were designed to produce photographs providing a broad view, GAMBIT was designed to allow the intelligence agencies to get a close-up look at a smaller area. The clearer and more detailed the picture, the more information NPIC's photo-interpreters could extract from the photograph.

One problem that would exist throughout the life of the KH-7 was that the camera could not easily be moved to cover targets located in different positions relative to the spacecraft. Thus, if one target was directly under the satellite it could be difficult or impossible to move the camera to a target off to the side in the limited time available to get a picture before the spacecraft was out of range. As a result, the KH-7 was "access-limited" and the satellites usually finished their missions with unused film because the consumables needed to keep the spacecraft in orbit ran out before the film.[48]

From its inception, the KH-7 GAMBIT was intended to work in tandem with the KH-4 CORONA. A CORONA mission would produce a broad view of portions of Soviet territory, returning its film at two intervals. Based in part on the examination of that photography, targets would be selected for the follow-up GAMBIT mission.

Among the prime targets for CORONA and GAMBIT were the new Soviet missiles being deployed throughout the Soviet Union. As noted earlier, in 1962 the Soviet strategic forces began deploying the SS-7 SADDLER ICBM. Ultimately, the Soviets were to deploy 197 of the 5900 mile range missiles, which carried a single warhead of either three or six megatons. The following year, deployments of the SS-8 SASIN, a 5400 mile range missile with a single 3-megaton warhead, began.[49]

Unfortunately, the initial GAMBIT missions were unsuccessful. The orbital control apparatus of the spacecraft continued to malfunction, preventing stabilization, but a solution was eventually discovered; the spacecraft remained attached to the Agena second stage.[50]

With that problem solved, CORONA and GAMBIT continued to operate throughout 1964 and 1965, surprisingly, without Soviet protest. While the Soviets had been protesting U.S. reconnaissance satellites they had been busy developing their own spy satellites, launching the first in April 1962. Almost a year later the Soviet position had apparently not changed, as the Soviets submitted an 11-point proposal to COPUOS, which included a ban on satellite reconnaissance. However, in his July 15, 1963 column in the *New York Times*, correspondent C. L. Sulzberger reported on a statement Khrushchev had recently made to a Western leader. On the subject of nuclear testing, Khrushchev argued that on-site inspection was not necessary: "that function can now be assumed by satellites. Maybe I'll let you see my photographs." Two months later, when the new Soviet delegate to the U.N. committee, Dr. Nikolai Fedorenko, addressed the committee he made no mention of satellite reconnaissance. When in December 1963 the United Nations adopted a "Declaration of Legal Principles Governing the Activities of

States in the Exploration and Use of Outer Space" the Soviet Union did not object to the provision which specified that "Ownership of objects launched into outer space, and of their component parts, is not affected by their passage through outer space or by their return to the earth. Such objects or component parts found beyond the limits of the State of registry shall be returned to that State."[51] Apparently, the Soviets realized that U.S. satellite reconnaissance was here to stay, that their own program made it more difficult for them to protest the U.S. program, and that the satellites might actually contribute to the avoidance of war.

CORONAs were launched about every six weeks, with at least one GAMBIT launched between CORONA missions. The CORONA lifetimes were regularly above 20 days.

During that period the Soviets continued to deploy their new ICBMs. In 1964, 100 SS-7 and SS-8s were added, mainly SS-7s. The SS-8 program probably concluded that year, since only 23 were ever deployed. Further ICBM deployments in 1965 brought the year-end total to 270 operational Soviet ICBMs.[52] CORONAs would periodically survey areas where missiles had been deployed and those areas that were logical candidates for the next set of deployments. GAMBIT missions would then produce photos to allow interpreters to make the best possible estimates of size, configuration, yield, and any other pertinent characteristics of Soviet arms.

Although the space segment of the U.S. reconnaissance program was functioning smoothly by 1965, there were serious problems back on earth. The partnership between Richard Bissell and Joseph Charyk, and hence between the CIA and Air Force, had functioned smoothly. Likewise, Herbert Scoville, the head of the CIA's Directorate for Research and Bissell's successor as the CIA's director of satellite reconnaissance efforts, also found Charyk a congenial partner.[53]

The situation changed dramatically however, when Charyk was succeeded as Under Secretary and NRO Director by Brockway McMillan in 1963. McMillan was a Bell Telephone Laboratories executive who had served with Scoville on the defense committee of the Killian Technological Capabilities Panel in 1955. McMillan battled with Bissell's two successors as head of the CIA satellite reconnaissance program, Scoville and Albert Wheelon. Many years later, with the battles long behind, Scoville referred to McMillan as an incompetent whose only talent was fighting organizational battles.[54]

Relations became even more acrimonious between Wheelon and McMillan when Wheelon replaced Scoville. When John McCone took office he was faced with recommendations from the President's Foreign Intelligence Advisory Board—specifically chairman James Killian and Edwin Land—and the Joint Study Group to establish a directorate to handle science and technology. Such a directorate would absorb both the technical collection operations of the Directorate of Plans (the U-2 and A-12* spy planes and CORONA) and the scientific intelligence production work of the Directorate of Intelligence. It was a proposal that had first been made in 1957, but Richard Bissell's insistence on maintaining control of the U-2 and Allen Dulles' traditionalist definition of intelligence prevented any change.[55]

McCone, who didn't like to be confronted without "outside demands" but felt the need to placate such influential advisors, first offered the job to Richard Bissell, who, after the Bay of Pigs fiasco, did not have the option of remaining as Deputy Director of Plans. But Bissell, who voiced opposition to the new directorate in a series of memos to McCone, declined. McCone then proceeded to establish a Directorate of Research. He named a Deputy Director for Research (DDR) he did not like and gave him inadequate administrative and political support and resources. The unlucky DDR was Herbert "Pete" Scoville. McCone and Scoville had been antagonists during the many years that the liberal Scoville was involved in the nuclear weapons program while the conservative McCone headed the Atomic Energy Commission.[56]

Trying to manage an inadequately staffed directorate with the less-than-full support of an individual he didn't get along with was difficult for Scoville. McCone did not provide Scoville with the political support required when Ray Cline, Deputy Director for Intelligence, refused to relinquish the Office of Scientific Intelligence, and the bureaucratic battles with McMillan made the situation intolerable for Scoville. McMillan wanted to seize control of the reconnaissance program for the Air Force. As Director of NRO he believed that he should be in full control of the satellite reconnaissance program and that the CIA should take orders from him, not be an equal partner. In the face of this situation Scoville

---

* The A-12 was the CIA's successor to the U-2. It began operations in 1962 but was superseded, largely for bureaucratic reasons, by the Air Force SR-71A, a modified version of the A-12.

headed off to his northwest Connecticut summer home, announcing that he had no intention of returning. To coax Scoville back to Washington, McCone sent a young scientist from the Office of Scientific Intelligence, Dr. Albert "Bud" Wheelon.[57]

Wheelon's mission was a failure: Scoville was adamant. Hearing Wheelon's report, McCone informed him that he was the new Deputy Director for Research. Wheelon was admirably qualified. He had received his doctorate from M.I.T. at the age of 21 and had reformulated the mathematical equations used to predict the trajectory of missiles and space vehicles. Wheelon had significant experience in government, weaponry, and scientific intelligence. After working for the technical research firm TRW, he joined the CIA's Development Projects Division in the late 1950s, and from there went to the Office of Scientific Intelligence (OSI), becoming its head. As OSI head he also served as chairman of the United States Intelligence Board's Guided Missiles and Astronautics Intelligence Committee (GMAIC). The GMAIC coordinated all research on Soviet and other nations' space and missile systems and was conversant with the U.S. reconnaissance assets that produced intelligence about those foreign systems. Wheelon agreed to take the job on the condition that he receive full backing from McCone and be given the necessary manpower and funds to do the job correctly. McCone agreed and Wheelon became the head of the most powerful component of the Agency, the Directorate of Science and Technology. By August 1963, Wheelon's directorate included the Data Processing Staff, Office of ELINT, Development Projects Division (which handled satellite development), and the Office of Research and Development. He also obtained control of the Office of Scientific Intelligence, leaving Ray Cline furious, and established a Foreign Missile and Space Analysis Center.[58]

Wheelon's empire-building put him on a collision course with McMillan, and it was a collision from which neither was going to shy away. Wheelon was described by one rather genteel former CIA official as the "most acerbic . . . son-of-a-bitch" that he had ever met. Wheelon resented McMillan on general principles, because although he himself had significant relevant government experience, McMillan had none and his service on the TCP hardly stood up to Wheelon's record. Wheelon did not consider him an equal. Beyond general principles, Wheelon had a specific, if petty, gripe: McMillan had, as editor of a physics journal, rejected an article submitted by Wheelon while he was at MIT.[59]

Nor was McMillan a gentlemen. The result was that there "soon developed a series of battles over turf that were so vituperative that they are still talked about by old hands who were involved in the agency's technical operations."[60]

The debate was not over what basic type of systems would be built, because everyone agreed that both general search and close-look capabilities were needed to satisfy both military and CIA requirements. It was instead over who controlled the development of those systems. Richard Bissell, though no longer a formal member of the intelligence community during the Wheelon-McMillan battles, had his sources of intelligence concerning the war over strategic reconnaissance. Many years later he told author William E. Burrows that

> Wheelon, essentially, was battling to maintain the agency's influence in the reconnaissance programs and also to have the agency designated by the NRO as the procurement agency for a lot of the payloads. The Air Force was battling for the exact opposite. They wanted to do as much as possible of the procurement and have as much influence as possible on the technical decisions and operational matters. And that was really the essence of Bud's continuing battles. What kind of programs will receive what kind of funding? Who will be the procurement agency for this or that? And they went on, and on, and on.[61]

The debate became so intense that DCI John McCone and Secretary of Defense Robert S. McNamara agreed in 1965 to create a National Reconnaissance Executive Committee (NREC), chaired by the DCI and reporting to the Secretary of Defense, to oversee NRO's budget, structure, and research and development activities.[62]

Initial membership in NREC consisted of the DCI, the President's Science Adviser, and a DOD representative. Decisions of the committee would be sent to the Secretary of Defense for approval or disapproval. If the DCI objected to the Secretary's decision he would be allowed to take his argument directly to the White House.[63]

Creation of the NREC did not eliminate all the battles, and the Air Force and CIA would soon face off over the question of manned space reconnaissance. At 10 A.M. on August 25, 1965, President Johnson strode into the East Room at the White House to announce that the United

States would build and deploy a Manned Orbiting Laboratory (MOL). The laboratory would test the military usefulness of a man in space and develop technology and equipment for manned and unmanned space activities. According to the President's announcement, MOL would perform "very new and rewarding experiments" with the first of two unmanned flights in early 1968 and at least five manned flights beginning in early 1968. All launches would be from Vandenberg AFB into a polar orbit.[64]

Of course, it was clear that reconnaissance was to be a major function of the MOL. When a reporter questioned a defense official at a background briefing about the purpose of MOLs polar orbit, the assembled press just laughed.[65]

Many believed that a manned reconnaissance vehicle offered some significant advantages over unmanned satellites, particularly with regard to target selection. Astronauts equipped with moderately powerful binoculars could easily pick out objects of potential interest and photograph them immediately. In addition, men could accurately shoot pictures through gaps in the clouds and maintain and repair equipment.[66]

The origins of the MOL concept dated back to at least 1960. In November 1960 the Aeronautical Systems Division of the Air Force Systems Command finished a study designated SR-178 and titled "Global Surveillance System." The objective of SR-178 was to determine the design characteristics of a manned, recoverable, reconnaissance satellite system.[67]

The study envisioned a fully recoverable vehicle in low earth orbit with a crew of three to six astronauts. Approximately four manned satellites would be in orbit simultaneously, each equipped with a variety of sensors. Sensors considered feasible included a high resolution camera with 3' resolution, an imaging infrared camera with 75' resolution, side-looking imaging radar with 25' resolution, and a communications intelligence–electronics intelligence antenna. Initial operations to identify exactly what role man could play in space reconnaissance would begin in the 1968–1970 period.[68]

In June 1962, the Air Force broached the idea of employing the Gemini vehicle from the NASA civilian program as a military spacecraft. Tests involving rendezvous, docking, and transfer would be the initial step in the Manned Orbital Development System (MODS). MODS would be a space station with a crew of at least four, a variety of equipment, a re-entry capsule based on either the Gemini or Apollo, and a supply module that would carry cargo and provide propulsion. The concept was further expanded in August 1962 into the BLUE GEMINI program, with a proposal to fly six Gemini missions with Air Force pilots for preliminary orientation and training.[69]

By January 1963 the BLUE GEMINI idea had been dropped. Defense Secretary McNamara had been so impressed by a NASA presentation that he suggested that NASA and Air Force efforts be combined and moved to the Department of Defense. NASA opposed the idea because of fear that it would interfere with Apollo, and the Air Force was afraid it would create competition for their Dynasoar space glider, so the idea was dropped.[70]

With the demise of BLUE GEMINI, the MODS effort continued and detailed studies were made to provide the cost and technical information necessary for full-scale development. At the same time, the Dynasoar program was approaching its end. It had been the victim of reorientations, delays, alterations in launch vehicles, and cost increases, and it seemed that Dynasoar did not offer substantial advantages over Gemini in terms of data or cost.[71]

No astute observers were shocked, then, when McNamara applied the coup de grace to Dynasoar at a December 10, 1963 press conference. In its place, he announced that the Air Force would begin work on a Manned Orbiting Laboratory. MOL was to be a modified NASA Gemini, called Gemini B, attached to a trailer-sized laboratory module. It would be placed in orbit as a single unit by a Titan 3C, avoiding the need for rendezvous and docking.[72]

McNamara's statement only indicated approval of an Air Force effort to do preliminary work on a MOL: it did not indicate a decision to proceed to an operational system. The Air Force had to demonstrate the system's value to receive final approval, and at the time of the press conference, U.S. astronauts had only spent 54 hours in orbit.[73]

Initially the Air Force proposed to launch MOL from Cape Kennedy with an orbit of 125 by 250 nautical miles and an inclination of less than 36 degrees. At that inclination it would overfly no territory north of the Middle East and China and no territory south of Africa and South America, completely avoiding the Soviet Union. There were no plans to take photos of reconnaissance quality.[74]

The Air Force concept was not favorably received when presented in January 1964 to higher authority, and revisions were quickly begun. After numerous rejections of its new ideas by McNamara, who didn't see that they had established the need for a man-in-space, the Air Force finally hit on a winning combination. Two experiments were added: the in-orbit assembly of a large radar antenna and a camera system to produce photos of reconnaissance quality. MOL was now part of the KEY-HOLE program. Its camera was given the designation KH-10 and the program to use the MOL for reconnaissance was codenamed DORIAN.[75]

In January 1965, the White House commissioned reports from both the Air Force and NASA while contractors were requested to submit proposals. On the basis of these reports, three options were prepared for President Johnson—approve MOL; combine MOL and the Extended Apollo System, Apollo X; or cancel MOL. It was the MOL's usefulness for collecting intelligence that induced many of those advising the President to advocate the first option because of the intelligence MOL could collect in support of arms control. An anonymous official told *The Washington Post* that, "If this does what we think it will do, MOL will be the greatest boon to arms control yet." Vice-president Hubert H. Humphrey agreed. At a July 9, 1965 meeting on MOL, Humphrey asked about 25 questions, the answers to which convinced him that any negative impact of having a manned military space presence was outweighed by its value for arms control.[76]

By the time of Humphrey's meeting not only had a justification for MOL been worked out but some additional evidence of the utility of manned reconnaissance had been accumulated. The NASA Gemini 4 and Gemini 5 missions of June and August 1965 constituted tests of manned space reconnaissance. One photograph taken when Gemini 4 passed over Cape Canaveral showed roads, buildings, and launch pads even when viewed with the unaided eye.[77]

The MOL was to be placed in orbit at an altitude of between 150 and 160 miles with an inclination between 85 and 92 degrees, and would revolve around the earth once every 90 minutes. As envisioned in August 1965, it was to be 54 feet long, of which the laboratory would make up 41 feet. The total weight was to be 25,000 pounds, including the 6000 pound Gemini B capsule and the 5000 pound reconnaissance payload. Though unmanned reconnaissance satellites had payloads greater than 5000 pounds, it was hoped that much of the automated equipment necessary for these satellites could be eliminated.[78] At the very end of the laboratory were living quarters for the astronauts, and in front of the astronauts' quarters would be the section that housed the telescope. A Gemini capsule would return astronauts from the laboratory and shuttle new astronauts to the laboratory.

Many were not convinced of the utility of the MOL. At the time, one Pentagon planner told *Newsweek* that "MOL is really an Air Force toy. It's fascinated with the idea of putting men in space, but there are still many unanswered questions as to whether we really have to put men there."[79] And for many in the CIA, space espionage and manned space flight were an unstable combination.

CHAPTER 4

New Programs, New Wars

Twice in the early 1960s the United States and Soviet Union found themselves in confrontations that threatened to explode into war—first, the 1961 Berlin crisis and then the even more serious Cuban missile crisis. The subsequent three years were relatively peaceful, with the superpowers reaching agreement to ban atmospheric testing.

The latter half of the 1960s also saw an absence of direct U.S.–Soviet confrontations. However, crises in Eastern Europe, in the Middle East, and between the Soviet Union and China were of great concern to those responsible for U.S. national security. The crises either threatened to severely damage U.S.–Soviet relations or to engulf the U.S. in regional conflicts. Each was the subject of close attention by America's intelligence establishment, including its sentries in space.

The United States entered 1966 with its tandem of area surveillance and close-look satellites, the KH-4A and KH-7, dropping their film at regular intervals. The KH-4A photographs had a resolution of approximately ten feet while the resolution of the KH-7 photos was 18 inches.[1] Six CORONA and six GAMBIT missions were launched between January 1 and June 21. Between mid-July and mid-August, CORONA and GAMBIT modifications would be introduced into the space reconnaissance fleet, although it would be almost another year before their predecessors would finally fade from the scene.

On July 29, 1966, a Titan 3B-Agena D blasted off from Vandenberg AFB with a new camera system, the KH-8, for the GAMBIT program. The rocket was a Titan 3 core without the usual two solid fuel strap-ons and transtage but with an Agena third stage, that could place 7500 pounds into polar orbit. The spacecraft, cylindrically shaped, 24 feet long, with a diameter of 4.5 feet and weight of 6600 pounds, was placed in an orbit with a 110 degree inclination, a perigee of 98 miles and an apogee of 155 miles. In addition to regular black and white film, color film was also carried on KH-8 missions, with frames of color film being spliced into the film reel. However, the resolution was far inferior to what was obtained using black and white photography. For the next seven days the first KH-8 proceeded on its mission, photographing selected targets in the Soviet Union and China. Its seven day lifetime equaled that of the KH-7 launched only 17 days earlier, which had the longest mission for a close-look satellite.[2]

Eleven days after the launch of the KH-8, on August 8, 1966, a new camera system for the CORONA program, the KH-4B, was launched into space. The KH-4B offered improved resolution over its predecessor—objects with any dimension of five feet or more could be spotted by the new camera. The satellite was also equipped with 18 small rockets to allow greater maneuverability, specifically to allow the spacecraft to dip down for a closer look. To place the new camera system in orbit, a new booster was employed, the Long Tank Thrust Augmented Thor-Agena D (LTTAT-Agena D). Weighing about 4200 pounds, of which 2600 pounds constituted the reconnaissance equipment, the initial KH-4B operated in a 120 by 178 mile orbit with an inclination of 100 degrees for the next 32 days.[3]

While the unmanned reconnaissance vehicles were operating in space, work on the proposed Manned Orbiting Laboratory was continuing back on earth. Because the existence of the MOL, if not its mission, was unclassified, the MOL became the only part of the KEYHOLE program ever to become a subject of extensive discussion by public officials outside the inner sanctum of the national security bureaucracy.

Much of this discussion came in the course of hearings held on the MOL program at the instigation of Senator Spessard L. Holland. Senator Holland was not concerned about the potential problems of manned space reconnaissance or the ability of the MOL to effectively perform its reconnaissance mission. Rather, he was concerned that the MOL would be launched from California rather than his home state of Florida, resulting in the loss of jobs and payroll.

Launches from Vandenberg AFB were traditionally used to place spacecraft in polar orbits. As Dr. John S. Foster, Jr., Director of Defense Research and Engineering, explained to the Senator, there were only two ways to achieve polar orbit from Cape Kennedy. One would involve trajectories passing directly over southern Florida and Miami and was therefore unacceptable because of the danger to the civilian population if any problems developed with the Titan. The alternative involved a "due east" launch, as if the spacecraft were being launched into a geosynchronous orbit, followed by turning the vehicle south in a "dog leg" maneuver. Dr. Foster mentioned two problems with such a launch strategy, which had been used on rare occasions previously. The Titan would pass over foreign territory including Jamaica, Cuba, and parts of Central and South America before it reached orbit, which created a problem under international law. But perhaps of more importance as the deciding factor was the fear that the MOL's reconnaissance system could fall into foreign hands if there was an aborted mission shortly after launch. A failure of equipment could bring the MOL down in the main street of some Latin town such as Havana.[4]

A third problem with a Florida launch was addressed in a Secret memo from Deputy Secretary of Defense Cyrus Vance to the president. The memo, possibly motivated by Holland's pressure, noted that launching the MOL into a polar orbit from Florida would require that the payload be reduced by 2500 to 5000 pounds from the 30,000 that could be lifted from Vandenberg and that *"With this loss in payload the primary intelligence mission simply could not be planned."* The memo also noted that the probability of impact and fatality would be 100 times greater when launching from Florida rather than California and that in the case of the MOL launch, "we would be very concerned with the probability of impact and recovery of any portion of the highly classified payload."[5]

Unable to budge Foster on the unacceptability of a polar launch from the Cape, Senator Holland shifted ground, questioning the necessity of a polar orbit. By this time the friction between Holland and Foster was evident. Holland pressed Foster as to "why the polar course is the sole and exclusively chosen one, under the thinking of the Air Force?" Foster's response, "I am sorry, I can only say that it is a requirement of the program," did not satisfy Holland. But Foster was clearly not willing to discuss the reconnaissance aspect of MOL operations in front of the committee. After several more unproductive rounds between the two antagonists the hearing was over.[6]

In addition to political difficulties, the MOL program was also having technical difficulties, which resulted in weight increases and a one-year slippage in the launch schedule. The first unmanned launch was scheduled for April 15, 1969; the first manned launch for December 15. MOL's weight grew to more than 30,000 pounds, beyond what the Titan 3C could handle. To handle the larger weight, a new version of the Titan was developed: the Titan 3M. Rather than having a mere five solid-fuel strap-ons, the M-version had seven and could generate a thrust at takeoff of three million pounds.[7]

Tests of the basic Titan-Gemini combination were conducted in November. The Air Force had cut a hatch into the old NASA Gemini 2 capsule to allow the crew to transfer from the Gemini back to the MOL. A modified Gemini 2 would be attached to a simulated MOL canister constructed from a Titan II first stage. A secondary objective was carrying out the nine on-board experiments carried by the canister. Although only 34 feet in length—shorter than a real MOL—the simulated MOL was sufficient to provide information on the launch stresses of the MOL/Titan 3M. On November 3, 1966, the Titan 3C was launched from Cape Kennedy. When the transtage and payload reached 125 miles, the transtage pitched down towards the earth and separated the Gemini 2. After making a severe re-entry that put maximum heating on the hatch, the capsule landed 5500 miles downrange, only seven miles from the recovery ship USS La Salle. After it had been pulled out of the ocean and subjected to a variety of tests, the news was good. The hatch had survived the 17,500 mile per hour re-entry without damage or loss of strength. In fact, it had welded shut, sealing itself.[8]

With the design of the Titan 3M, the basic MOL mission profile was complete. The Titan 3M would place the MOL into a polar orbit from the Western Test Range. The two-man crew would arrive for their 30-day missions and then depart using the Gemini B vehicle. Ground control would be the responsibility of the Air Force Satellite Control Facility.[9]

Once in orbit the crew would first activate the MOL's laboratory systems and then put the Gemini B in storage. Next, they would open the 26-inch diameter hatch and float through the tunnel into the laboratory. After having verified that the life support system was functioning properly, the crew would be free to remove their space suits and begin work.[10]

The planned laboratory was 10 feet wide, 41 feet long, and divided into two parts. The experiments and work area were located in the forward section. The rear section made up the living area and contained the crew's sleeping bags, an exercise bicycle, and the toilet. A wall divided the two areas and a floor ran the length of the laboratory, an up-and-down reference requested by the astronauts. Velcro carpets were tested as a way to defeat the zero gravity atmosphere and hold the crew in place. The outer skin of the laboratory bulged slightly, apparently as a meteorite shield.[11]

The MOL camera was to have a lens measuring six feet across that would have a theoretical resolution of four inches and an actual resolution, allowing for atmospheric distortion, of nine inches.[12]

The primary purpose of MOL was strategic reconnaissance, but the claim that the MOL would test man's military usefulness in space was also true. Fifteen primary and 10 secondary experiments were scheduled. The primary experiments included the assembly, erection, and alignment of large structures such as radar antennas, tracking ground and space targets, spotting targets of opportunity, electronic reconnaissance, ocean surveillance, post-attack bomb damage assessment, extra-vehicular tests, in-space maintenance, navigation tests, biomedical and sociological experiments, and general performance in the military area.[13]

Secondary experiments included communications propagation, air-glow analysis and photography, recovery of space objects, and the multi-band spectral analysis of planets.[14]

It was hoped that putting a man in the loop and in space would heighten the value of the photos that would be produced. Detection and high-resolution photography of a new target might take weeks or months using unmanned satellites, but men could spot and photograph a new target without delay. In addition, rather than wasting film where there was clearly no activity of interest, certain areas could be bypassed. Thus, the astronauts would serve as part-time photo-interpreters as well as determining some of the targets for the MOL camera.[15]

The Navy hoped that the MOL astronauts would prove proficient at ocean surveillance. Each day, the laboratory would pass over all the world's oceans. Earlier manned missions had already demonstrated that it was possible in clear weather to spot the wake of ships with an unaided eye. Under poor weather conditions the MOL would use radar. The crew would detect shipping, locate it, identify and classify the ships according to type, and track their course. They would also separate warships from commercial vessels and Allied from Soviet bloc shipping. Some Navy officials hoped that if the MOL astronauts were to prove adept at the tasks, the fourth MOL would be specifically equipped for ocean surveillance and anti-submarine warfare and would have an all-Navy crew.[16]

After the crew's 30-day mission was completed they would shut down the laboratory systems, transfer back to the Gemini B, and then separate. The retro-fire and splashdown would be the same as if it were a NASA mission. The laboratory would be left to burn up on re-entry.[17]

Plans for recovering the capsule and the astronauts included provisions for the standard recovery scenarios, for recovery after a launch abort, or for recovery from hostile territory. During the pre-launch period, surface and air-recovery forces would be stationed at close intervals along the first 1000 miles of the launch trajectory in the event of a launch abort. Once the MOL had attained orbit, the launch recovery forces would then be dispersed. It was anticipated that under normal circumstances the Gemini B would re-enter into one of three recovery zones located in the Atlantic, Pacific, and Indian Oceans, with recovery forces on station during the length of the mission. If the Gemini B was required to re-enter during orbits which would land the capsule outside the three zones, then the World Wide Air Rescue Service (ARS) would be utilized to rescue the crewmen.[18]

If the crew were to land outside the designated recovery zones it might be hours or days before they could be located and recovered. To help the astronauts prepare for such an event, they would be sent to jungle survival training at the USAF Tropical Survival School in the Panama Canal Zone and to ocean survival training off the coast of Florida.[19]

There was a possibility that the astronauts would not only land outside of the designated recovery zones but in hostile territory. Provisions for possible forceful recovery of the astronauts and spacecraft, in the event that they were forcibly detained by rebel or guerilla groups or by unfriendly tribal elements were incorporated in the "Conceptual Global Air Recovery Plan for Projects GEMINI, MOL, and APOLLO."[20]

After four years of operation, the KH-4A and KH-7 programs were terminated in 1967, with the KH-4B and KH-8 becoming the sole photographic satellites. The KH-4A program concluded on March 30, 1967, with its 46th successful launch in 51 attempts. Those 45 spacecraft had orbited with a mean inclination of 78.7 degrees, perigee of 107 miles, apogee of 242 miles and a mean lifetime of 23.6 days.

The KH-7 program concluded with the successful launch of June 4, 1967. Thirty-six of 38 KH-7 launches had successfully placed the close-look satellite into orbit. The satellites operated at an average inclination of 97.2 degrees, with a mean perigee of 92 miles, a mean apogee of 187 miles, and a mean lifetime of 5.5 days.

The value of those programs was enthusiastically affirmed by none other than President Lyndon Baines Johnson. Johnson's praise did not come at a meeting of the National Security Council or some other classified forum. Rather, it was during a report on the contribution of space to national security before educators in Tennessee that Johnson made his oft-quoted statement. Speaking to 125 educators, wives, and reporters in a drawing room of the Governor's Mansion in Nashville, Johnson dramatically told his audience that "I wouldn't want to be quoted on this but we've spent 35 or 40 billion dollars on the space program. And if nothing else had come out of it except the knowledge we've gained from space photography, it would be worth 10 times what the whole program has cost. Because tonight we know how many missiles the enemy has and, it turned out, our guesses were way off. We were doing things we didn't need to do. We were building things we didn't need to build. We were harboring fears we didn't need to harbor."[21]

Indeed, while Johnson was praising the value of U.S. reconnaissance satellites, the satellites were busy keeping him apprised of key developments in Soviet offensive and defensive forces.

Soviet offensive deployments included the continued deployment of the single warhead SS-11 SEGO and the initial deployment of the SS-9 SCARP and SS-13 SAVAGE ICBMs. Their deployment in underground, rather than on above-ground silo pads, meant that the CORONAs and GAMBITs generally produced pictures of different types of silos being constructed rather than of the missiles themselves. While the SS-13 deployment never exceeded 60, the SS-9 and SS-11 missiles became mainstays of the Soviet arsenal. The SS-11 force eventually reached 1030, some of which had as many as six independently targetable warheads. The SS-9 had the most dramatic impact. Its approximately 10,000 pound throw-weight allowed it to carry a 20 megaton warhead or three 5 megaton warheads. Whether those warheads were independently targetable and targeted on U.S. ICBM silos became a subject of intense debate within the intelligence and defense communities during 1969.[22]

Keeping track of those developments required the integrated operation of both the CORONA and GAMBIT systems. The SS-9 missiles were placed in newly constructed silos in the south central Soviet Union. Before GAMBIT satellites could produce their high-resolution photos of the new missile fields, those fields had to be found. CORONA surveys of regions where other intelligence or logic indicated the possible construction of new silos were required to confirm the existence of the new missile fields.

At the same time, the Soviet Union was pressing ahead with its development of an anti-ballistic missile system to degrade any attack from the U.S. strategic missiles, which had reached a ceiling of 1000 Minuteman, 54 Titan II ICBMs, and 656 Polaris SLBMs on 41 submarines. Soviet engineers and work crews were busy at several ABM sites outside of Moscow where construction had begun in 1964. The interceptor was designated the ABM-1 Galosh, a nuclear tipped two-stage missile with a 250-mile range. The ABM network was to consist of four complexes of 16 launchers (divided into batteries of eight) on soft sites, along with target acquisition and tracking radars. Construction was sporadic during 1964 and 1965, possibly due to design problems, but by the second half of 1966, the project accelerated. Land was cleared for eight complexes, mostly at old SA-1 sites, deployed in a circle about 45 miles from Moscow. Construction work went ahead at six sites, with no construction at the two southernmost sites.[23]

The operations of the CORONA and GAMBIT programs allowed Robert McNamara to provide a detailed description of the Soviet activity to Congress. The Defense Secretary described three large radars and "bulldozers clearing launching pads, excavating shovels digging trenches for cables and deep holes for launchers, concrete pourers laying out pads and access roads." Each complex had 16 launchers and two radar "triads" (two sets of one large and two small radars). These were supposed to track the incoming warheads and guide the interceptor missiles toward them. The command center tying them together was a "multi-level structure built entirely underground."[24]

Before McNamara and Johnson left office they were to learn the limitations of the satellites that were so effective in monitoring the deployment of Soviet missiles. The first lesson began on June 5, 1967, the morning after the launch of the final KH-7 mission, when Israel launched a devastating air strike at the end of the early morning alert of the Egyptian air force. Since May 16, the Israeli cabinet had seen the withdrawal of U.N. troops from the Sinai and Gaza, the closure of the Suez Canal to Israeli shipping, and the blockade of the Straits of Tiran. A cable came from Meir Amit, the head of the Mossad (the Israeli secret service), who was in Washington, indicated that the United States no longer thought a serious international effort to open the canal likely. On June 4 the Cabinet voted to go to war.[25]

Over the next six days Israeli forces racked up devastating victories on three fronts. By noon on June 5, Egypt had lost 309 of its 340 serviceable aircraft, including all 30 of its TU-16 bombers that could be used against cities. Three Israeli Defense Force armored corps broke into Egyptian territory, took the Gaza strip, and penetrated to the heart of the Sinai.[26]

To the east, Israeli forces took on the Jordanian army. In response to the Jordanian strafing of a small Israeli airfield, Brigadier General Mordechai Hod, the Air Force commander, struck back. The Israeli Air Force caught 30 Jordanian planes on the ground, delivering a crippling blow. Nor did the Israeli ground forces find the Jordanian army a difficult adversary, rolling through the West Bank in a matter of days.[27]

Syria had also struck against Israel on the opening day of the war. Syrian planes had bombed an oil refinery, Israeli positions at the Sea of Galilee, and an air base. The Israeli response, an air strike, led to the loss of 57 Syrian planes, all but eliminating that branch of the Syrian armed forces. Ground combat, however, did not begin until after the Egyptian and Jordanian campaigns were virtually completed. On June 9, Minister of Defense Moshe Dayan instructed the IDF to seize the Golan Heights, from which Syria had been conducting artillery attacks in peacetime. During the first day of fighting, four breaches were made in the Syrian line. At dawn the next day Israeli pressure increased. By noon the town of Kuneitra had fallen into IDF hands and the road to Damascus was open.[28]

From the beginning of the war, the United States was monitoring events as closely as possible. Washington's first notification that war had broken out came at 2:38 A.M. from an Israeli Defense Ministry announcement. President Johnson was wakened by a call from his national security adviser, Walt Rostow, at 4:30 A.M., to announce the beginning of the war.[29]

The events were of the greatest concern to the President and his advisers. It was not just that Israel was an important ally or had intense support in the United States. Given the heavy involvement of both the Soviet Union and the United States in the region, it was also a source of danger. Continued fighting might either threaten Israel's existence, which could be devastating for the United States, or more likely result in pressure for Soviet intervention to prevent Israeli troops from marching into Cairo and Damascus.

One means of monitoring events was communications intelligence, the intelligence derived from intercepting the communications of the governments involved in the conflict. The U.S.S. Liberty was stationed off the Israeli coast to provide such intelligence, which it did before it was bombed by Israeli planes. U.S. spy planes, the U-2 and SR-71, could also be dispatched from Cyprus to photograph battle areas, but neither CORONA nor GAMBIT made a contribution. As noted earlier, a KH-7 had been launched on the day before the war. In addition, a CORONA mission that began on May 9 continued for 64 days, including the entire period of the war. In an attempt to get better coverage, the orbit of one of the satellites was altered, but apparently the film returned was of poor resolution. It is not surprising, therefore, that former Defense Secretary Robert McNamara does not recall satellite reconnaissance as having played any role in U.S. intelligence-gathering during the war.[30]

At the same time that KEYHOLE satellites were busy monitoring military activity, the intelligence community was trying to restore some order to the exploitation of the imagery they were producing. When the U-2 aircraft had been approaching operational status, an Ad Hoc Requirements Committee of the DCI's Intelligence Advisory Committee (IAC) was established. Chaired by CIA official James Q. Reber, the committee's function was to determine and prioritize the requirements for photography from the different consumers. As CORONA and SAMOS approached operational status a Satellite Intelligence Requirements Committee was also established, under the DCI's United States Intelligence Board, as IAC had become known.[31]

The two committees were subsequently merged into the Committee on Overhead Reconnaissance (COMOR), chaired by Reber. COMOR only had authority over satellite targeting—deciding which particular satellite would photograph a specific target, when, and how often. So although a rational plan could be developed to select targets, none existed for photo-interpretation. The CIA, DIA, Army, Navy, and Air Force elements at NPIC as well as many of the service intelligence agencies had their own photo-interpretation capability. In the absence of a coordination mechanism, there was no means of ensuring that a photo-interpreter at the Air Force Systems Command's Foreign Technology Division at

Wright-Patterson AFB in Ohio was not investing significant amounts of time studying the same picture of Soviet MiGs that an analyst at DIA was scrutinizing. Even the CIA and DIA analysts at NPIC might be duplicating each others' work. To resolve the problem, on July 1, 1967, Director of Central Intelligence Directive 1/13 transformed COMOR into COMIREX, the Committee on Imagery Requirements and Exploitation. DCID 1/13 specified that COMIREX's members were to include representatives from all the departments and agencies represented on the United States Intelligence Board plus representatives from the Army, Navy, and Air Force.[32] COMIREX was to be responsible not only for targeting the satellites but also for determining which agency's interpreters would have primary responsibility for interpretation of the imagery being returned by the satellites.[33]

Thus, by 1968 the institutional infrastructure for managing the KEY-HOLE program had been established. And 1968 was a busy year for the program. Eight KH-8 satellites were launched along with eight KH-4Bs, one less of each type than the previous year. The lifetime of the KH-8 satellites was extended somewhat, varying from 17 days (the first launch of 1968) to as low as eight days (the last launch). Most of the satellites orbited for 11 to 15 days.

Even though the KH-8 launch vehicle performed flawlessly, placing every satellite in its proper orbit, there were serious problems actually getting useful imagery. These problems were particularly frustrating because they occurred as the crisis in Czechoslovakia came to a climax.

The crisis dated back to late 1967 when Soviet leaders acquiesced in the Czech Communist Party removal of hard-line Antonin Novotny as First Secretary and his replacement by Alexander Dubcek. Dubcek's ascent to power fully unleashed reformist sentiment within the party and Czech society. The new Czech leaders' concept of "socialism with a human face" was incorporated into their Action Program of April 1968. The program called for allowing greater intraparty democracy, grant-ing more autonomy to other political parties and parliament, restoring basic civil rights like freedom of assembly, vigorously continuing polit-ical rehabilitation, and instituting economic reforms. Dubcek also per-mitted the establishment of several new political clubs and abolished censorship.[34]

These developments alarmed those who ran the Soviet Union and other Warsaw Pact countries on two grounds. The poison of liberalization, they feared, might spread to their countries and make it harder to maintain their repressive regimes. Walter Ulbricht, the East German communist party leader, and Pyotr Shelest, Soviet Politburo member and Ukrainian party boss, were particularly concerned. Such a dramatic shift in domestic policy was also seen as a serious threat to the unity of the Warsaw Pact. In light of internal liberalization, the Czech commitment to following the Soviet line in foreign policy and being a loyal member of the Warsaw Pact was considered questionable.[35]

That the Czech experience was perceived as a threat by the Soviet Union, East Germany, and others in the Warsaw Pact was evident early on from public statements by Soviet leaders. Attention in the West turned to the question of whether the Soviets would use brute force again as they had in Hungary in 1956. Among those most concerned were the top national security officials of the United States and the U.S. intelligence community was tasked with monitoring Soviet/Warsaw Pact activity and estimating likely reactions.

In addition to the data that could be obtained from the Soviet press, diplomatic reports, clandestine agents, and signals intelligence, the KEYHOLE satellites also could provide important data. Signs of impending invasion that might show up in satellite photography were increased activities at airfields, troop departures, extensive logistics activities, and, most dramatically, the massing of troops near the Czech border. Such activities would be detected by U.S. satellites, but the intelligence would be too late to even attempt to forestall an invasion.

A KH-8 launched on August 6 performed poorly and was deorbited after nine days, so the CIA was forced to rely solely on the KH-4B launched on August 7. A film package returned before August 21 proved reassuring because it showed no indications of Soviet preparations for an invasion. But on August 21, Warsaw Pact troops led by troops from the Soviet Union stormed into Czechoslovakia and ended the Prague Spring.[36]

The Czech invasion was the second major military event in two years to highlight the limitations of satellite photography that was considerably less than instantaneous. When the second and last of the CORONA film buckets was recovered and analyzed, it clearly showed Soviet preparations—including massing troops—for an invasion. By then, however, those photos were only of historical interest.[37] The experience was not forgotten by many in the photo reconnaissance program. One former CIA official recalls that people were "still talking about it years later."

According to the official, "a lot of good work was done in retrospect" because the photo intelligence did prove valuable in developing warning indicators to be used in conjunction with the KH-11.[38]

That real-time photography could have substantially enhanced U.S. ability to follow developments leading up to the invasion was made evident by the report of the House Select Committee on Intelligence chaired by Otis Pike. The committee report indicated that revelatory signals intelligence had not arrived in Washington until after the Czech radio had announced the invasion, and human source reporting had been "so slow to arrive it proved of little value to current intelligence publications."[39] Near instantaneous imagery could have provided warning.

Though the Czech army did not resist the overwhelming Soviet forces in 1968, a year later the army of another troublesome socialist state—the People's Republic of China (PRC)—was considerably more willing to engage in combat with the Red Army. The conflict was a vestige of the Czarist period, when Russia had seized land previously controlled by the Chinese emperor.

In 1963, for the first time since the establishment of the PRC, Mao Zedong began to question the border treaties signed by Czarist Russia and the Ch'ing Dynasty, calling them unequal and subject to revision. In a July 1963 meeting with members of the Japanese Socialist Party, Mao noted that "about a hundred years ago, the area to the east of Baikal became Russian territory, and since then Vladivostok, Khabarovsk, Kamchatka and other areas have been Soviet territory. We have not yet presented our account for this list."[40]

In secret conversations in February 1964, China informed the Soviets that they were not demanding the return of the territory but wanted a Soviet admission that the treaties were unequal. After such a Soviet admission, the treaties could be renegotiated. The Soviets were afraid, however, that such an admission would establish a precedent with regard to Eastern Europe, that the Chinese might fail to negotiate, and by the Chinese insistence that most of several hundred islands in the Amur and Ussuri rivers belonged to China.[41]

The Soviets had much at stake in the region of the dispute. Two of Siberia's major cities, Vladivostok and Khabarovsk were located there, and the Trans-Siberian Railroad passed only 20 miles from the Chinese border. The railroad was the only lateral land route from Central Russia to the Pacific and a crucial means of shipping supplies to East Siberia, including oil for the Pacific Fleet.[42]

Given the Chinese claims and the general drift of Chinese foreign policy, the Soviet Union began to improve its Far East defenses, beginning by signing a new defense treaty with Mongolia in January 1966. Soviet troops, which had not been garrisoned in Mongolia since 1957, began to return. By early 1967 nearly 100,000 had arrived.[43]

The defense treaty with Mongolia was not the only 1966 initiative. Moscow began supplying its Far Eastern forces with nuclear warheads and transferred at least seven Soviet divisions from central Asia to positions east of Lake Baikal, reinforcing the 15 to 17 divisions already there.[44]

Tensions between the Soviet Union and China continued to build over the next two years, but it was not until 1969 that serious shooting began. On March 2, Soviet troops on Damansky Island in the Ussuri, about 250 miles from Vladivostok, were attacked by machine-gun–firing Chinese troops. Over 20 Soviet soldiers died that day and the Soviets reacted two weeks later when a Soviet force of approximately battalion strength, armed with tanks and artillery, attacked a Chinese force on the same island.[45]

Fighting and threatening gestures continued over the next two months, and in May, Soviet and Chinese forces exchanged fire along the Amur River. The Soviets also began serious preparations for air raids inside China. Late-model Soviet planes were transferred from Eastern Europe to newly enlarged airfields in Outer Mongolia. In June, Soviet bomber units flew mock exercises against targets in northwest China.[46]

While the Soviets and Chinese were shooting at each other, the United States was listening and watching. Using signals intelligence stations in Asia, and possibly a newly launched COMINT satellite codenamed CANYON, the National Security Agency was intercepting massive amounts of Soviet and Chinese communications. The Asian antennae of the CIA's Foreign Broadcast Information Service were busy receiving the broadcasts of Soviet and Chinese regional radio. And flying overhead were the KH-4B and KH-8 cameras, spotting, among other things, Soviet installation of missiles across from Manchuria.[47]

During the three months in which the two communist giants were on the brink of war, three KH-8 and two KH-4B systems were lofted into space. Increased coverage of key border areas, troop barracks, and airfields kept the spaceborne cameras busy. A great deal of effort was made to acquire data and some was acquired, but there was also much frustration. The problem was not timeliness, as was the case with the Six-Day War and Czechoslovakia. Rather, the swath width of the CORONA satellites—the amount of territory they could photograph with a single frame—was relatively small. As a result there was less chance of spotting an event of significance or of obtaining repeat coverage than with a satellite with a greater swath width. Such a satellite would not be in operation for another two years.[48]

The year 1969 saw the demise of the MOL. Increasing costs and diminished political support pushed the MOL into an early grave. The MOL's demise was unexpected given that its budget for 1968 was greater than for any previous year, which gave the impression that it had more than adequate support. The 1968 budget included structural testing and fabrication of the first three flight MOLs (two unmanned test vehicles and the first manned spacecraft), development of the seven-segment solid-fuel strap-ons, and installation of equipment at the launch pad at Vandenberg. Six-hundred million dollars would be required to cover the cost and keep the program on schedule.[49]

In quieter times such a request might have been approved without much difficulty, but the times were anything but quiet. The war in Vietnam was escalating, with repeated air strikes on Vietnam. As a result, the war was absorbing an increasing proportion of the defense budget. Military programs not related to the war effort were reduced or simply cancelled outright to provide more money for the war. The MOL, the largest non-war item in the Air Force research and development budget, made an inviting target.[50]

Even with the Administration's requested defense budget, MOL would have been in trouble. Congress expressed no enthusiasm for the project and directed that a $6 billion cut be made in DOD funding. It was feared that the MOL might be cut by as much $100 million, which would result in a major delay. Congress finally directed an $85 million cut, which affected the program's timetable. The first unmanned launch was still scheduled for late 1970, but the first manned MOL, which had been scheduled for the summer of 1971, was pushed back three months. The delay in the schedule, necessitated by the $85 million cut and technical changes, sent the MOLs' total cost up to $2.2 billion from $1.5 billion.[51]

Initially, Richard Nixon's election seemed to mean increased support for the MOL because the new administration's budget included $576 million for MOL. Though not quite the $700 million that some people believed necessary, it was still substantial. It would pay for thermal and dynamic testing, delivery of the first flight MOL and the start of construction on those to follow, completion of the launch pad, pilot training, and work on booster components.[52]

But Vietnam proved to be a budgetary black hole, absorbing funds without anything coming back. The new Secretary of Defense Melvin Laird found that the outgoing Johnson administration had been too optimistic, as it had continually been concerning Vietnam. Laird foresaw high war costs for at least another two more years, and to further complicate matters, several defense projects including the C-5A transport aircraft had major cost overruns. By cancelling the fifth manned MOL mission, $20 million was saved for these other projects. It was argued that because of improvements in technology, the test goals could be accomplished in four manned flights.[53]

Soon another $31 million was cut, again delaying the launch schedule. The first unmanned MOL flight would not be until early 1971 and the first manned launch would have to wait until mid-1972. The delays raised the total projected cost to $3 billion—twice the original estimate.[54]

Apparently, the actual decision to terminate the MOL program came at a White House meeting between President Nixon, national security adviser Henry Kissinger, and Budget Bureau director Robert Mayo. Mayo suggested eliminating MOL and Nixon and Kissinger approved. Only after the decision was made was Defense Secretary Laird informed that the MOL was even in trouble. Mayo called Laird, who apparently became very upset and objected strongly. In subsequent congressional testimony Laird told Congress: "I fully advised the President . . .that the Joint Chiefs of Staff supported this program. I asked for the opportunity for an appeal to be made directly to him and an appeal was directly made to him but the decision was made not withstanding the position of the Joint Chiefs and the position which was taken by the Department of Defense." The public announcement was made on June 10, 1969. A total of $1.4 billion had been spent.[55]

Another factor in the cancellation of the MOL was the lack of support from CIA Director Richard Helms. Without Helms' endorsement it was harder to convince the president of the importance of the MOL. Some Air Force officials apparently viewed Helms' lack of support as retaliation for the Air Force takeover of the aerial reconnaissance mission, specifically for the demise of the CIA A-12 reconnaissance plane in favor of the Air Force SR-71. But it was more likely a CIA concern about the advisability of a manned space reconnaissance platform. Some feared that such manned spy vehicles might be considered intolerable by the Soviets and that they would take action, possibly using their Galosh ABM interceptors to destroy the MOL as it passed over Moscow. But more important, they feared that an accident that cost the life of a single astronaut might ground the program for an extended period of time and cripple the reconnaissance program.[56]

The MOL left the United States with two legacies. The launch pad for the MOL that had been constructed at Vandenberg, SLC-6, had been completed and would eventually be assigned as the launch pad for space shuttle launches from Vandenberg.[57] Also, the astronauts who had been selected for the MOL program went on to some high profile positions in the Air Force and space program. The MOL astronauts included James A. Abrahamson, later director of the Strategic Defense Initiative Organization (SDIO); Robert Herres, future Air Force Chief of Staff; and Richard Truly, NASA Administrator.[58]

In August 1968, when Warsaw Pact forces invaded Czechoslovakia, the United States and Soviet Union were about to announce their plans to begin arms control negotiations. The impact of the invasion on U.S.– Soviet relations plus the change of administrations meant a substantial delay before the talk could be rescheduled. It was not until November 17, 1969 that the United States and Soviet Union began the Strategic Arms Limitation Talks (SALT) in Helsinki.

In addition to the communications exchanged between U.S. and Soviet delegations, the CORONA and GAMBIT satellites were able to pick up signals that involved actions, not words. From late 1969 to early 1970 the Soviets started only 10 new SS-11 and SS-13 silos. Of even greater significance was the total absence of SS-9 starts made after August 1969, which was interpreted as a signal of Soviet good faith in the SALT negotiations.[59]

At the same time, construction of existing silos continued. By April 1970, 222 SS-9s were operational while another 60 were still under construction. The SS-11s continued to be deployed and the fifth Moscow ABM site began operation.[60]

The Soviet moratorium on new silos held until May 1970, when the Soviets began construction of 24 new SS-9 silos. Six were in an established field and the remainder in a new field in the south central

U.S.S.R. On June 25, a KH-8 began an 11-day mission, photographing Soviet missile installations. Within 48 hours of the end of the mission on July 5, the film was flown to Washington, delivered to NPIC, processed, and analyzed. The following day, July 8, Defense Secretary Melvin Laird announced that the Soviets had broken the construction moratorium. Observation of the new silos continued and revealed that the Soviets had apparently reacted to Laird's announcement. A KH-8, which blasted off on October 23, rephotographed the new missile sites. Again the Soviets were relying on America's secret eyes to detect and transmit a signal. The photography from the mission, which concluded on November 10, indicated that the Soviets had dismantled the 18 most recent silos and had slowed work at the other more complete ones.[61]

CHAPTER 5

The Big
Bird Arrives

By the late 1960s, the Air Force began to consider successors to the CORONA system. An Air Force initiated study, codenamed VALLEY, examined the possibilities and concentrated on incremental improvements in the KH-4B camera system. Two space tests of a 20-foot diameter antenna were also conducted. Such an antenna could be used in another attempt to return imagery by scanning the film and converting that data into electronic signals.[1]

Others had more radical changes in mind. Edwin Land had expressed his preference for a system that would have the 18-inch resolution of the KH-7 and the swath width of the KH-4B, but "Bud" Wheelon had an even more radical vision. Wheelon envisaged putting a camera system in space that could produce photographs covering a much greater area than the KH-4B but still have the 18-inch resolution of a KH-7.[2]

Years earlier when the Air Force had proposed a high-resolution close-look system, Wheelon had prepared an elaborate technical analysis demonstrating that the system would not work. Wheelon was proven wrong, for the system that he had demonstrated would not work turned out to be the highly-successful KH-8. Now it was Wheelon's turn to hear from the skeptics. An advisory panel reviewed Wheelon's proposal and concluded that his revolutionary system would not work, but Wheelon was far too convinced of his own expertise to be discouraged. Before he left the CIA in 1967 his vision appeared to be well on its way to realization.[3]

Initial work on the system began in 1965 and continued for a year or so. Not surprisingly, Lockheed's Space and Missile Division was again selected the prime contractor and the satellite was given the Lockheed designation "Program 612." In 1968 the program number became 467, with development well underway. The program had also been assigned a BYEMAN codename, HEXAGON, although it was to become much better known by the name it received in the press—"Big Bird."[4]

But in 1969 it appeared that HEXAGON would never see the darkness of space when the program was abruptly cancelled. The same budget crunch created by the Vietnam war and the Great Society that shot down the MOL also threatened to end the HEXAGON program before it reached the launching pad. As a substitute, plans codenamed HIGHER BOY were developed to put some KH-8 spacecraft into an area surveillance orbit. Such ad hoc substitutes were not acceptable to those running the reconnaissance programs. In an attempt to restart the flow of money, COMIREX chairman Roland Inlow was sent to talk to James Schlesinger, Assistant Director of the Bureau of the Budget and the person responsible for national security programs. Inlow explained that the KH-9 was essential to arms control verification and the money began to flow again.[5]

The spacecraft that lifted off a Vandenberg AFB launch pad on June 15, 1971 was a tribute to Wheelon's ruthlessness in pursuing his vision. Launched into space by a Titan 3D with three million pounds of thrust, the newest addition to America's reconnaissance arsenal was a 30,000 pound cylinder, 40 feet long and 10 feet in diameter. Its size would allow it to host a variety of other projects. Thus, the new generation of surveillance satellite would often carry antennae to collect signals intelligence and to relay messages from U.S. covert agents in the Soviet Union and elsewhere. The additional missions would often lead NRO to keep the spacecraft in orbit even after all the film had been returned to earth. HEXAGON launches would also often carry a second payload: "ferret" spacecraft designed to detect and record signals from Soviet and other radar systems.[6]

The U.S. had begun orbiting ferret satellites in 1962. From that time until 1972 one class of ferrets was the sole payload carried on their boosters. Generally operating in orbits of 300 miles, they would collect the signals emitted by the air defense, ABM, and early warning radars of the Soviet Union, China, Vietnam and other nations. A second class of ferrets was put into operation beginning in August 1963. This second class was launched as a secondary payload on numerous CORONA missions. By 1972 they were also operating in 300 mile circular orbits. By 1973 only this second class continued to operate and HEXAGON missions carried these electronic hitchhikers into orbit.[7]

The camera system, designated the KH-9, consisted of two cameras with 60-inch lenses. The cameras could operate individually or could obtain overlapping photos of a target. The overlapping photos could then be used with a stereoscope to extract additional information about the target's dimensions. The cameras were able to produce images covering a much wider area than the KH-4B but with a resolution of two feet, almost as good as the 18-inch resolution of the KH-7. Whereas the KH-4B camera system had a swath width of 40 by 180 miles, the KH-9 system was twice that, 80 by 360, resulting in a four-fold increase in the territory that could be covered by a single photo. And while the KH-4B returned two film capsules, the KH-9 returned four.[8]

As a supplement to the standard film designed for visible-light photography, both black and white and false-color infra-red film was spliced into some KH-9 film reels. Such film is sensitive to the near-infrared portion of the electromagnetic spectrum. Because such film records the radiation (usually sunlight) that is reflected from objects, it requires daylight conditions. Near-infrared photographs taken on a hazy day will show distant objects with more clarity and contrast than visible-light photos. Greater contrast between land and water can also be obtained using infrared film. Most significantly, film sensitive in the infrared can often expose attempts at camouflage. Black and white infrared film shows healthy vegetation in a light grey tone while ordinary green paint and cut vegetation, either of which might be used for camouflage purposes, appear in a darker grey. Color infrared film produces false-color photos of its targets—dying vegetation appears as blue or cyan but healthy vegetation as red.[9]

A third camera would be carried on five HEXAGON missions. This 12-inch mapping camera would have its film fed into a fifth film capsule and be returned at the end of the mission.[10]

A secondary experimental system was used to transmit pictures by radio signals—essentially the same system that had failed when it operated on SAMOS. The results were no better this time and the system was eventually jettisoned.[11]

The KH-9 represented a major advance in U.S. reconnaissance capabilities. The greater film capacity meant longer lifetimes and that film could be returned as frequently as with the KH-4B in normal circumstances. In the early days of the program the KH-9 would return film capsules every three or four days, and in emergencies, an incomplete reel could be returned without drastic damage to the overall mission. But most importantly, the tremendous swath width of the KH-9 meant that true wide-area searches could be conducted to locate new missile fields, test ranges, and nuclear facilities. It also meant that a greater number of requests for photography could be accommodated because of the ability to incorporate a wider area in a single scene. Thus, lower priority targets had a better chance of being photographed.[12]

The ability of the KH-9 to photograph huge chunks of territory was a delight to the mappers of the newly created Defense Mapping Agency (DMA), which was established in 1971 to consolidate the mapping activities of the military services. The fewer photos needed to cover a part of the world, the easier it was to construct an accurate map. When the KH-9 program was terminated, the mappers at DMA "wept blood" according to one intelligence official.[13]

The first KH-9, also designated 1901, operated in an elliptical 114 by 186 mile orbit, with a 96.4 degree inclination. Its inclination ensured that 1901 would not only cover the entire earth from pole to pole in the course of its operations but also that its orbit was sun-synchronous. A sun-synchronous orbit ensures that each daylight pass over an area is made at an identical sun-angle, avoiding differences in pictures of the same area that result from different sun-angles. Each ground track would repeat every 3.5 days. On the days that a particular area was overflown it would be overflown twice—once in daylight and once in darkness.[14]

Between its launch on June 15 and its destructive re-entry 52 days later on July 6, 1901's operators checked out its imaging, communications, and propulsion systems and began operations. They also tested it extensively to determine how well the new camera system held up to its theoretical promise. As with all new satellites, photographs were taken of a variety of locations in the United States where the dimensions of the target and energy emissions could be precisely determined.[15]

From that point on, both the KH-4B and KH-9 were employed until the supply of KH-4B spacecraft was exhausted. As with the transition from the KH-7 to KH-8 close-look spacecraft, the new spacecraft was first deployed while several older-generation spacecraft remained, to provide a fallback if the new generation spacecraft was found to be flawed. Once the new generation proved successful it was alternated with the older generation until the supply of the latter was exhausted.

The several months bracketing the launch of the first KH-9 were notable for the intelligence produced by the KH-4B and KH-8 systems.

On February 8, 1971, a KH-8 satellite launched on January 21 returned a film pack that was analyzed by the start of March. The pictures revealed 10 new Soviet missiles in SS-9 fields in central Russia. The holes were larger than those normally associated with SS-9 deployments, and the Soviets were employing a digging technique that did not allow for the determination of the likely dimension of the missiles intended for the hole. The Soviet practice, when starting work on new SS-9 silos, had been to clear a wide area of land and then to start digging a hole for the silo in the center. It was possible to estimate the size and make a reasonable guess at the likely dimension of the missile intended for the hole. With the new holes the Soviets had used a "funnel technique." That is they started with a very large hole, and then narrowed it down inside. The cameras could not see how much narrowing was being done.[16]

The data were passed to Senator Henry Jackson, who claimed that the data presaged the deployment of "a new generation . . . of offensive systems . . . big or bigger than the SS-9s," a claim that did not help the SALT I negotiating atmosphere. These concerns were apparently raised by Henry Kissinger in discussion with Soviet Ambassador Anatoly Dobrynin, who offered verbal assurances that no "monster missiles" were being deployed. However, the Soviets offered more than verbal assurances. The pictures that came back from a KH-4B satellite on April 12 showed in great detail the contents of the new "holes." The photographs revealed next to several holes the entire set of liners for the silos, laid out in rank order of emplacement and face upward, so they could be properly photographed. Even missile canisters were provided so that the diameter of the "new" missiles could be assessed.[17]

The detail provided by the photographs enabled the well-connected columnist Stewart Alsop to write a detailed description of the new silos on May 10:

> First they would build two fences, sometimes three, around a 100 acre site. Then they would dig a big flat hole, around 100 feet across and 25 feet deep. . . .
>
> Inside the first hole the Russians would then dig another, deeper hole, about 30 feet across and 120 feet down. They would then line the hole with concrete, put a steel liner inside that and then lower the big missile into the liner. In the remaining empty space of the first big hole, they would build

a complex of workrooms, generators, fuel pumps and so on, and cover the whole thing with a thick sliding door.[18]

The Soviets had obviously decided to send a message via spy satellite to the United States, for the photos indicated that the alleged monster missiles would be no larger than the SS-9. That in itself was only partially reassuring because silo construction was proceeding rapidly, with approximately 40 silos at six different locations being identified by the end of April. The figure was to grow to 60 in May, almost 80 by June and reach its maximum of 91 in October. But the KH-4B photographs from April indicated that the silos would not be housing 91 of the three-warhead SS-9s, for the new silos were shown to be under construction at both SS-9 and SS-11 missile fields. The photos also showed different-sized protective concrete liners, which indicated to most analysts that silo improvement programs were underway for both the SS-9 and SS-11 missiles. That conclusion was comforting because the SS-11 was considered far less threatening to the American ICBM force than the SS-9, being less accurate and having only a single warhead mounted on its nose.[19]

The issue was officially resolved with the statement of Pentagon spokesman Jerry W. Friedheim that "new information now available to us leads us to conclude the Soviets may be involved in two separate silo improvement programs."[20]

By January 1972, U.S.–Soviet negotiations towards an arms limitation agreement had been in progress for over two years. The sixth of an ultimate seven rounds of negotiations, alternating between Helsinki and Vienna, was wrapping up. In addition to the negotiations conducted at those European capitals by Ambassador Gerard Smith and his SALT delegation, Henry Kissinger and Anatoly Dobrynin were negotiating via their "backchannel," much to the annoyance of Smith.

As the ultimate meeting at Moscow approached, a second KH-9 (1902) was launched on January 20, 1972 and remained in its 92 by 213 mile orbit for 40 days. The launch, part of a step-up in U.S. reconnaissance activities, was designed to provide an updated survey of Soviet strategic forces in preparation for the final SALT negotiations. Between January and May 1972, one KH-8 and two KH-4B spacecraft were launched in addition to 1902.

It was quite logical that such satellites would not only provide intelligence support to arms control negotiators but would also be crucial in verifying compliance. In testimony to Congress in 1970, Herbert Scoville noted that fixed land-based ICBMs "require extensive launch-site construction in order to provide the necessary hardening to make them resistant to blast from a nuclear explosion. This construction requires many months and therefore ample time is available to permit its detection." In addition, Scoville noted that "after . . . submarines are launched they require many months for fitting out, during all of which they are subject to observation."[21]

But two years earlier the intelligence community had argued that its sensitive overhead systems could not be employed for monitoring compliance. The community had argued that use of those systems for verification purposes would require the United States to make its capabilities public to establish that it could verify compliance. It was also argued that signing an agreement to be monitored by intelligence systems would reveal the capabilities of those systems, and that charges of violations would have to be backed up by revelation of the data indicating such violations. Thus, on-site inspection and other sources of information would be required. To explore the possibility of verification without the use of reconnaissance satellites, the Air Force conducted a study codenamed CLOUD GAP.[22]

Before the Johnson administration left office it overruled the objections of the intelligence community. But a plan to declassify "the fact of" U.S. satellite photography met opposition from NRO, which stalled until the Johnson administration was out of office.[23]

But the Johnson administration's decision to employ the NRO's space assets for verification held up. After some last-minute snags that threatened to delay the signing of the treaty, U.S. and Soviet negotiators agreed to limits on both offensive and defensive forces. At 11:00 P.M., at Spasso House, Leonid Brezhnev and Richard Nixon signed the "Interim Agreement Between the United States of America and the Union of Soviet Socialist Republics on Certain Measures with Respect to the Limitation of Strategic Offensive Arms" and the "Treaty Between the United States of America and the Union of Soviet Socialist Republics on the Limitation of Anti-Ballistic Missile Systems." The Interim Agreement set limits on the number of launchers each side was allowed — 1710 for the United States (1054 ICBM + 656 SLBM), and 2358 for the Soviet Union (1618 + 740). No more than 313 ICBMs could be heavy missiles such as the SS-9. The Soviets could increase the number of submarine launched missiles to 950. To do this, they would have to scrap one of the older SS-7 or SS-8 ICBMs for each new SLBM (16 ICBMs for each Y-class submarine).[24]

While Nixon and Brezhnev were signing the historic treaties, one U.S. photo reconnaissance satellite, launched the day before, was circling the globe. That the last CORONA should watch over that signing and the beginning of a new era in arms control was appropriate, as if it were taking a bow and handing over the mantle to the following generation. For without CORONA and its successors there would have been no possibility of stabilizing the arms race and reaching agreements such as those signed at Spasso House. In the treaties there were no provisions for on-site inspection. Rather, "national technical means of verification" were to be used, a term which most emphatically included CORONA, GAMBIT, HEXAGON, and their Soviet counterparts.

That final KH-4B was deorbited on June 3. The program showed 33 successful launches out of 34 attempts. The satellites had operated at a mean inclination of 83.5 degrees, had a mean perigee of 105.9 miles, a mean apogee of 198 miles, and a mean lifetime of 23.4 days.

When the third KH-9 was launched on July 7, 1972, it had a dual and overlapping function: monitoring Soviet compliance with the newly signed treaty as well as Soviet and Chinese military activities. For the next 68 days it operated in its 96 degree, 108 by 156 mile orbit. The third KH-9 also established a pattern with the KH-8 satellite launches. On September 1, two weeks before 1903 re-entered, a Titan 3B-Agena D placed a KH-8 into a 110 degree, 87 by 236 mile orbit for the next 29 days. The pattern of launching a KH-8 satellite during the latter part of a KH-9 mission would continue for the rest of the 1970s.[25]

One more pair of launches took place in 1972. On October 10, ten days after the end of the KH-8 mission, the fourth KH-9 began a landmark mission. The KH-9 operated in its normal orbit of 102 by 166 miles with a 96 degree inclination. Most notably, 1904 remained in orbit for a full 90 days, surpassing by 22 days the lifetime of the previous KH-9 and establishing a pattern of its own. With some rare exceptions each KH-9 mission would have a longer lifetime than the previous one and in some cases a much greater lifetime. Before the HEXAGON program was terminated some KH-9 missions would seem to last forever.*

---

* As might be expected a KH-8 was launched on Dec 1 on a 33 day mission, with a 110 degree inclination, 83 mile perigee and 251 mile apogee.

One apparent exception occurred in 1973, however, with the launching of 1905 on March 9. The fifth KH-9 remained in its 94 by 167 mile, 96 degree orbit for only 71 days. Ironically, the failure was apparently due to the intervention of new DCI James Schlesinger, the same James Schlesinger whom Roland Inlow had convinced of the need for the KH-9. Schlesinger, an economist and long-time student of strategic questions, had worked at the RAND Corporation in the 1960s. During that time he developed the persona of a strategic analyst—cool, slightly jaded, arrogant, and tough-minded.[26] In 1970, while at the Office of Management and Budget, Schlesinger had studied the intelligence community for President Nixon and produced "A Review of the Intelligence Community." In his report Schlesinger noted a significant rise in costs for intelligence activities that was not matched, in his judgment, by a significant increase in the value of those activities. Causes of the problem included the continuation of old technical collection programs even as new programs began to operate as well as the overlapping collection activities of different agencies. Schlesinger also noted that "How much power the leader [DCI] can exercise, particularly over collection programs, will determine the size of the economies that can be achieved."[27]

According to a former intelligence official, in an attempt to exert such power, Schlesinger ordered the virtually immediate launch of a KEYHOLE satellite whose launch had been delayed. But launching a satellite is not a process that can be rushed. According to that same former intelligence official, almost every instance in which a reconnaissance satellite has been launched in a rush has ended with a failure. Preparations for launching include readying the launch pad, checking tracking and communications equipment, and programming the satellite. In addition, each satellite requires thermal preparation. Given the orbit the satellite flies, it must be prepared so that it absorbs sufficient but not excess heat when it faces the sun and retains sufficient heat when it faces away from the sun. Covering the satellite with tape is the means of ensuring that the correct amount of heat is retained, and given the orbital parameters and time of year it may be necessary to remove some of the tape. However, Schlesinger's rush order left no time to calculate such matters. The consequence was launching a satellite that returned only 10–15 percent of the data that would normally be returned.[28]

Fortunately, the remaining KH-9 and KH-8 spacecraft launched in 1973—on May 16 (KH-8), July 13 (KH-9), September 27 (KH-8), and November 10 (KH-9)—were properly prepared, for 1972 and 1973 would be busy years for the satellites as new generations of Soviet missiles reached the testing or deployment stage. New and more accurate versions of the SS-11 and SS-13 missiles were deployed in 1972 and 1973. Testing began in 1972 of the SS-16 SINNER, a three stage, solid-fuel missile with a range of 5000 nautical miles and a single 650-kiloton nuclear warhead that was meant to destroy soft targets. More importantly, the SS-17 SPANKER, SS-18 SATAN, and SS-19 STILETTO were also undergoing testing. Unlike the SS-16, which would never be deployed, the other three missiles would become the nucleus of the Soviet ICBM force for the next 15 years. Estimating their capabilities would be of tremendous significance in preparing for SALT I, monitoring Soviet compliance, and determining U.S. force requirements.[29]

KH-8 and KH-9 cameras would frequently provide pictures of the Tyuratam space launch and missile test facility from which the liquid-fueled SS-17, SS-18, and SS-19 missiles would be launched toward their impact area on Kamchatka. After the conclusion of the test they would train their cameras on the impact area to locate the craters created by the impact of the dummy warheads. Although the primary means of determining the accuracy of the missile warheads was the analysis of telemetry, specifically of the signals triggering mid-course path corrections, analysis of the distribution of craters would be used as a supplementary tool.

The photographs produced by America's reconnaissance satellites also showed that there were three new types of silos under construction. The largest were referred to as III-X silos, one of which was being dug at each of several complexes. During the spring of 1973 it was discovered that 150 III-X silos were under construction. If they were ICBM silos, they would constitute a massive SALT violation. In June 1973 the United States vigorously protested to the Soviet representative to the Standing Consultative Commission, which had been established in December 1972 to address compliance issues. In response the Soviets claimed that they were command and control centers and that this would become apparent as construction progressed.[30]

The Soviet explanation was met with substantial skepticism since the doors on the III-X silos were identical to those on normal silos: they could be blown clear moments before a missile was to be launched. Why command and control centers would need such a capability was not apparent. That the III-X silos might later be converted to missile silos was not considered beyond the realm of possibility. However, U.S. intelligence, probably via signals intelligence, was able to verify the Soviet claim.[31]

Events in the fall of 1973 reinforced the lessons of 1967 and 1968, that monitoring missile deployments was far easier than monitoring wars on a timely basis. A missile build-up proceeds slowly. Silos have to be excavated, concrete poured, and missiles transported. International crises and battles proceed quickly. Information that is hours old, much less days or weeks old, may be of little use. In 1973 it was the Yom Kippur War that served to illustrate that fact.

On October 6, 1973, the Jewish Day of Atonement, Israel was surprised by a unified attack by Egypt and Syria. Despite various items of intelligence that indicated such an attack was likely, the adamant refusal of the Israeli chief of military intelligence, General Eli Zeira, to credit such information prevented Israel from taking countermeasures until shortly before the attack.[32]

Three weeks before the war began, Israeli intelligence noticed that Syria had begun amassing its forces and was erecting an extremely dense network of anti-aircraft missiles along its border with Israel. Similar reports indicated that Egypt was starting to move troops toward the Suez Canal. By the end of September there was clear evidence that SA-6 missiles had been distributed among the armored divisions of the Egyptian Army. There had also been reports of Egypt employing excavations for earth-removing operations along the northern section of the Suez Canal. Additional evidence of an impending attack included the mobilization of Egyptian civil defense, the declaration of blackouts in Egyptian cities, and an appeal for blood donors.[33]

Further indication came when Soviet advisers in Syria made a sudden departure on the night of October 4. However, given his adamant belief that Egypt would not attack unless it could win, it was not until 3:00 A.M. on October 6 that Zeira believed Egypt would attack. An agent in the Canal Zone radioed information that Egyptian troops had been ordered to attack at 6:00 that evening. In fact, the attack was launched earlier, at 2:05 P.M.[34]

The value of surprise was evident during the first days of the war. After heavy but brief shelling and an air attack on various objectives in the Sinai, the land assault began. Ten thousand Egyptian infantrymen crossed the canal in small boats, establishing the bridgeheads that would allow another 30,000 soldiers to cross by nightfall. Egyptian forces soon made a dramatic breakthrough, taking control of large areas of the Suez Canal and numerous Israeli strongholds that made up the Bar-Lev line.[35]

Syrian forces also scored initial triumphs. Included among their triumphs was the conquest of the Israeli position on Mt. Hermon, where Israeli military intelligence maintained an eavesdropping post. Indeed the Israeli position was perceived to be so grave at one point that Prime Minister Golda Meir gave orders that Israel's Jericho missiles be armed with nuclear warheads.[36]

But within a week the situation had changed dramatically. On October 12, Israeli Defense Force units came within artillery range of Damascus. The threat to their Syrian ally was sufficient to stimulate the Soviet Union into issuing warnings to deter a further Israeli advance. In addition, Leonid Brezhnev and the Kremlin leadership informed Washington that two Soviet paratroop divisions were being placed on alert.[37]

It took Israeli forces a little longer to bring the Egyptian capital under the threat of siege. Still, Israeli forces turned back the Egyptian troops and broke through to the central sector of the canal, occupied some 720 square miles of Egyptian territory, isolated Egypt's Third Army, and advanced to within 60 miles of Cairo by October 24.[38]

That night Henry Kissinger received a call from Soviet ambassador Anatoly Dobrynin who proposed that the United States and Soviet Union "urgently dispatch" troops to implement a cease-fire. General Secretary Brezhnev raised the stakes by threatening unilateral action if the United States did not comply. At 10:30 that night Henry Kissinger convened a meeting of senior officials, including Secretary of Defense James Schlesinger, Admiral Thomas Moorer, CIA Director William Colby, and three senior NSC staff members. Within an hour this group, without consulting with the Watergate-obsessed President Nixon, raised the alert level of U.S. forces from DEFCON 5 (the lowest) to DEFCON 3. All Strategic Air Command training missions were canceled, some refueling tankers were dispersed, certain airborne command posts were readied for takeoff, and some B-52s were placed on heightened ground alert. Fortunately, by October 26 a cease-fire was in effect.[39]

As with other major crisis situations, the United States employed whatever means it could to obtain information concerning developments. Diplomats from U.S. embassies filed dispatches, and CIA officers queried their agents for any information that might be of use. Allied intelligence services were undoubtedly asked for all relevant information, and America's technical intelligence establishment was targeted, to an even greater extent than before, on the region. The United States Intelligence Board approved a DIA request to change the targeting for one SIGINT collection system.[40]

The Pike committee, which investigated intelligence performance in its 1975 hearings, concluded that "technical intelligence-gathering was untimely as well as indiscriminate." It went on to note that an intelligence community post-mortem indicated that U.S. national technical means of overhead coverage of the Middle East were of no practical value. Aircraft missions flown on October 13 and 25 were of little use because they "straddled the most critical phase of the war." The two KEYHOLE satellites in orbit—the KH-8 launched on September 27, which operated throughout the entire war, and possibly the KH-9 launched on July 13, which completed its mission after the first week of the war—did return photographs of the war. But by the time the photographs were in the hands of the analysts, they represented only history. One CIA analyst recalls that "we had wonderful coverage but we didn't get the pictures until the war was over." As a result he had a portfolio of photos of the situation that had existed right after the war began, but "a portfolio no one wanted to see."[41]

This lack of timely intelligence represented to the Pike committee a serious threat to the United States. Because of CIA and DIA reliance on overly-optimistic Israeli battle reports, "the U.S. clashed with the better-informed Soviets on the latter's strong reaction to Israeli cease-fire violations. Soviet threats to intervene militarily were met with a worldwide U.S. troop alert. Poor intelligence had brought America to the brink of war."[42]

The HEXAGON mission that had begun on November 10, 1973 did not conclude until 123 days later on March 12, 1974, setting a new endurance record. The next HEXAGON satellite, which blasted off on April 10, lasted 109 days and the final mission of 1974, which began on October 29, extended well into 1975 for a total of 141 days. The GAMBIT missions also showed extended lifetimes. The first GAMBIT satellite of 1974 was commanded to re-enter 32 days after its launching on February 13, representing the third occasion that a GAMBIT mission had exceeded 30 days. The next two KH-8 satellites, which were launched on June 6 and August 14 into orbits that were virtually identical with previous KH-8 orbits, had significantly longer lifetimes of 47 and 46 days respectively.

The increasing lifetime of the CORONA and HEXAGON missions meant that on most days the United States would have at least one photo-reconnaissance satellite in orbit. In 1974, there were only 53 days in which the United States did not have at least one photo-reconnaissance satellite in orbit. In 1973, there were only 33 days without some coverage. This compared to 226 days with at least one satellite in orbit in 1972, 158 in 1971, and 138 in 1970.

There would be plenty to keep the KH-8 and KH-9 cameras occupied over the next few years. During the summer of 1974, KEYHOLE satellites observed the Soviets placing huge canvas covers over the Delta submarine construction yards at Severomorsk. The covers made it impossible for U.S. intelligence analysts to determine the number of Delta submarines under construction or what missiles were being loaded aboard them. Whether such concealment represented a violation of the SALT I treaty became a subject of debate within the State Department and intelligence community. Since the treaty prohibited concealment only if it went beyond previous levels, and the canvas covers had been put up intermittently before the treaty was signed, Defense Secretary James R. Schlesinger and others argued that the covers were permitted.[43]

Satellite photography in 1974 also sparked concern over Soviet compliance with the ABM treaty. Satellite photos showed several Square Pair radars at the Sary Shagan ABM test center. The photos and electronic intelligence, probably both obtained from the RHYOLITE signals intelligence satellite and Iranian ground stations, suggested that the Square Pair was being used to track ICBM warheads in conjunction with the operation of the SA-5 surface-to-air missile. Since the ABM treaty banned testing of SAM radars "in an ABM mode" such an activity would represent a violation. Extensive testing, it was feared, would permit the Soviets to upgrade their 1800 SA-5s to ABMs to give them a nationwide ABM defense, which the treaty prohibited.[44]

During the summer months of 1974, the KH-9 also detected the development of a phased-array radar at the Sary Shagan ABM test center. Although this did not violate any treaty commitments, it certainly concerned U.S. defense officials. The radar detected was probably the prototype for the series of Pechora early-warning radars the Soviets began installing along the Soviet periphery in the late 1970s as replacements for the older Hen House radars.[45]

Later that year there was another Soviet strategic development for America's spy satellites to monitor—the initial deployment of Soviet SS-18 missiles in the fields in the south-central Soviet Union. The initial SS-18s were tipped by a single 24-megaton warhead, the largest in either superpower's arsenal. The missiles had a range of 6500 miles, and at the end of their journey, the warheads would have a 50 percent chance of getting within 1380 feet of their targets. Most impressive to some was the huge size of the new missile. With a throw-weight of 16,000 pounds, 5000 pounds greater than the SS-9, future SS-18s could be fitted with a large number of independently targetable warheads. With improvements in warhead accuracy, the SS-18 force alone could threaten the survivability of all U.S. ICBMs.[46]

Soviet deployment of the new generation of ICBMs continued in 1975. In addition to continued deployment of the single-warhead SS-18 Mod 1, the Soviets also began deployment of SS-17 and SS-19 missiles. The SS-17, of which only 150 were ever deployed, carried four independently targetable warheads, each with a yield of 750 kilotons and a 50 percent chance of getting within 1440 feet of their targets. Like the SS-18, the SS-17 was "cold-launched," meaning that it was ejected from its silo by compressed gases before its main engines were ignited. Cold-launching allowed silos to be reused in short periods of time. The SS-19, of which 300 would be deployed, carried six 550-kiloton warheads. It was originally estimated by the intelligence community that each of those warheads had a 50 percent chance of getting closer than 979 feet to their targets, which translated into a 75 percent chance of destroying a U.S. ICBM in its silo with two warheads.[47]

To visually monitor deployments, construction of silos for future deployments, and the Soviet silo upgrade program that increased the ability of the Soviet silos to resist nuclear blasts, the CIA, DIA, and other interested intelligence units relied on three KH-9 and two KH-8 satellites. Until its re-entry on March 18, the KH-9 launched on October 29, 1974 continued to operate, dropping its film canisters at regular intervals. Exactly a month later, on April 18, a KH-8 was launched into a standard GAMBIT orbit of 83 by 249 miles, inclined 110 degrees. For the next 48 days the KH-8 trained its high resolution camera on Soviet missile fields and other targets. By the time its mission concluded, U.S. photo-interpreters had identified 10 operational SS-18 silos, 10 operational SS-17 silos, and 50 operational SS-19 silos.[48]

Among the other targets was a major Soviet naval exercise—Okean 75, which involved 120 ships. The maneuvers included anti-submarine, anti-aircraft carrier, sea lines interdiction, convoy escort, and amphibious landing operations. Land-based naval aircraft as well as planes from the Soviet Air Defense forces participated in the exercises.[49] The imagery obtained by the KH-8, along with signals intelligence and ocean surveillance data, would be used by analysts at the various naval intelligence units in Washington and attached to the unified and specified commands.

Four days after the KH-8 burned up in the atmosphere yet another KH-9 lifted off from its Vandenberg launch pad. From June 8 until November 4 the KH-9 would circle the globe, establishing a new KH-9 endurance record of 150 days.

The KH-8 launched on October 9 before the HEXAGON mission had concluded would also establish a new record for the longest KH-8 lifetime—52 days. It would also be notable in another way. Only four of the previous 40 launches had inclinations of less than 106 degrees (all four less than 96 degrees) while the previous 11 launches all had 110 degree inclination. At that later inclination the satellite could not observe events north of 70 degrees, an area which included the Novaya Zemlya nuclear testing area and much of the Barents Sea where the new Soviet Delta submarines could fire their 4200-mile range missiles against targets in the United States. The spacecraft launched on October 9 was inclined 96 degrees, as were all subsequent KH-8 launches.

One of the provisions of a 1974 protocol to the SALT agreements specified that as older SS-7 and SS-8 missile launchers were replaced by modern SLBM launchers, the SS-7 and SS-8 launchers would be dismantled within four months after the new submarines entered "sea trials"—four months after they sailed out to sea. That meant when the Soviets first sent such replacement submarines to sea around mid-September, about 20 launchers would have to be dismantled by mid-January, 1976. The actual missiles had been removed some months earlier.[50]

By late January it appeared that the Soviets had not made the deadline, but the evidence was inconclusive. By March, it was conclusive. It is likely that one element of the conclusive evidence was imagery returned from a KH-9 launched on December 4, 1975, which remained in operation through the end of March. By March it had probably returned three film capsules.[51]

As a result, the United States prepared to raise the issue at the March 29 meeting of the Standing Consultative Commission. Sidney Graybeal, the U.S. Commissioner, sat down with the Soviet commissioner, General Dimitri Ustinov. Ustinov began his report by acknowledging that the Soviet Union had failed to meet the required deadline for dismantling the older launchers and by presenting numbers concerning the remaining launchers which corresponded precisely with U.S. intelligence estimates.[52]

By the time the KH-9 of December 4, 1975 was launched, the KEYHOLE program had been in operation for over 15 years. In that time approximately 200 photo-reconnaissance satellites had been orbited, including the KH-9 launched on December 4, 1975. The lifetime of close-look missions had increased from a few days to over 50 days. The lifetime of area surveillance missions had leapt from less than a month to five months. Whereas in 1961 there were fewer than 50 days when the U.S. had at least one operational photographic reconnaissance satellite in orbit, in 1975 there were 332 such days. In addition, the quality of the product improved dramatically. The area covered in a single KH-9 image far surpassed that in a KH-4B photo. There was also a marked improvement in resolution. The KH-9's two foot resolution was almost the equivalent of that of the KH-7 close-look system, and the KH-8 camera system was capable of producing even greater detail than that of the KH-7. The photos were so detailed and crisp that they appeared to have been taken from 85 feet above the target, not 85 miles. But the biggest improvement was yet to come.

CHAPTER 6
Quantum Leap

Sitting on a launch pad at Vandenberg AFB on the morning of December 19, 1976 was a Titan 3D booster. Its scheduled mission was to place yet another KEYHOLE satellite into orbit.

Three camera-carrying satellites had been launched earlier in the year, which began with the CIA and other interested parties relying on a KH-9, launched on December 4, 1975, to provide the required photography. From the beginning of the year until April 1, when it was deorbited, the KH-9 camera was busy producing its extremely wide-area and very detailed photos.

Ten days before the KH-9 mission concluded, a Titan 3B-Agena D shot a KH-8 spacecraft into orbit on a 57-day mission that ended on May 17. It was to be 51 days before another KEYHOLE satellite was to lift off. That gap was the longest since a 58-day gap early in 1973. And it made some people nervous, including Major General George Keegan, the Air Force's Assistant Chief of Intelligence.[1] A lot could happen in 51 days—troops could be mobilized and deployed, aircraft redeployed, nations invaded. Even those less suspicious of the Soviet Union than Keegan would have felt better without such gaps.

Coverage resumed on July 8 with the launch of a KH-9 and continued through December 13, when the satellite was deorbited. The 158-day lifetime set a new HEXAGON endurance record. In addition, less than halfway through that mission, on September 15, a new KH-8 was launched into an 84 by 205 mile orbit. Until its fiery return on November 5, the NRO operated both spacecraft simultaneously. Although simultaneous operation of a KH-8 and KH-9 was not unprecedented, the KH-8 mission represented the first time that a KH-8 spacecraft was orbited and deorbited during the same KH-9 mission.

Up until December 19 there had been 296 days of coverage out of a total of 353. Although there had been no crises or wars equivalent to the Middle East wars, the Soviet invasion of Czechoslovakia, or the Cuban missile crisis to monitor, the spacecraft certainly did not lack targets to photograph.

Among the targets were the old standby, Soviet ICBM installations. In 1976 the Soviet Strategic Rocket Forces began to deploy the SS-18 Mod 2. Unlike the single warhead Mod 1, the Mod 2 carried between 8 and 10 independently targetable warheads, significantly expanding the Soviets' overall ICBM warhead arsenal. In addition, the Delta II submarine, with 16 launch tubes, each loaded with a single warhead SLBM, began operation that year. On the research and development side, it was a year in which the Soviets began flight-testing three missiles—the SS-17 Mod 2 ICBM, the SS-N-17 SNIPE, and SS-N-18 STINGRAY SLBMs—requiring monitoring of launch sites at Nenonska and Tyuratam and the impact zone on Kamchatka.[2]

The launch on December 19, 1976 went smoothly, with the Titan 3D sending its payload into a polar orbit. Seemingly, yet another KH-9 was lofted into orbit. But close observers noted at least one difference. The new satellite had a markedly higher perigee and apogee than previous KH-9 satellites, coming no closer than 164 miles to earth's surface. The normal KH-9s had perigees of a little over 100 miles. Thus, in an article published in early 1978, space expert Anthony Kenden, noted that "A Big Bird was launched on 19 December 1976 into an unusually high orbit. . . . This new type of orbit may indicate that it was the first test of a Program 1010 vehicle."[3]

In fact, the payload launched on December 19 was the very first launch of a new type of reconnaissance spacecraft. The new satellite was known by three designations—the KH-11, for its optical system; KENNAN, for the program name; and 5501, to designate the specific satellite and mission number.[4]

The KH-11 represented a personal triumph for the CIA's Deputy Director of Science and Technology, Leslie Dirks, and for Dirks' predecessor, Carl Duckett. Dirks had come to the CIA after obtaining a B.S. at M.I.T. in 1958 and a Research Degree from Oxford University in 1960. In 1962 Dirks and colleagues in the CIA science and technology directorate, began pondering whether the United States could launch a truly secret reconnaissance satellite, one that could be kept secret not only from the American public but from the Soviet Union as well. Obviously, if the Soviets did not know that such a satellite existed they would take no steps to hide weapons or activities from it as it came overhead. But Dirks and his colleagues quickly concluded that a secret satellite in low-earth orbit was not feasible. The Soviet space detection and tracking network would easily pick up the launch and then the orbit of the satellite. An alternative was to place the satellite in a much higher parking orbit, bringing it down only when needed. Possibly the Soviets would miss or be confused by this unusual maneuver. But this strategy also had a fatal flaw. As the film sat in space, unused, it would begin to degrade, and by the time the secret satellite received NRO's call the entire film supply might be worthless.[5]

The alternative to film brought Dirks and his colleagues back full circle to RAND's initial concept of a television-type imagery return system. The desirability of such a system had not been forgotten, despite CORONA's success and SAMOS' failure. Whether or not such a satellite could be kept secret from Soviet space watchers, it could send back timely data. While the technology was no more at hand in 1962 than it was in 1952, Dirks realized that it might be at hand in 1972. For the next 10 years he kept the project alive, looking for advances in technology that would permit such a system, seeking support for research into areas relevant to the development of such a system.[6]

By 1969 the Cuban missile crisis, Six-Day War, and Soviet invasion of Czechoslovakia had dramatically indicated the potential value of a real-time system. Of course, prior to any action being taken, a full-scale interagency study would be necessary—because it was bureaucratic ritual and because it would provide an opportunity to systematically consider the pros and cons of such a system and to give all interested parties an opportunity to express their views.

The study was managed by Roland S. Inlow, who became the new COMIREX chairman in 1969. Inlow had joined the agency in 1951 with an academic background in Russian area studies and public administration. Because of his Russian language training, Inlow was assigned to be an analyst in the Directorate of Intelligence office responsible for Soviet affairs. In 1956 he became head of a new unit responsible for analyzing information on the Soviet guided missile program. When the CIA created the Office of Strategic Research in 1967 and gave it responsibility for analyzing Soviet and other military activity, Inlow became deputy director, and a year later he was given the additional task of being the CIA member of COMIREX. As a COMIREX member he worked on various studies concerning the utility of overhead photography.[7]

One of Inlow's first actions as COMIREX Chairman was to direct the study to evaluate the utility of near-real-time satellite imagery to potential users. The study group examined three cases—the Cuban missile crisis, the Six-Day War, and the Prague spring invasion. The group explored how the intelligence community, the president, and other high-level officials could make use of such data in crises. Specifically, it focused on what information could have been obtained in each situation, how it might have changed perceptions of the crisis, and what the utility of such information would have been. It attempted to determine how different degrees of timeliness could have aided decision makers.[8]

The study's conclusions were sufficiently positive to encourage the CIA's Directorate of Science and Technology to begin a full-scale effort to develop a real-time system. Not surprisingly, the CIA and Air Force were soon in competition. As had been the case for many years, the Air Force sought incremental improvements to currently operating systems rather than radical improvements. Thus, the Air Force Office of Special Projects, the Air Force component of NRO, proposed development of FROG, Film-Readout GAMBIT. As its name indicated, FROG would take the film-return KH-8 satellite and add a film-scanning capability in the manner of SAMOS. The Office of Special Projects advertised the system as providing the timeliness that the Inlow study recommended. What the special projects office did not advertise was that the resolution of the imagery produced by the film-scanning system would be inferior to the resolution of the film-return KH-8, and the area covered by a single photographer would be limited.[9]

Despite its limitations, the FROG system was initially selected by the Secretary of Defense, Melvin Laird, as the next generation KEYHOLE system. FROG had the advantage of being a modification of an existing system. It would take less time to achieve the FROG capability than a more revolutionary approach. But Laird's decision, if not reversed, would probably mean that it would be a long while before any revolutionary change was made. That was a prospect that did not sit well with the CIA's Deputy Director for Science and Technology, Carl Duckett, or with many of the eminent scientists who served as advisers to the CIA and NRO.[10]

Duckett differed from many of those who rose to high levels in the CIA. He did not have an Ivy League background. Rather, he grew up in rural North Carolina, and when he was 17 his mother presented him with a new pair of jeans, some money, and instructions to get a job at the mill down the road. Duckett left and didn't stop for 200 miles, until he got a job at a radio station. He was drafted for service in World War II, and when the results of his IQ tests came in, it was apparent that the military had a genius on its hands. Duckett was then sent to study radio at Johns Hopkins University.[11]

After the war, Duckett was assigned to White Sands Proving Ground and became involved in missile testing and telemetry analysis. From White Sands he moved on to Huntsville, Alabama to become head of the Army missile intelligence unit that is now known as the Army Missile and Space Intelligence Center. At Huntsville, Duckett directed his organization to focus on Chinese missiles. In 1957 he was brought to Washington to advise Art Lundahl and his photo-interpreters how to identify missile sites from overhead photos. Five years later he so impressed DCI John McCone at a meeting of the U.S. Intelligence Board that McCone ordered him hired. Within a year Duckett was head of the Foreign Missiles and Space Activities Center within the Directorate of Science and Technology.[12]

In addition to his technical skills, Duckett was also a skilled salesman and became part of the campaign to reverse the decision in favor of FROG. He journeyed to Capitol Hill to talk to Senator Allen Ellender, the powerful Louisiana Democrat and chairman of the Appropriations Committee and persuasively explained the need for a more revolutionary system than the Air Force was intending to build.[13]

When a panel headed by Dr. Eugene Fubini concluded that the CIA's advanced concept was not feasible, a member of that panel who strongly disagreed, Richard Garwin of IBM, convened a meeting of the advisory Reconnaissance Panel, which he and Edwin Land chaired. The panel concluded that the CIA's concept was quite feasible.[14]

Garwin, along with Stanford physicist Sidney Drell, journeyed to the White House to talk to Henry Kissinger. And Edwin Land talked to the president, advising him that there was nothing simpler than a tube with a mirror in front of it, which was the essence of the CIA approach.[15]

Further consideration took place at a meeting of the President's Foreign Intelligence Advisory Board, known for short as the PFIAB (pronounced piff–ee–ab). PFIAB's members included Edwin Land, William Baker of Bell Labs, Nelson Rockefeller, Gordon Gray, John Connally, and Maxwell Taylor. Usually Henry Kissinger and his deputy attended, representing the president. But this meeting also drew the president himself, along with James Schlesinger, then of the Office of Management and Budget. Schlesinger supported the FROG concept.[16]

Nixon's decision to approve the CIA approach was, according to an individual present at the meeting, "a direct consequence" of the meeting. It was a decision that pleased Duckett, Helms, and most especially Leslie Dirks. It was one, however, which greatly displeased Colonel Ralph Jacobsen, of the Office of Special Projects, who saw what would have been a $2 billion program (FROG) snatched from his hands.[17]

Naturally, it was a decision that led to another round of studies. Now the focus was not on how such real-time information would be useful in a theoretical sense, but how the new system would be incorporated into the intelligence community's operations. A special study group was established to determine how the intelligence community would exploit the KH-11 as a source of information and how it would be used in the analytic product.[18]

The decision also became known, relatively rapidly, to the public. Newspaper and trade magazine stories appearing in *Aerospace Daily*, the *Los Angeles Times*, the *Washington Post*, and *Aviation Week and Space Technology* all noted that a Project 1010 was underway to develop a real-time photographic-reconnaissance satellite. The *Washington Post* article noted that the satellites "would constantly be on call while in orbit," and *Aviation Week* indicated that the system would become operational in the 1976-1977 period.[19]

When it was launched on December 19, 1976, the KH-11 was within two months of the date originally projected for its initial launch. And despite reports to the contrary, it also came in substantially under budget. However, subsequent versions of the satellite proved to be excessively costly.[20]

Although no two versions of any imaging satellite are necessarily identical, with modifications being made to sensors and other equipment, the basic dimensions of the KH-11 have stayed the same from the initial launch—the cylindrical spacecraft are about 64 feet long, 10 feet in diameter, and 30,000 pounds.[21]

The basic means by which the KH-11s have collected and transmitted their imagery in real-time, the charge-coupled device or CCD, has also remained constant. The CCD originated at Bell Telephone Laboratories in the late 1960s when two researchers, William S. Boyle and George E. Smith, sought to invent a new type of memory circuit. It rapidly became apparent to the researchers that the tiny chip of semiconducting silicon they first demonstrated in 1970 had other applications, including signal processing and imaging (the latter because silicon responds to visible light). By 1975 scientists from the California Institute of Technology's Jet Propulsion Laboratory and the University of Arizona were using a CCD with a 61-inch telescope to produce a picture of Uranus, approximately two billion miles from Earth. Today CCDs are in widespread use in hundreds of civilian and military programs, including medical imaging, plasma physics, ground astronomy, the Galileo spacecraft, the Hubble space telescope, and home video cameras.[22]

The KH-11's optical system scans its target in long, narrow strips and focuses the light onto an array of charge-coupled devices with several thousand elements. The light falling on each CCD during a short, fixed period of time is then transformed into a proportional amount of electric charge. In turn, the electrical charge is read off and fed into an amplifier, which converts the current into a whole number between 0 and 256 that represents a shade of color ranging from pure black to pure white. Thus, each picture is transmitted as a string of numbers, one from each element.[23]

More specifically, the CCD captures particles of energy, including visible light, in an array of picture elements known as pixels. The pixels automatically measure the intensity of the particles and then "send them on their way in orderly rows until they are electronically stacked up to form a kind of mosaic." The standard CCD used in the Space Telescope has a total of 640,000 pixels arranged in an 800 by 800 format and occupies less than half of a square inch.[24]

One analogy used to illustrate the working of CCDs involves 640,000 buckets, an open field, and plenty of rain.

> Each of the 640,000 buckets, all arranged in neat rows to form
> a square, represents a pixel. The rain, heavy in some places
> and light in others, is the equivalent of incident radiation

(photons, electrons, neutrons, or protons, for example). Since different amounts of rain are coming down on the various parts of the field, the buckets catch varying amounts of water and therefore fill to different levels. As the buckets catch the water they move in vertical rows, called channels, toward a conveyor belt—a line transport register—which in turn carries them, a line at a time, off to the side where each of the eight hundred buckets in each line is simultaneously measured for its water content. If the amount of water in each bucket represents a different color or tone—say black, white, and several shades of gray—then the entire string of eight hundred would take on the appearance of a line that is darker in some spots and lighter in others. Eight hundred pieces of precisely quantified information, constituting a line in a picture, would have been created. No sooner has the last of the eight hundred buckets moved off the conveyor than a second line gets on, and it too is measured. Then a third line follows, and so on, until all eight hundred lines of eight hundred buckets have been measured for their water content, or tonality. The net effect would be a highly detailed "picture" composed of 640,000 dots of varying tone. But the process is instantaneous and continuous, so that the picture moves.[25]

Of course, a CCD collects particles of radiated energy rather than water. In addition, pixels, unlike the buckets of water in the example, do not move. Instead, the photons that strike the pixels generate charges that move from one pixel to the next along a series of horizontal "gates." The result is that the CCD acts as an extremely small, extremely precise light meter, capturing radiated energy emissions across the visible and invisible bands of the electromagnetic spectrum that can then be amplified and converted into photographs.[26]

The KH-11's charge-coupled devices cannot, however, do the job alone. Without a good mirror system in front of them, even the best CCDs would produce photographs with poor resolution. But the KH-11 mirror, which is essentially the mirror developed for the MOL, is quite good. Several years ago, the Air Force donated six 71-inch mirrors to the University of Arizona and to Harvard's Smithsonian Observatory. It has been inferred that the KH-11's mirror is wider than that, and indeed, it is much wider. The primary mirror for 5501 was 92 inches wide. Since then mirror size has increased. It would follow that its secondary mirror would be greater than one foot in diameter. The secondary mirror would narrow the image coming off the primary mirror and sharply focus it.[27]

Another key to the KH-11's ability to produce high-quality photographs is its computer. About the size of a sleek VCR, the computer is the key to maintaining the KH-11 in a stable position, pointing the mirror, and obtaining photographs of the desired targets.[28]

Once the visible light has been collected and transformed into an electrical charge, the signals would then be transmitted to one of the two or three Satellite Data System (SDS) spacecraft that relay the signals. SDS spacecraft, the first two of which were launched in 1976, have a variety of functions. In addition to transmitting the KH-11 digital signals, they also relay communications to any B-52s flying on a polar route and serve as communications links between the various parts of the Air Force Satellite Control Facility. The SDS orbit, which comes as close as 250 miles when it's passing over the southern hemisphere and goes as far as 24,000 miles as it passes over the northern hemisphere, makes it ideal to relay KH-11 imagery when the satellite is transmitting imagery of the Soviet Union. The SDS takes eight to nine hours to pass over Soviet territory, leaving it available to receive and transmit KH-11 imagery for long stretches of time.[29]

The SDS satellite then transmits the KH-11 signals for initial processing to a ground station at Fort Belvoir, Virginia, about 20 miles south of Washington. The Mission Ground Site—a large, windowless, two-story concrete building, is officially the Defense Communications Electronics Evaluation and Testing Activity.[30]

The other part of the KH-11 ground network is the Air Force Satellite Control Facility. Between its initial operation at the beginning of the CORONA program and the initial KH-11 launch it underwent substantial changes. Most significant was the reduction of U.S. sites in favor of overseas sites. Thus, as of 1976 (and 1989) the Remote Tracking Stations that made up the Air Force Satellite Control Facility (renamed HQ, Consolidated Satellite Test Center in 1987) were located at Vandenberg Air Force Base, California; New Boston Air Force Station, New Hampshire; Thule Air Base, Greenland; Mahe Islands, Seychelles; Anderson AFB, Guam; and Kaena Point, Hawaii. Oakhanger, England serves as an operating location.[31]

As a result of its electronic method of transmitting data, the KH-11 launched on December 19 would remain in orbit for over two years, or

770 days to be exact. The KH-11 was not limited, as were the KH-8 and KH-9, by the amount of film that could be carried on board. An additional factor was the KH-11's higher orbit, approximately 150 by 250 miles, which reduced atmospheric drag on the spacecraft.

Only a select group of government officials was permitted to know of the KH-11's existence or even see its product. The KH-11 was treated with even a greater degree of secrecy than usual in the black world of reconnaissance satellites, and the photographs and data derived from those photographs were not incorporated with data from the KH-8 and KH-9 systems. The decision to restrict the data to a very small group of individuals was taken at the urging of senior CIA officials but opposed by military officers who wanted the information to be more widely distributed throughout the armed forces.[32]

But that decision was almost made irrelevant by *Aviation Week and Space Technology*. Writer Clarence A. Robinson, Jr., according to one of his colleagues at the time, had "sources who would blow your mind" who would provide him with information that "people should be shot for [leaking]." The launching of the KH-11 was one bit of information that his sources provided for him. Robinson prepared to write a story announcing the existence of the new satellite as well as discussing the operation of the KH-8 and KH-9 spacecraft, but before the story could appear Robinson met with a senior intelligence official (possibly then DCI George Bush) to discuss the story. The official persuaded him to sit on the story because the CIA believed that the Soviets had misidentified the function of the new satellite. It would be more than a year before the public and the vast majority of the government would first become acquainted with the KH-11.[33]

Among the officials who did know about KENNAN in early 1977 was Jimmy Carter, who had been briefed extensively on the new system after his election in November. In the month between its launching and the new administration's taking office, the KH-11 underwent checkout and testing. It was only on the day of Carter's inauguration that the first operational photos were transmitted.[34]

The occupant of the DCI position at that time was Mr. Enno Henry "Hank" Knoche, although he was Acting Director of Central Intelligence pending confirmation of the new president's choice as DCI. Knoche was a career analyst who had been promoted to the Deputy Director of Central Intelligence position in 1976 by DCI George Bush. Bush had replaced Colby, who had been fired by Gerald Ford for having been too forthcoming with the committees then investigating CIA and intelligence community activities.[35]

Bush had wanted to remain as DCI but Carter was not interested in having the Republican stalwart remain, not even until his choice could be confirmed. Until that happened, and it was to take longer than usual because Carter's first nominee withdrew when it became clear he could not win approval, Knoche was head of the world's largest and most expensive intelligence establishment. One of Knoche's first roles was to show the new president an example of the capabilities of America's newest spy satellite as one way of demonstrating the CIA's value to the new president.[36]

It would have been most dramatic for Carter to see the photos within moments of their coming down, but the new president was rather busy on Inauguration Day—taking the oath of office, strolling down Pennsylvania Avenue in subfreezing weather, and attending the various traditional celebrations. Also, the most dramatic effect would have required either Carter to make the trip to Ft. Belvoir or Ft. Belvoir to come to Carter, neither of which was feasible. As a result, Knoche decided he would wait a day before visiting the new president.[37]

So it was 3:15 P.M. on January 21 when Knoche and Admiral Daniel J. Murphy, the DCI's deputy for intelligence community affairs, began a 15-minute meeting with President Carter and national security adviser Zbigniew Brzezinski in the White House's second-floor Map Room. Knoche had a handful of six-inch square black-and-white photos with him. Knoche can no longer remember what those photographs showed, apparently because the photos were rather mundane.[38]

As noted above, Carter had been briefed in some detail during the transition period. Now he was to have his first chance to see the KH-11's product. Carter looked down and examined the photographs that Knoche spread on the map table. After peering at the photos for a few moments, Carter looked up at Knoche, grinned, and then laughed appreciatively. He congratulated Knoche and Murphy over the apparent quality of their new spy satellite and requested Knoche to send over some more samples for the next day's National Security Council meeting, his first as president. "Of course," Carter said as he turned to Brzezinski, "this will also be of value in our arms control work."[39]

Carter and Brzezinski knew that they had something of immense value. Indeed, it was more obvious to them than to some in the CIA, for it was the president and his advisers who might be pressed to make crucial crisis decisions that could dramatically affect the fate of the United States. Now they would be able to make those decisions with timely information.[40]

There was one serious initial limitation on the new system. While it could transmit its data instantaneously, it could only do so for two to four hours per day. The power required to transmit the data to the relay satellite using the KH-11's Traveling Wave Tube Amplifier was so great that it drained power far faster than it could be replaced by the satellite's solar panels. Thus, the new model of the spy satellite fleet could only be used sparingly at first.[41]

As a result the KH-11 would operate along with KH-8 and KH-9 satellites for several years before becoming America's lone model of photographic reconnaissance satellite. On March 13 a KH-8 was placed into a 96 degree inclination, 77 by 216 mile orbit. The KH-8 continued to operate for another 74 days, establishing a new record for KH-8 lifetimes, beating the past high by 17 days. On June 27, 33 days after the KH-8 was deorbited, a KH-9 was launched into a 97 degree, 96 by 148 mile orbit. For the next six months (179 days to be exact), the KH-9 stayed in orbit, taking its area surveillance pictures and periodically dropping its buckets of film. In the middle of the its lifetime, a KH-8 joined it in orbit when a Titan 3B–Agena D placed 1747 into a 96 degree, 78 by 218 mile orbit on September 23. For the next 76 days, when the KH-8 deorbited, the United States simultaneously had all three models of photo reconnaissance satellite in orbit.

Just as the KH-11 was to have a dramatic impact on space reconnaissance operations, it also was to significantly influence the nature of photo-interpretation.*

As it had been since 1961, the National Photographic Interpretation Center was the focus of the interpretation effort. Still operating from Building 213 in the Washington Navy Yard at 1st and M Streets, it

---

* Strictly speaking, there are photo-interpreters and photogrammetrists. The former determine what is shown in a satellite photo, and the latter determine the dimensions of what is shown. Of course, the two tasks may be performed by a single individual. Often, determining the dimensions of an object may be the key to determining what it is — for example, the only difference between an intermediate range missile (the SS-20) and an intercontinental range missile (SS-16) may be the length, because the latter has an additional stage.

is identified by a blue and white sign on its beige exterior. It is also noticeable for its lack of windows, most having been cemented over to keep outside light from interfering with the interpretation task.[42]

Within the building are photo-interpreters assigned to NPIC from the CIA, DIA, and several military service intelligence units. In addition, the DIA, Air Force Intelligence Agency, Navy Intelligence Support Center, and Army Intelligence and Threat Analysis Center all maintain photo interpretation components under their own command—either located at NPIC or in the Washington area. Other major intelligence units—the Air Force System Command's Foreign Technology Division at Wright-Patterson AFB, Ohio; the Army's Missile and Space Intelligence Center in Huntsville, Alabama and Foreign Science and Technology Center, Charlottesville, Virginia; and the Strategic Air Command's 544th Strategic Intelligence Wing at Offutt AFB, Nebraska—have imagery interpreters working at their headquarters. Even the CIA has within its Directorate of Intelligence an Office of Imagery Analysis, established when NPIC became a "service of common concern." Such diversification was sanctioned by NSCID No. 8, which established NPIC, when it specified that "Departments and agencies represented on the United States Intelligence Board shall continue to be individually responsible for photographic interpretation and the production of photographic intelligence in support of established department or agency responsibilities."[43]

Wherever they are located, the photo-interpreters have all experienced the revolution brought about by digital imagery. Traditionally, photo-interpreters have worked with light tables and magnification equipment to analyze aerial and satellite photos. Among the magnification equipment of great value has been the stereoscope, which allows photo-interpreters to superimpose photos taken from different angles. Such superimposition can yield a three-dimensional effect that makes it easy to determine the height and length of weapons bunkers, space launch vehicles, or other objects of interest.[44]

While the light table and stereoscope are by no means defunct, the computer has come to occupy a central place in the interpretation process, although it is not a substitute for the interpretation process. It is the computer that is vital for the processing of the enormous volume of data flowing into NPIC and other interpretation centers. Sitting at a computer terminal an interpreter may face the task of examining recently obtained photographs or older photographs which have taken on new significance. In the aftermath of the 1988 discovery that China was supplying Saudi Arabia with CSS-2 IRBMs, analysts at NPIC began poring over past KH-11 photos of Chinese railyards and port areas.[45]

Recovery of such photos is also done by computer. Former NPIC director Arthur Lundahl recalls that during his tenure the CIA had a major filing and recall problem. But use of a computerized data base now allows for easier retrieval. An analyst can now specify earth coordinates along with a time period and the computer does the rest—it searches and produces a printout that lists all photographic references on the subject, whether they are classified or not, and the location where they are stored. It also gives the angle of the sun when the photos were taken, because as former NPIC official Dino Brugioni has explained, "In some cases you might want long shadows for measurements; in others you want very little shadows for greater detail."[46]

Once they have the image on their computer screen, analysts have a large variety of options. According to Kevin Hussey of the California Institute of Technology's Jet Propulsion Laboratory, the image enhancement computer program at JPL has 400 algorithms for image enhancement and allows for "anything you want to do to a picture."[47]

One computer option is the manipulation of contrast to increase the visibility of objects in shadows, obscured by haze or thin cloud cover, or photographed with too much or too little exposure. Adaptive filtering is one of the methods of manipulation, used when thin cloud cover is a problem. Thin cloud cover partially obstructs the transmission of light from the ground to the camera as well as reflecting considerable amounts of light directly back to the camera, resulting in the overexposure of the haze relative to that of the ground. The adaptive filtering process first computes the average brightness and the local contrast values for all regions of the picture and then reduces the average brightness and increases the variations in such a way that contrast is enhanced.[48]

Computers can also perform the task of object detection, which may be particularly useful in situations where the resolution of a particular image has been degraded either by mechanical problems with the satellite or environmental conditions. One object detection technique seeks to enhance anomalies in the photographs. In this process the expected intensity of each pixel is predicted from the intensity values of a large number of pixels in its neighborhood. A statistical test is then applied to determine if the pixel intensity differs significantly from the predicted value, in which case it is classified as an anomaly or object.[49]

Changes in a particular target area can also be determined by computer, using a technique known as electronic optical-subtraction. According to Dino Brugioni, the CIA has used computers to spot changes automatically since at least 1983. Research since then has been directed toward developing systems that distinguish between significant alterations like new structures and superfluous changes like the length or darkness of a shadow.[50]

Image resolution can be improved by computer. One technique compensates for known shortcomings in the sensors or for atmospheric characteristics by conducting tests to distinguish between the actual and theoretically attainable resolution. Special algorithms can then adjust the photograph to produce the theoretically attainable result, although there is the possibility that artifacts will be introduced into the photo.[51]

It is also possible to improve resolution by the computer-assisted combination of multiple images of a scene. Overlaying the photos compensates for deficiencies in the sensor. "You can add these things up, and if you work really hard you get up to a factor of 2 improvement in resolution," the JPL's James L. Anderson has observed, no matter how good the resolution was to begin with.[52]

Image restoration can also be found in the computer's bag of tricks. All photographic systems introduce some distortion into the images they create. The distortion can be analyzed mathematically, relying both on basic optical theory and empirical measurements on the actual optical system. The analysis can then be translated into a computer program which can be applied to any image produced by the system to remove the distortions. Blurring caused by the relative motion of the camera and subject can also be corrected by the same computer routine.[53]

Once computers have restored, enhanced, or detected, the photo-interpreter must then make sense of what is revealed by computers. Brugioni believes that the ultimate judgment must remain with man and not machine—"If I'm monitoring a treaty, you can bet your bottom line that I'm not going to let a machine take my place."[54]

By 1977 it was clear to the United States that several smaller nations were interested or potentially interested in developing nuclear weapons. The first hint that South Africa was one of those nations came in 1965, when one of the members of South Africa's Atomic Energy Board, Dr. Andres Visser, advocated building a nuclear arsenal for "prestige purposes" and reportedly said "We should have the bomb to prevent aggression from loud-mouth Afro-Asiatic states."[55]

In subsequent years, even higher ranking officials hinted at South African interest in nuclear weapons. Later in 1965, Prime Minister H. F. Verwoerd urged, during the inauguration ceremony of the Safari I nuclear reactor, that it was "the duty of South Africa to consider not only the military uses of the material but also to do all in its power to direct its uses to peaceful purposes"—implying that South Africa was already pondering the military uses of the material. In December of 1968, General H. J. Martin, Army Chief of Staff, reportedly stated that South Africa was ready to make its own nuclear weapons. In addition, South Africa refused to sign the just-completed Non-Proliferation Treaty.[56]

A 1974 CIA memorandum, "Prospects for Further Proliferation of Nuclear Weapons," noted that South Africa

> . . . apparently has developed a technology for enriching uranium that could be used for producing weapons-grade material. South Africa would probably go forward with a nuclear weapons program if it saw a serious threat from African neighbors beginning to emerge.[57]

While the estimate concluded that such a serious threat was unlikely in the 1970s, the South Africans may have disagreed. Independence for Angola and Mozambique in 1974 was followed by the movement toward black rule in Rhodesia and internal violence in 1976. Thus, Information and Interior Minister Connie Mulder claimed that South Africa would "use all means at our disposal, whatever they may be. It is true that we have just completed our own plant [the Valindaba uranium enrichment plant] that uses very advanced technology, and we have major uranium resources." In addition, the election of Jimmy Carter in November 1976 made it doubtful that the South Africans could rely on sympathy from Washington.[58]

In 1976, U.S. intelligence picked up definite indications that South Africa had embarked on a program to develop nuclear weapons. Soviet attention was also focused on the potentially explosive situation. On August 6, 1977, the Soviet Union notified the United States that one of its Cosmos reconnaissance satellites had detected what appeared to be a nuclear test site in the Kalahari Desert. The initial reaction within the CIA was one of skepticism—the responsible analysts did not believe the Soviet report. Despite such skepticism, COMIREX decided to reprogram one of the two operating KEYHOLE satellites to survey the area reported to contain the nuclear test site.[59]

When the satellite returned its imagery it confirmed the Soviet claims, so the United States began working with France, Great Britain, and West Germany to develop a concerted response to the impending test. The concerted pressure by the United States and other governments, which reportedly included the threat to break diplomatic relations and a warning by France that a test might cause France to terminate its aid in the construction of the nuclear power plants that South Africa had purchased from France, led to an apparent South African decision to forgo any testing. On August 23, President Carter announced that "South Africa has informed us that they do not intend to develop nuclear explosive devices for any purpose, either peaceful or as a weapon, that the Kalahari test site which has been in question is not designed for the use of nuclear explosives, and that no nuclear explosive test will be taken in South Africa now or in the future."*[60]

Jimmy Carter arrived in Washington promising a more open government than the previous Republican administration, with greater accountability in the area of intelligence activities. But although part of his October 1, 1978 speech at Cape Canaveral touched on intelligence activities, his statements were motivated not by those promises so much as the need to sell what was to be known as SALT II, the second Strategic Arms Limitation Treaty. In the first part of 1978 it appeared that the U.S. and Soviet negotiators were on the verge of agreement on the specifics of a new and far more complex arms limitation agreement.

It was apparent to everyone involved that the struggle for ratification would not be easy. There were formidable forces likely to oppose any treaty with the Soviets on the grounds that the United States needed to undertake a massive build-up to counter what they saw as the tremendous Soviet advances in strategic hardware that had occurred since 1972. One key element in the administration's strategy to obtain public and Senate approval was to make it clear that the United States had the capability to verify Soviet compliance with the prospective treaty.

But the major means of verification involved the KEYHOLE satellites, the very existence of which was still considered Top Secret. It

---

* U.S. pressure may ultimately have influenced South Africa to test covertly. South Africa may have conducted such a test in the area of Prince Edward Island in September of 1979. A U.S. nuclear detonation detection satellite detected signals that some U.S. agencies (the Defense Intelligence Agency and Naval Research Laboratory among them) believe indicated a test. A presidential panel concluded that the evidence was ambiguous.

did not matter to many within the Defense Department and intelligence community that it was an open secret. Nor did it matter that Lyndon Johnson's 1967 off-the-record comments on the value of satellite photography had been widely quoted. Even the public statements of Secretary of State William Rogers (in 1972), of former DCI William Colby in his CIA-approved 1978 memoirs, or of Carter himself, who had told listeners to a radio call-in program in March 1977 that "As you probably know, with space satellite photography we . . . guarantee the security of our country," did not constitute official declassification.*[61]

But when Jimmy Carter stated at the Kennedy Space Center medals ceremony that "Photo-reconnaissance satellites have become an important stabilizing factor in world affairs in the monitoring of arms control agreements. They make an immense contribution the security of all nations. We shall continue to develop them," it did represent formal declassification. It was the president's intent that U.S. representatives and legislative supporters of SALT be able to talk about their use in assuring the United States that the Soviets were not cheating on the provisions of the treaty.[62]

Carter's acknowledgment of the U.S. photo-reconnaissance satellite program provoked virtually no reaction, domestically or internationally. Virtually every American who had even minimal interest in arms control and whether the United States could verify compliance was aware of such satellites. The Soviet Union had been consulted in advance and had no objection to the statement. Nor did other nations protest.[63]

But while Carter's speech caused no ripples afterwards, it caused a modest degree of debate within the administration before it was given. The possibility of relaxing the secrecy surrounding the KEYHOLE program had been raised in several earlier administrations. Robert McNamara recalls both his desire to release information on Soviet strategic forces based on satellite photography and his inability to obtain the agreement necessary to implement such a plan. The issue came up again as SALT I negotiations approached their conclusion. Some believed that the Soviets would inevitably cheat on such an agreement if they could get away with such conduct. Just as Carter wished to assure the public

---

* In all likelihood, the lawyers were relying on a doctrine presented several years later in an FOIA case concerning U.S. covert operations in Nicaragua. The doctrine asserted that no matter how many times U.S. officials might mention a "covert" activity or operation in public, such mentions did not represent formal declassification unless they had the specific intent of declassifying the activity or operation.

that such cheating could be detected, so did Richard Nixon. But ultimately, no statement was made. Partially, it was a result of intelligence
community opposition, who continued to argue that the Soviets would
not admit that satellites were over their territory and that they would be
pressed into more extensive denial and deception. Perhaps an even more
important factor was the political situation in 1972. In that era, before
the dimensions of Watergate were fully known, there was far greater
trust in government by the general public. Both the Joint Chiefs of Staff
and CIA were solidly behind the treaty, and Richard Nixon had a record
as a hard-line anti-communist. There was no perception, except perhaps
by the extreme right, that he would be taken in by the Soviets.[64]

In the spring of 1978 it appeared to many that a SALT II agreement
was approaching and that verification would be a major issue. A public
statement concerning the ability of the United States to verify compliance via photo-reconnaissance satellites was considered to be important,
especially since many of the provisions would be written on the basis of
photo-reconnaissance capabilities. In addition, it would allow pro-treaty
Congressmen and Senators to tell their constituents that "we've got spy
satellites that can tell if they turn the screws the wrong way."[65]

What many would have expected to be a source of a tremendous
debate was not. Paul Warnke recalls that it was "not a spirited debate."
A NSC staff member, Roger Molander, recalls the issue more as not
"that big a deal at the time," but as a "medium deal."[66]

The exchange of memos that were part of the debate indicates the
positions staked out by the various participants. The National Security
Council, assigned as mediator between policymakers and the intelligence
community, prepared an eight-page summary of the case on August 31,
1978. The NSC memo began by acknowledging that the "secret" which
Carter planned to divulge was "already widely known." The main argument for disclosure was to "enable government spokesmen to make a
more effective case for a SALT II agreement." At the same time, the
NSC moderators noted that classification of the existence of the reconnaissance program "has served as the first line of defense for the security
of overhead space intelligence programs. After declassification, U.S.
agencies and officials would be under considerable pressure to provide
more information." There would be, the NSC predicted, annoying questions from reporters, Freedom of Information Act requests from the public, objections from the Soviets, and meddling from the United Nations
or Third World countries.[67]

Over the next two weeks, the debate was conducted with the classified memoranda circulating around Washington. Both Secretary of State Cyrus Vance and Arms Control and Disarmament Agency director Paul Warnke argued that declassification was vital to the ratification of SALT II since it answered doubts about U.S. verification capabilities. Stansfield Turner and the Central Intelligence Agency leaned toward declassification.[68]

The military intelligence agencies formed the opposition. As Eugene Tighe, the director of the Defense Intelligence Agency, wrote in a memo to Turner: "I do not find the alleged benefits of [declassification] compelling." Secretary of Defense Harold Brown and the Joint Chiefs of Staff, in a four-page Top Secret "Joint Talking Paper," spelled out seven risks of the proposed disclosure, essentially the same risks noted by the NSC. Brown tried to bury the proposal by agreeing to declassification "but only after careful preparation WHICH HAS NOT YET OCCURRED."[69]

The stratagem failed when the eight Cabinet-level members of the NSC's "Space Policy Review Committee" met on September 13 and voted for declassification. Anticipating a deluge of press inquiries after disclosure, national security adviser Zbigniew Brzezinski prepared a Confidential advisory memo to the appropriate agencies, complete with 19 sample questions and the approved answers.[70]

It was a "decision whose time had come," according to Carter administration ACDA official Spurgeon Keeny. It was also a truly minimalist decision—it was only agreed to acknowledge the "fact of" satellite photography. But that was as far as the decision went. Paul Warnke wanted to go much further. The ACDA director advocated releasing some photos of less-than-highest quality that would demonstrate U.S. ability to verify Soviet compliance with the ICBM and submarine limitations. In addition to being a means of obtaining public support, Warnke felt that "the better the Soviets know of our photo-reconnaissance capability the more they would be inhibited" about committing violations.[71]

Warnke was not alone. Bobby Inman, then Director of the National Security Agency, responsible for signals intelligence collection, and Hans Mark, Director of the National Reconnaissance Office, also favored release of satellite photographs. Inman recalls that his objective was to smooth the way for ratification of the SALT agreement by convincing pessimists that the United States could indeed verify any agreement with the Soviets. Mark's motivation was somewhat different. He writes in his memoirs that "I had long felt that we should provide more information to the American people about what the Russians were doing in terms of building up their strategic nuclear forces in order to generate more public support for our own strategic modernization program."[72]

Proposals to release actual photos ran up against a variety of objections. It was feared that the release of photographs might allow the Soviets or other nations to identify weaknesses in U.S. reconnaissance capabilities that they could exploit.[73] Another issue was whether nations, including the Soviet Union, who did not protest the acknowledgment of U.S. satellite spying might find release of actual photographs of their territory as going too far. Not only might they be acutely embarrassed that their secret installations were put on display, but they also might fear that information contained in such photos would be exploited by third parties.[74]

But the objection that was decisive in preventing the release of any photos was one raised by Department of Defense lawyers. They pointed out that release of a single satellite photo could lead to a flood of requests for other photographs under the provisions of the Freedom of Information Act. As long as it was policy that all such photos were classified, any requests for satellite photography could be rejected without searching for and examining the photography. Once one photo was released every request would require a time-consuming response in which the photo would have to be found and then examined before a declassification decision could be made.[75]

This argument proved decisive. According to Bobby Inman the United States had gone down a long road in reducing intelligence manpower from its peak in the 1958-to-1964 time period to 40 percent below the peak figure by 1978 and 1979. The manpower decrease was used to pay for the new technical collection systems, particularly satellites. In view of that reduction there was no enthusiasm for a course of action that had the potential to significantly divert personnel to the FOIA search-and-review process. Even the need to publish overhead photographs in pursuit of foreign policy objectives did not change the decision. When the United States wanted to provide evidence of a Soviet naval base at Berbera, Somalia, a SR-71 reconnaissance aircraft was used to rephotograph the target already photographed by KEYHOLE satellites.[76]

1. President Eisenhower views the capsule returned by the Discoverer XIII satellite during a White House ceremony.

Photo provided by the U.S. Air Force.

2. A Hercules JC-130B in the process of snatching a satellite capsule. The planes carried heavy duty winches that were used to pull the parachutes and capsules inside.

Photo provided by Lockheed–Georgia, Courtesy of William E. Burrows.

3. Headquarters of the Central Intelligence Agency. The CIA developed the first operational photographic reconnaissance satellite, CORONA. The photos it sent back drastically altered U.S. perceptions of Soviet strategic strength.

Photo provided by the Central Intelligence Agency.

4. The headquarters of the National Reconnaissance Office. Established on August 25, 1960, NRO has played a central role in photographic and signals intelligence satellite operations.

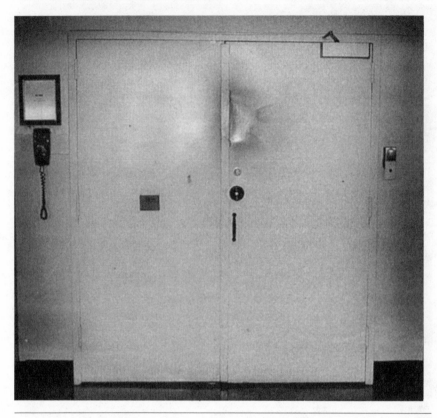

5. A Titan 34D seconds after launch. Titan 34Ds were used to launch KH-9 and KH-11 satellites in the later years of those programs. Malfunctions resulted in the loss of a KH-11 in 1985 and of a KH-9 the following year.

Photo provided by Martin Marietta.

6. The shuttle orbiter Atlantis. In a mission that began on December 2, 1988, Atlantis placed the first LACROSSE radar-imaging satellite into orbit. LACROSSE's radar-imaging capability allows the production of imagery in darkness and in the presence of cloud cover.

Photo provided by the National Aeronautics and Space Administration.

7. An aerial view of Space Launch Complex 6 (SLC-6) at Vandenberg AFB. SLC-6 was originally to be the launch pad for the MOL Titan 3M booster, subsequently for shuttle missions from Vandenberg.

Photo provided by the U.S. Air Force.

151

8. The Defense Communications Electronics Evaluation and Test Activity at Fort Belvoir, Virginia, the Washington area receiving station for KH-11 and Advanced KENNAN imagery. Other receiving stations are probably located in West Germany and at Buckley ANG Base, Colorado.

Photo provided by Robert Windrem.

9. A deliberately distorted KH-11 photo of a Soviet Blackjack bomber on the runway at the Ramenskoye test center near Moscow. The photo was taken on November 25, 1981 and appeared in the December 14, 1981 issue of *Aviation Week and Space Technology*. It was the first leaked U.S. KEYHOLE photo.

Photo provided by *Aviation Week and Space Technology*.

10. Two satellite photos, not produced by the KH-11. The photos were accidentally not excised from 1984 Congressional hearings before release to the public. The photos show two Soviet aircraft—a MiG-29 Fulcrum (left) and a SU-27 Flanker.

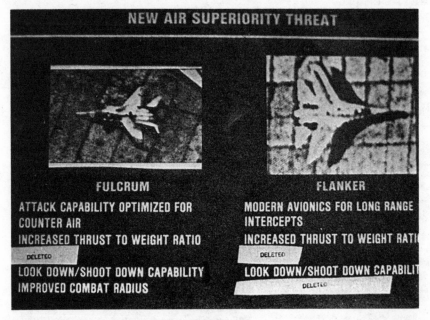

Photo provided by William E. Burrows.

11. A KH-11 photo, taken in July 1984, of the first Soviet nuclear-powered aircraft carrier under construction at the Black Sea shipyard of Nikolaev. The photo and two others were leaked to *Jane's Defence Weekly* by naval intelligence analyst Samuel Loring Morison. Morison was brought to trial, convicted, and sent to prison in a controversial case.

Photo provided by AP / Wide World Photos.

12. The Krasnoyarsk radar was first spotted by a **KEYHOLE** satellite in 1983. Built 400 miles from the nearest Soviet border and oriented in the opposite direction, across the vast expanse of Siberia, it was a clear violation of the ABM treaty. Under U.S. pressure the Soviet Union agreed, in 1989, to dismantle the radar.

Photo provided by William J. Broad / *The New York Times*.

CHAPTER 7

Betrayal

Although Jimmy Carter raised the curtain a bit on the highly secretive KEYHOLE program, it was William Kampiles who ripped it away. While the Carter administration was debating whether to acknowledge the existence of the KEYHOLE satellites and release so much as a single photograph, Soviet intelligence analysts, thanks to Kampiles, were poring over a detailed Top Secret manual describing the capabilities of the KH-11.

Kampiles, the son of Greek immigrants, was born in 1955 and grew up in the well-scrubbed Hegewisch section of southeast Chicago. His father died of cancer in 1964, so his mother, Nicoleta, supported the family by working in the cafeteria at Ford Motor Company's assembly plant in the neighborhood.[1]

Kampiles worked as a paper boy and a grocery boy in his childhood as well as serving as an altar boy at the neighborhood Catholic Church. He put himself through college by delivering groceries, driving a cab, and working at a steel mill. After graduating from Indiana University in 1976, he got a job as a hospital supply salesman and quickly became "No. 1 in the territory."[2]

But being a number one hospital supply salesman was not exactly what Kampiles had in mind as a way of life. In 1975, during his senior year of college, Kampiles had interviewed with a CIA recruiter, but heard nothing by the time of graduation. However, about a year after his interview, Kampiles received an offer from the agency and he accepted it, despite the fact that it meant a 30 percent cut in pay from $16,500 to $11,523 a year.[3]

After a three-week training period, Kampiles began, in March 1977, his job as one of about 65 watch analysts in the seventh-floor CIA Operations Center. The Operations Center served as a training ground for inexperienced junior officers and offered an alternative to the analyst and covert operators training programs. Yet watch analysts like Kampiles received information from the full range of American intelligence community sources: CIA agent reports, NSA communications intercepts, Defense Attaché reports, KEYHOLE photographs, and Foreign Service officer reports. Kampiles and the other officers monitored the incoming intelligence reports from around the world and routed them inside the agency.[4]

Since CIA personnel might need to be located in an emergency, watch analysts also had access to the CEMLOC—the Central Emergency Locator System. The CEMLOC was a six-volume computer print-out used to identify agency employees in an emergency. For each employee it specified name, home address, telephone, whether they were overt or covert employees, whether they were assigned abroad, and if so, where they were assigned.[5]

So that the watch analysts could understand the significance of the incoming data and take appropriate action, they had access to a variety of documents. One set included a large number of unclassified newspapers and magazines, but far more sensitive were the classified analytical reports they were cleared to read. There were also documents that provided insight into the satellites and other technical collection systems that were providing data. Among those documents was Copy 155 (of 350) of the *KH-11 System Technical Manual*, written under the direction of COMIREX chairman Roland S. Inlow. Once copy 155 was within the "vaulted" confines of the Watch Office, there was no further restriction on access. The 65 watch officers could use the manual at will, for it was not kept locked up. Its normal resting spot was on a shelf alongside an almanac in an unlocked cabinet known as the CONSERVA file, located beneath a standard copying machine.[6]

By the end of April, Kampiles had been formally indoctrinated into the KEYHOLE system. On April 19, 1977, Kampiles signed an "Indoctrination and Secrecy Oath Talent–Keyhole Clearance." The oath began with the explanation that

> A special document and material control system known as the
> "TALENT" Control System with a compartmentation called
> "KEYHOLE" has been established by the Director of Central
> Intelligence under the authority of the National Security Act of
> 1947 for the purpose of controlling the intelligence products
> of extremely sensitive intelligence sources.[7]

The document closed with the stipulation that

> I do solemnly swear or affirm that I will never divulge, pub-
> lish or reveal either by word, conduct or any other means
> any classified information relative to 'TALENT–KEYHOLE'
> material or sources except in the performance of my offi-
> cial duties and in accordance with requirements set forth in
> the 'TALENT' Control System Manual, unless specifically
> authorized in each case by the Director of Central Intelligence
> or his designee.[8]

Despite the secrecy oaths, codewords, and access to highly sensitive
intelligence, Kampiles found the work tedious—even though he occa-
sionally delivered late-night intelligence to DCI Stansfield Turner at his
home, minutes from the agency. Indeed, Kampiles was Turner's favorite
courier—for the simple reason that he was on good terms with the Admiral's dog. He had envisioned using the fluent Greek he had learned from
his parents to work for the CIA overseas, but instead found life in the
CIA to be a drab procession of 12-hour shifts in a single room. Five to
10 times a day he filled the burn bag, dutifully stapled it, and took it
down the hall to the incinerator chute.[9]

After a time, Kampiles began to express his dissatisfaction with standing unglamorous watches at all hours of the day and night. He pleaded
that he always wanted to be a case officer in the Directorate of Operations, which ran the overseas spy networks, but his record in the center was not sufficiently impressive to stimulate interest in Operations.
Initially told that his desire for a transfer was premature, he continued
to make requests for a transfer. His performance deteriorated, however,
and any chance of a future transfer vanished. Indeed, one of his supervisors recommended that he be fired before his probationary period was
up. The former supervisor recalls that Kampiles pawed the women in the
office, bragged of his sexual exploits, and "stood out markedly from the
rest of the people; he was a bullshitter." In November 1977, after only
eight months in the CIA, and after receiving a formal letter indicating
dissatisfaction with his performance, he resigned.[10]

Before departing for good, however, Kampiles pilfered copy 155 of
the KH-11 manual, which was numbered but not regularly inventoried.
He stuffed a copy of the manual into his sport jacket one day on the way
out of the building. Stealing a document was not particularly difficult.

Removal of classified documents from the agency by employees who wanted to do some work at home had reached such proportions that Director Stansfield Turner had addressed a memo to employees warning against such activity.[11]

That Kampiles picked the KH-11 manual was not simply chance. During his tenure at the Operations Center, Kampiles was encouraged to study the manual in detail so he could give briefings that included information on the KH-11 and its products. After purloining the manual, Kampiles took it to his apartment in Vienna, Virginia and put it in a dresser drawer. When he quit the CIA, he carried the manual to his mother's home in Chicago and again put it in a dresser drawer. Several weeks later, with a plan to take it to Greece, Kampiles cut off the classification markings. Cutting off the markings would prevent his mother from getting upset if she saw it before he left for Greece and would avoid any unpleasantness with Greek customs when he arrived.[12]

With his hopes of being a CIA operative in Greece shattered and his employment at the CIA terminated, Kampiles journeyed to Greece on his own. Kampiles left for Greece on February 19, 1978. He brought the CIA manual along, and not for vacation reading. On February 22 or 23, he went to the Soviet Embassy and told "an older, balding, fat Soviet official" he could provide information on a long-term basis. The security officer demanded that Kampiles provide documents, claiming to be concerned that Kampiles might be a U.S. plant.[13]

A day or so later Kampiles came back prepared and told a clerk there that he had previously spoken with the security officer. The clerk told him to wait. Within a few minutes he was greeted by a man who identified himself as Michael. Michael was Major Michael Zavali, a military attaché and officer in the GRU—the Chief Intelligence Administration of the Soviet General Staff.[14]

Almost immediately Michael took Kampiles for a walk outside the embassy grounds during which Michael identified himself as a Soviet attaché and asked to see identification from Kampiles. Kampiles refused, telling the GRU officer that his name was Robert Jackson. Naturally, Michael didn't believe him and pressed him further. Kampiles claimed he countered by telling him that his first name was Michael but refusing to give a last name.[15]

Before the meeting was over, Kampiles gave Zavali two or three pages of the KH-11 manual—its table of contents, summary, and an artist's conception of what the satellite looked like. But Zavali refused to part with any money until his experts examined those pages and arranged a meeting with Kampiles for a week later.[16]

The two met for the second time on March 2 in the vicinity of the Greek National Stadium. Michael told Kampiles he was interested in the manual but needed the rest of the document to find out how much he could pay him. Kampiles requested $10,000 and turned over the rest of the document. In return for the entire document, Kampiles received a mere $3000.[17]

Zavali expressed interest in obtaining additional documents from Kampiles and requested that the ex-CIA man photograph them since it was easier to get exposed film through Greek customs. Kampiles should give priority, Michael instructed, to information on U.S. military capabilities, military installations, and CIA personnel abroad rather than on what the U.S. knew of Soviet military capabilities.[18]

The GRU officer also gave Kampiles elaborate instructions for meeting again in August or September. Zavali jotted down an address on a 50-drachma ticket stub that tourists get on visits to the Acropolis. Kampiles was to send a "Happy Birthday . . . I am well" note to that address to signify he was returning to Athens. Once back in Athens, he was to go to the Athens stadium, make his way up a cobblestone path to a certain telephone pole, and stick a thumb tack in it. That was to be a signal for Michael to meet Kampiles the following Saturday night at Seranos, an Athens pizzeria.[19]

Upon returning to the United States, Kampiles was anything but discreet. Perhaps out of a need to be caught or a need to convince himself that his fantasies had come true and that he had conned the Russians, he began telling friends his version of his Greek encounters.

From Chicago Kampiles called Anastasia Thanakos in Washington, a friend and protégé who had worked briefly as a clerk at the CIA. Kampiles told her of meeting a "foreigner" on his February flight to Greece. In his account, Kampiles told the foreigner that he was hoping to get a job in Greece, and the foreigner in turn told Kampiles to go to the Soviet embassy. Kampiles did so, and as he told it, gave the Soviets "information you could pick up anywhere—such as in *Newsweek* magazine."[20]

The story did not surprise Thanakos. She also knew him as a bullshitter and Thanakos questioned him about his latest tale. Kampiles insisted that he was telling the truth and that he had received $3000. At his request, Thanakos conveyed Kampiles' story to a mutual friend, CIA officer George Joannides. Kampiles wanted to return to the CIA as an operative.[21]

On April 29, 1978, Kampiles and Joannides sat on a bench outside CIA headquarters. Kampiles told Joannides about the trip and his "conning" the Soviets in Athens. Kampiles said he crashed a garden party at the Soviet embassy one night and struck up an acquaintance with a Russian named Michael. Kampiles said he convinced "Michael" that he was still with the CIA by showing him some sort of "identity documents" he had used as a watch officer. Kampiles also told Joannides of his plans to return to Greece in the late summer. He told Joannides that Michael had given him an address in Athens where he could be contacted, that he was "playing the Soviets along" and that he had received $3000 from them.[22]

Joannides, who worked in the General Counsel's office, told Kampiles he should talk to someone from the Soviet section. While Kampiles sat on a bench near a statue of Nathan Hale, Joannides went in search of someone in the Operation Directorate's Soviet Bloc Division. Joannides talked to a Soviet Bloc expert who refused to meet Kampiles or pass him on to a colleague in the mistaken belief that President Carter's Executive Order on intelligence activities prohibited him from contacting a U.S. person.* Joannides argued unsuccessfully with the CIA officer and suggested that Kampiles write a letter to the agency, a suggestion to which the officer agreed.[23]

Kampiles wrote to Joannides and mailed his letter in May. Joannides, who was recovering from a heart attack, placed the letter in a zippered briefcase that he took to the office with doctor bills and insurance forms. Because of changes in assignments and in his status at the CIA, Joannides didn't get to the briefcase until July.[24]

In his letter, Kampiles wrote that in his meeting with Michael they discussed "the use of a camera and my role in delivering the information." He enclosed a 50-drachma ticket, apparently to a Greek sport-

---

* Executive Order 12036, "United States Intelligence Activities," of January 24, 1978, permitted collection of information on U.S. persons under a variety of circumstances, including—if a person gave consent; if the information was relevant to a lawful counterintelligence, personnel, physical, or communications security investigation; or if that person was a former intelligence agency employee and the information was required to protect intelligence sources and methods. All three applied in the case of Kampiles.

ing event. On the ticket, ostensibly in the handwriting of the Soviet intelligence officer who was trying to recruit him, was the address of a proposed rendezvous in Athens. Kampiles wrote that "if you think there might be agency interest, I would be willing to discuss the experience in full detail."[25]

Joannides turned the letter over to the Soviet Bloc Division while Kampiles settled into a new apartment in Munster, Indiana and began his new job at Bristol Meyers.[26] Soviet Bloc official Vivian Psachos soon requested that Kampiles journey back to Washington to discuss the matter. Kampiles agreed and said he would try to make arrangements to come to Washington before his scheduled trip to Athens on August 18. When he did so and phoned Joannides, it was once again Friday afternoon and Joannides' secretary "erroneously refused to put it through, saying [he] was in conference." Joannides called Kampiles back and arranged for a subsequent visit to Washington for interviews with CIA and FBI officials.[27]

The officials were certainly interested, but not because they believed Kampiles' story. The reference to money piqued their interest because it is a known Soviet intelligence technique never to pay anyone without receiving documents or information in return. In addition, the FBI was directly involved in investigating a perceived leakage of KH-11 secrets. The CIA had already concluded that the satellite had been compromised because of Soviet ground movements that seemed designed to thwart specific KH-11 capabilities.[28]

On August 14, 1978, at a suburban Washington motel, Kampiles met with Vivian Psachos, FBI agents Donald E. Stukey and John Denton, and Bruce Solie of the CIA Office of Security. At the meeting with Stukey, Kampiles told him that he gave "Michael" a forged CIA identification card and that "Michael" had responded with $3000 in a brown envelope. Kampiles said in this first interview that the $3000 was payment for expenses but that he was promised $10,000 for secrets on long and short range missiles, the USAF/Rockwell International B-1 bomber, cruise missiles, reconnaissance satellites, and information on CIA agents abroad.[29]

"I said I didn't believe him, because the Soviets don't give money for nothing," Stukey later testified. There were also several other holes in Kampiles' story. Kampiles claimed he didn't want the Soviets to know his real identity. According to Kampiles, it was just a lark and he wanted to be able to end it when the time came and not become subject to blackmail in the future. But Kampiles said he gave Zavali his ID card bearing his photo and true name. Kampiles also first said he had signed a receipt using the name Robert Jackson, although that was obviously not the name on the ID card.[30]

Stukey arranged for a second interview with Kampiles at the Washington Field Office on August 15. In the FBI interrogation room in Washington, FBI agent James Murphy told Kampiles he didn't believe his story. Murphy later recounted that "I told him that if I didn't believe it, nobody in the whole world would believe it." Kampiles, who had failed two polygraphs, slumped down in his chair and buried his head in his hands. After several seconds, Kampiles looked up at Murphy. "You're right," he said. "I didn't get the $3000 for nothing, I sold them the document."[31]

The document, of course, was the KH-11 manual. Kampiles told Murphy that he had taken the manual from the Operations Center and stuffed it into the left-hand pocket of his sport jacket with the vague idea of selling it to the Soviets. FBI agents asked the CIA if the manual was missing only to discover that no one knew. Kevin J. Donoghue, one of the Operations Center supervisors, was dispatched to find out. After combing the Watch Office, he reported that he could not locate Copy 155.[32]

The next day Kampiles identified a *KH-11 System Technical Manual* as the document he sold the Soviets. Pointing to the copy held by Agent Stukey he said, "That's it." He was then shown the document at which time he looked through it, flipped pages, and said, "Yes, that's it."[33]

The decision to bring Kampiles to trial was only made because of a change in policy initiated by the Ford and Carter administrations and only after strenuous objections by the Defense Department. Previously, the government would often refuse to prosecute those who had committed espionage for fear that a trial would compound the damage by the disclosure of additional secrets.

Eventually, the CIA cooperated fully with attorneys at the Department of Justice in the prosecution of Kampiles, but the Department of Defense remained obdurate. The problem was the obligation to introduce the KH-11 manual as evidence at the trial in a way that would keep its contents secret. An exchange of letters between DCI Stansfield Turner and Attorney General Griffin Bell should, theoretically, have settled matters, because the CIA had developed the KH-11 and CIA employee and COMIREX chairman Roland Inlow had directed its production.[34]

However, the Secretary of Defense held day-to-day management responsibility for the National Reconnaissance Office, and the manual was "published" under NRO auspices. Exactly who held ultimate authority for security concerning satellite reconnaissance systems had long been a sore point between the CIA and NRO. Thus, one month after Kampiles' arrest and six weeks before the trial was to begin, lawyers from the Department of Defense objected to the procedures that the Justice Department and CIA had worked out. They contended that the government couldn't acknowledge at the trial that the United States was conducting overhead reconnaissance of the Soviet Union. At the time they did not realize that President Carter was already planning to acknowledge the program in his October 1 speech.[35]

Pentagon lawyers also argued that the government couldn't refer in the courtroom to the NRO, whose initials were on the cover of the KH-11 manual. No one present realized that the CIA had already authorized the publication of a book by its former director, William Colby, in which he discussed both the overhead reconnaissance program and NRO. But the Pentagon lawyers probably would have objected anyway.[36]

Pentagon lawyers presented four "theories" that prosecutors could choose from at the trial to avoid revealing the program or the office. One theory was that the KH-11 documents had actually never been given to the Soviets. A second was that the KH-11 system had been proposed but never implemented. A third theory postulated that the satellite system had been constructed but never launched, and the final theory was that the system had been launched but never placed in operation.[37]

One problem with any of the "theories" was that Kampiles knew the Soviets had the manual in their hands and that the KH-11 was an operational system. Furthermore, the first falsehood would obviously lead to a dismissal of the charges. The Pentagon approach, according to Bell, "showed no respect for the integrity of our criminal justice system. All of the proposals would have required making material misrepresentations to the court in violation of Section 161 of Title 18 of the U.S. Criminal Code and several provisions of the Code of Professional Responsibility of the American Bar Association."[38]

When John Martin of the prosecution team made those points in a meeting with Pentagon lawyers, they replied that "conversations are now taking place at higher levels," and that the prosecutors would be "receiving instructions." The following day, when Martin and others on the team told the Attorney General that the Defense Department was exerting extreme pressure, Bell reaffirmed that Justice, not the Pentagon, was in charge of the case.[39]

The instructions from higher authority never came, but the Pentagon lawyers did propose to solve the problem by conducting a closed trial. Apparently, in the years that the Pentagon lawyers had immersed themselves in national security law, they had grown rusty in constitutional law. The Justice Department prosecutors politely informed them that the Sixth Amendment to the Constitution provided that "In all criminal prosecutions, the accused shall enjoy the right to a speedy and public trial."[40]

The Defense and Justice departments did agree that two elements had to be kept secret: the KH-11 manual and the method used in establishing the KH-11's compromise. The trial judge, Phil M. McNagny, Jr., solved the problem by ruling that only the attorneys, jury, and expert witnesses—not the press—could see the manual with some "extraordinarily sensitive numerical tables and geographical references excluded and that the document would be kept out of the public record." In addition, he instructed defense lawyers not to ask in open court how the government learned that the KH-11 was not a secure system.[41]

Michael Monico, Kampiles' lawyer, attacked the government's case against the former CIA employee in a variety of ways. Kampiles had recanted his confession and Monico claimed that Kampiles' original story was the true one—that he had simply been acting out a dream that the CIA denied him, seizing on an unexpected opportunity to play double agent and beat the Soviets at their own game.[42]

As Monico told it, Kampiles had walked around Athens' Constitution Square after a night on the town and noticed what seemed to be a party in a nearby building, the Soviet embassy. "He had a crazy notion, a wild and unbelievable notion," Monico declared. "He wanted to test himself, to test his fantasies and dreams, so he walked in. It was the worst decision he had ever made in his life up to that point."[43]

He marched up to the embassy's bar, and tried to convince those present that he was still with the CIA and had secrets they would want to have. "His brain was pounding," Monico explained. "Here was Bill Kampiles in the Russian embassy doing exactly what he'd always wanted to do . . . He thought he could return to the United States, to be accepted [by the CIA] and return to Greece as an undercover agent." After three visits and much talk about strategic missiles and the B-1 bomber, Kampiles got $3000 in exchange for nothing but promises, the lawyer said. Monico denied any actual sale: "Perhaps if we wait long enough, they'll find No. 155 and then we can all go home."[44]

Another theme of Kampiles' defense was to suggest that he may have been set up for some unknown master spy or "mole" in a key position at the CIA or another strategically placed government agency. U.S. Attorney David T. Ready had admitted in pretrial proceedings that "other suspects were developed in this investigation," but he said that the investigation of those suspects proved fruitless. He refused to give the defense any more information concerning that aspect of the investigation on grounds of privacy and potential compromise of "sensitive national security information."[45]

One other suspect whose name Monico introduced in court was John Paisley, a CIA official who had died (or according to some, had escaped) in somewhat bizarre circumstances in the Chesapeake Bay in 1978. Paisley had served in the CIA's Office of Strategic Research, the office responsible for the assessment of Soviet strategic forces, and had served as a CIA liaison to Team B in the controversial Team A–Team B competitive analysis of Soviet strategic forces. Some accounts have suggested that he served the Soviets as a mole and that the body recovered from the Chesapeake Bay in 1978 was not really Paisley's.*[46]

Monico also recounted Kampiles' discussions with George Joannides and the events that followed. But he claimed that the FBI man who broke Kampiles, James Murphy, had exploited the fact that Kampiles put the $3000 into a joint account with his mother. According to

---

* In a 1979 *Reader's Digest* article, journalist Henry Hurt reported that the KH-11 was first known to be compromised in August 1977, rather than in July 1978 as CIA Director Stansfield Turner testified before the Senate Intelligence Committee. Richard Helms was quoted as claiming that "the Kampiles case raises the question of whether or not there has been infiltration of the United States intelligence community at a significant level." Another possibility, assuming for the sake of argument that the KH-11 was compromised in 1977, is that Christopher Boyce had given information on the KH-11. Boyce, who worked at the TRW defense contracting firm, had provided the Soviets, with the aid of friend Andrew Daulton Lee, documents concerning the RHYOLITE signals intelligence satellite and the planned PYRAMIDER agents communications satellite. Hurt reported that in his confession Lee stated that he "turned over to the Soviets five to ten typed pages dealing with the [cryptonym omitted] communications satellite . . . the type that flies daily over Russia taking photographs."[47]

Monico, Murphy "told Bill if he didn't change his story, his mother would be implicated. . . . Bill Kampiles is a young, patriotic boy that got involved in a bizarre notion. . . . They turned him into a criminal."[48]

Two other themes concerned the thousands of CIA employees who had access to the manual, including a summer intern, and the number of manuals that were unaccounted for. Ray Hart, a security officer who briefed CIA employees on the importance of maintaining document security, testified that he had sworn "nearly a thousand" CIA employees, including Kampiles, to a special secrecy oath concerning the KH-11 documents in a two-year period.[49]

A November 1, 1978 inventory of 350 technical manuals, reported in a Memorandum for the Director of Security entitled "Status Report—KH-11 System Manual Accountability," showed that the whereabouts of 17 manuals remained unknown. Thirteen of the missing copies had been assigned to the CIA, two to the DIA, and two to the Army. The CIA's unaccounted for copies were signed out to COMIREX (4), the DCI (1), National Foreign Assessments Center (2), and the Directorate of Science and Technology's Office of Development and Engineering (6). DIA subsequently reported prior to the trial that it had located one copy.[50]

On the stand Kampiles was asked by Monico if he had ever taken a document out of CIA headquarters.

"Yes sir, I did. It was a blank ID card."

"Anything else?" Monico asked.

"No sir, I did not," Kampiles said.

"Did you ever give any Soviet, any person, a CIA document?"

"I did not."[51]

Monico went to the evidence table, retrieved hardbound and softbound covers of the *KH-11 System Technical Manual*, and held them aloft. Had Kampiles ever given any Soviet or any person a copy of the manual?

"No sir, I swear it."[52]

In addition to the testimony of agents Murphy and Stukey, who recounted their interrogation of Kampiles and his ultimate confession, the prosecution relied heavily on the testimony of Leslie Dirks. Dirks told the jury that U.S. national defense could be "seriously harmed" if the Soviet Union had access to the KH-11 manual. He said knowledge of the manual would suggest ways that the Soviet Union could hide its nuclear and military capabilities from the satellite.[53]

Dirks explained that the "KH-11 system is a photographic satellite with associated ground facilities for controlling [the] satellite and distributing its products" and "one of the principal intelligence collection sources used to verify that the Soviet Union is indeed living up to the terms of their [SALT] agreement with the United States."[54]

Disclosure of the Top Secret manual to a hostile foreign power would, Dirks testified, do serious harm to national defense. The manual described "the characteristics of the system, its limitations, and its capabilities" and described the "process of photography employed by the KH-11 system and illustrates the quality of the photos and the process used in passing the product along to the users of the system." He also detailed the "responsiveness and timeliness in the delivery of the 'product'." Page 8 of the 64-page manual described the satellite's "limitations in geographic coverage."[55]

Possession of the manual would, said the Deputy Director for Science and Technology, "put the Soviet Union in a position to avoid coverage from this system. For example, by rolling . . . new aircraft under test into hangars when the system passes overhead, thereby preventing photographs of the new airplanes." In addition, knowledge of the quality of the photographs could enable the Soviets to devise effective camouflage.[56]

Compromise of the KH-11 system would, according to Dirks, cause the U.S. to lose the advantage of being able to produce "accurate and current information" on Soviet capability for the U.S. President "in a time of crisis."[57]

In his summation, Monico urged the jury to disregard the confession as "untrue, unreliable and incredible" and that the government was in "such a rush to prosecute this case" that no effort was made to find out whether anyone at the CIA operations center had seen the manual after Kampiles quit the agency in November 1977. He went on, "Just because you get a kid to admit to something he didn't do, that doesn't make a case. He wants to scream out his innocence. Don't convict him on statements that came out of his own mouth that day [Aug 15]. He told them it was not true on the 17th. They paid no attention. Please! Pay attention! End the nightmare!"[58]

The jury deliberated for over 10 hours before returning the verdict of guilty on all counts. Two counts dealt with espionage and the defendant's intent to injure the U.S. and carried the possibility of a life sentence. Other counts, carrying lesser penalties, concerned passing U.S. documents to unauthorized persons and the sale of U.S. documents valued at more than $100.[59]

At his December 22 sentencing, Kampiles said "First of all, Your Honor, I'm sorry for everything that has happened. Not at any time did I want to injure my country in any way. I only wanted to serve my country." Attorney Ready suggested a "substantial sentence" because Kampiles "chose to casually disregard the safety and well-being of 200 million Americans."[60]

Judge McNagny sentenced Kampiles to 40 years in prison, stating that "This case is a complete tragedy for a young man who has never been in trouble before" but that "the United States has suffered a severe setback because of the sale to the Russians."[61]

After Kampiles' conviction was upheld by the appellate court, Richard J. Stone, the Pentagon's deputy assistant general counsel for intelligence, international, and investigative programs, excoriated the prosecution in a letter to Philip Heyman, assistant attorney general in charge of the Criminal Division. Stone claimed that many of the government's problems in sensitive national security cases "stem from a curious perspective evidently entertained by lawyers of the Criminal Division's Internal Security Branch that Defense Department and CIA were not 'clients' but 'victims.' "[62]

The exact extent of damage done by Kampiles may never be known with certitude. It has generally been asserted that not only were the Soviets previously unaware of the specific capabilities of the satellite, but that they were unaware that it was taking pictures at all. According to one account,

> The Soviets knew the schedule of the United States' KH-9 spy satellite to the minute, and when it flew over the Uzbekistan missile center everything was tucked out of sight. But a few hours later, another U.S. satellite, the KH-11, passed over the same field and caught an aerospace glider out in plain view—giving U.S. intelligence its first evidence that the Soviets were making a craft similar to the U.S. space shuttle.
>
> The Soviets had been *tricked* into believing the second satellite was electronically "dead."[63]

One aerospace expert has suggested that the Soviets believed the KH-11 was a KH-8 or KH-9 that was no longer operating and that the United States had engaged in some very specific deception to convince them. Without such deception it would have taken a monumentally bad analysis by those monitoring the U.S. reconnaissance program, given the 1972 stories in the media and the only moderately higher KH-11 perigee. Former DCI Stansfield Turner, who believes the Soviets did not understand that the KH-11 was taking photographs, cautions that "one should never underestimate the role of stupidity."[64]

Whatever the Soviets knew before Kampiles' sale, they certainly knew the KH-11's true mission after they received the manual. And one thing they undoubtedly did not know until they received the manual was how easy it was for the KH-11 to maneuver its optical system in order to obtain coverage of targets in front, to the side, and under the spacecraft. Within the course of a minute the KH-11 could take 8 to 12 pictures in any sequence of directions.* But even though the knowledge the Soviets gained from Kampiles' betrayal allowed them to decrease the satellite's effectiveness, it by no means allowed them to significantly lower its value. One official, quoted by the *Washington Post*, noted that "The Russians know the satellite has been in orbit taking pictures of their country for some time. Getting their hands on the manual didn't stop the satellites taking pictures."[65]

The countries that the KH-11 and its predecessors continued to take pictures of included Cuba, the Soviet Union, and Iran. Jimmy Carter entered office with a desire to reduce the level of hostility between the United States and Castro's Cuba. Eighteen years of enmity had not changed the Castro regime and the new President was ready to try a different course. As a conciliatory gesture, Carter halted, in early 1977, direct overflights by SR-71 aircraft. Of course, the U.S. did not totally suspend overhead coverage: approximately twice a month, a KEYHOLE satellite would be instructed to send back imagery of selected targets on the island.[66]

A KH-11 image sent back to earth in January 1978 showed a Soviet freighter unloading crates at Havana. The dimensions of the crates, calculated from the photos, led to the conclusion that the crates carried

---

* John Pike, a space policy analyst for the Federation of American Scientists in Washington, D.C., described the properties of the KH-11 mirror in testimony at the trial of Samuel Loring Morison. Based on his knowledge of how the civilian LANDSAT system worked, Pike proceeded to describe the "mirror-flipping" capabilities of the KH-11. Representatives of COMIREX immediately began rushing about leaving notes on the prosecutors' table. I have independently verified Pike's description.

MiG-23s. Before the year was out, additional photographs showed 15 MiG-23s at a nearby airbase.*

What was disturbing to some administration officials was that the MiG-23 came in two versions, both of which the Soviets had been supplying to allies. The MiG-23E was an interceptor and Cuban possession of such planes was not of great concern, but the MiG-23F, also known as the MiG-27, was another matter. It was a ground attack aircraft capable of delivering nuclear weapons. Supplying such aircraft would represent a violation of the 1962 pledge made by the Soviet Union not to introduce offensive nuclear delivery vehicles into Cuba.[67]

The diverse assets of the U.S. intelligence community were undoubtedly exploited to help determine the precise nature of the 15 new MiG-23s. To help provide the best possible data, President Carter revoked his suspension of overflights on November 16, 1978, authorizing an SR-71 mission to target the MiG-23s. The ultimate result of the collection and analysis efforts was, to the relief of the administration, the conclusion that the MiG-23 crates first spotted by 5501 contained the interceptor version of the MiG-23.[68]

But the MiG-23 flap was the precursor of a more serious crisis involving Cuba, although the crisis was largely a U.S. domestic political crisis. Events such as the MiG incident and turmoil in Central America (in which some saw Cuban and Soviet hands) had stimulated concern in the administration. In March 1979, in an attempt to get thorough assessment of Soviet forces in Cuba, national security adviser Zbigniew Brzezinski directed DCI Stansfield Turner to assess the size, location, capabilities, and purposes of Soviet ground forces in Cuba.[69]

The first step in the process was the evaluation of older data. Beginning in the late 1960s, if not before, U.S. intelligence received periodic and fragmentary reports of a Soviet ground force unit of a few thousand men in Cuba. Apparently, the reports were not taken at face value by the analysts. And it was doubtful, according to CIA officials, that higher ranking officials, given the preoccupation with Vietnam, ever saw such reports.[70]

Such initial reports were written in apparent ignorance of the Soviet military forces that remained in Cuba after the 1962 missile crisis.

---

* These photographs might have been produced by 5501, the original KH-11, which operated throughout the year; by the KH-9 that operated from March 16 to September 10; or by 5502, launched on June 14, 1978.

Initially, 17,000 Soviet troops and technicians remained on the island. Included were four heavily armed combat groups totaling about 5000 men. The groups had distinct responsibilities, such as air defense of the Soviet missile installations, and were located in different camps.[71]

In early 1963 Secretary of Defense McNamara noted that some of the military equipment associated with the combat units had been removed from Cuba. In April of that year President Kennedy announced that another 4000 troops had departed, leaving 13,000–14,000 men. In June the State Department lowered the estimate to 12,500 and the *New York Times* reported that it was expected that a residual force of 10,000 would remain indefinitely.[72]

The intelligence community, however, was not certain of the number of Soviet troops remaining in Cuba or their precise mission. Several events proceeded to further confuse intelligence analysts. With the crisis over, the Soviets proceeded to set up a military training mission consisting of several thousand advisors. The mission had two purposes — to affirm the Soviet-Cuban military relationship and to train Cuban forces to use the equipment that the Soviets had provided. In addition, the Soviets consolidated the combat forces. As the distinct units disappeared from view it was assumed that they had returned to the Soviet Union. By the late 1960s those in the intelligence community who monitored Cuba had come to assume that the only Soviet forces in Cuba were the military advisors.[73]

Sometime in 1975 or 1976, the National Security Agency intercepted communications that referred to a Soviet "brigada." Again the information was ignored, this time because a brigade is an aberrational unit in the Soviet military, which is largely organized into regiments and divisions. The Soviets went to great lengths to mask their presence, and the colonel who led the brigade was never mentioned in public either in Havana or Moscow. Rather than being camped together in recognizable Soviet style, the brigade was split between separate locations to resemble Cuban camps a few kilometers from each other near Los Palacios, 60 miles west of Havana. Also the brigade only rarely conducted maneuvers and maintained a high degree of radio silence to hide its presence from NSA listening antennas in Florida.[74]

Additional information became available in 1978. The KH-11 returned photographs of modern Soviet military equipment deployed in camps near Los Palacios and photographs of a Soviet training mission at a Cuban gunnery range in western Cuba, but again the wrong conclusion was drawn. CIA and DIA officials concluded that the military equipment belonged to Cuban and not Soviet forces and that the brigade's bivouac areas were Cuban camps.[75]

Additional KH-11 photos revealed tents placed in the troop encampments near Lourdes in patterns standard for Soviet, but not Cuban, troop deployments. Communications intercepts of the administrative channels of Soviet and Cuban military units from April through June 1979 produced references to a Soviet unit stationed at the two sites southwest of Havana as a "brigada."[76]

A study of this data, conducted by an NSA analyst, led to the conclusion that there was at least a Soviet brigade headquarters in Cuba. A fierce interagency debate ensued. Army intelligence sided with NSA, but the remainder of the intelligence community was arrayed against them. The chiefs of the CIA, DIA, State Department, Air Force, and Navy intelligence components all rejected the Soviet brigade theory. On July 12 the intelligence community released a coordinated interim report stating that a Soviet force, separate from the Soviet advisory group, existed in Cuba. However, no agreement could be reached on the size, organization, and mission of the Soviet force.[77]

In light of the inconclusive report and because talk of a Soviet brigade in Cuba could threaten ratification of the SALT II Treaty, President Carter ordered a step-up in surveillance. In early August, communications intercepts indicated that a Soviet training exercise was planned in mid-August at the San Pedro maneuvers grounds, a few miles west of Havana's Jose Marti International Airport. With such advance knowledge it was easy to assure that at least one KEYHOLE satellite would be ready, for although 5501 had been deorbited in January, NRO had three KEYHOLE satellites in orbit from May 28 until August 25. In addition to 5502, a KH-9 had been launched on May 16 on what would be a 190-day mission and a KH-8 went into orbit on May 28, where it would remain until August 25.[78]

The available KH-11 was certainly employed to monitor the maneuvers. Not only was the maneuver area photographed, but also the area believed to contain the headquarters of the Soviet brigade. Those photographs showed that the units maneuvering with tanks, personnel carriers, and mechanized infantry at San Pedro were from the suspected Soviet base, and photographs on later days demonstrated that the equipment had been returned to that area. Such photographs represented conclusive proof to the entire intelligence community of the presence of a Soviet military unit. The unit was estimated to have between 2000 and 3000 men equipped with tanks.[79]

NSA declared it to be a combat brigade and the National Intelligence Daily, an above-Top Secret report available only to about 200 individuals, carried notification of the brigade's detection and also referred to it as a combat brigade. The term "combat brigade" was chosen by default for lack of a better designation. A high-ranking intelligence official, who normally would have had the opportunity to review and revise the NID but was out of town, is reported to have said he could have edited out or qualified the combat designation.[80]

But the combat designation did stand and based on intelligence about Soviet brigades in East Germany and the Soviet Far East, a size of 2000–3000 was estimated. After being informed of the administration's conclusions Senator Frank Church, whose Foreign Relations committee had issued a statement a month earlier that there was no buildup in Cuba, called a news conference to announce the findings and call for "the immediate removal of all Russian combat units from Cuba."

Once the brigade issue became public knowledge, the result was a political headache of massive proportions for the Carter administration. Senator Richard J. Stone of Florida, with a militantly anti-Castro constituency, decided to exploit the news politically. It was one means of recovering the ground he lost with his conservative constituents when he had voted for the Panama Canal Treaty a year earlier. Likewise, Church, facing a tough re-election campaign in conservative Iowa, sought to gain a new lease on political life by his vocal opposition to the brigade.[81]

The Cuban brigade issue inevitably became intertwined with the SALT II issue, as Church proceeded to suspend the hearings on the treaty, announcing that "I see no likelihood that the Senate would ratify the SALT II treaty as long as Soviet combat troops remain stationed in Cuba."[82]

Those who opposed the treaty used it as another example of Soviet "treachery." Whether they could have used the information to defeat the treaty will never be known, because the Soviet invasion of Afghanistan ended all debate over the treaty.*

---

* The true purpose of the brigade has never been conclusively established. At the time (1979) the Army and NSA offered the controversial theory that the brigade might be responsible for the protection of nuclear weapons. One former NSC official believes the brigade's purpose is to protect the Soviet's massive signals intelligence complex at Lourdes. Others believe that the brigade is the Soviet equivalent of the U.S. 7th Army in Europe—a trip wire to deter attack.

*　　　*　　　*　　　*

The SALT II Treaty that Jimmy Carter and Leonid Brezhnev signed at Vienna on June 18, 1979 was the product of six and a half years of negotiations. Far more complex than SALT I, the new treaty provided for equal ceilings on the number of strategic nuclear delivery vehicles (2400, to be further reduced to 2250 by the end of 1981), the total number of launchers of MIRVs and heavy bombers with long-range cruise missiles, the number of MIRVed ballistic missiles (1200), and the number of MIRVed ICBM launchers (820). It also imposed limits on the number of warheads specific missiles could carry and bans on the construction of additional fixed ICBM launchers and on increases in the number of fixed heavy ICBM launchers.

Another treaty provision specified that if a launcher was employed for testing a MIRVed ICBM or SLBM, then all launchers of that type would be counted as launchers for MIRVed systems. Thus, any SS-19, which came in single warhead and six warhead versions, would count against the limit for MIRVed missiles.[83] For although the United States could estimate the maximum number of warheads on a particular type of Soviet missile by monitoring its initial flight tests, the KEYHOLE satellites had no capability to see through missile nosecones to see how many warheads were deployed on any individual missile.

A related problem involved the launch sites located near two small towns, Derazhnya and Pervomaisk in the Ukraine. The solution that was incorporated in SALT II was a reflection of the strengths and limitations of the KEYHOLE satellites. Long before the signing of SALT I there had been 60 silos at each location. The silos contained SS-11 ICBMs, an older (third-generation) single-warhead ICBM that was more powerful than the Minuteman II but far less accurate. Then, in early 1972 the Soviets began excavating 30 new silos at each site. By 1978 the two sites had a total of 180 silos, of which 120 were estimated to contain SS-11s while 60 awaited new tenants. While the 60 holes lay empty, the Soviet Strategic Rocket Forces were busy testing the SS-19, potentially their most lethal MIRVed ICBM.

The KEYHOLE satellites observed the next step. First, the Soviets started to lower SS-19s into the 60 empty silos at Derazhnya and Pervomaisk. Next, they remodeled the 120 older silos but left the SS-11s in place. Since the KEYHOLE cameras could not penetrate concrete or any other opaque material and there was no outward distinction between SS-11 and SS-19 silos, photos provided no basis for distinguishing between silos housing SS-11s and those with SS-19s. There was a difference, however, in the nearby underground command-and-control silos, because SS-19 command-and-control silos were marked by a domed antenna that some officials at the State Department and ACDA dubbed the "midget."[84]

A negotiating conflict arose over whether just the 60 Derazhnya and Pervomaisk SS-19s should count as MIRVed ICBMs. Naturally, the Soviets argued that, since it was agreed that the SS-11 and SS-19 were different rockets, only the SS-19s should count toward the limit on MIRVed ICBMs. Although there was no intelligence, despite vigilant monitoring, to suggest that the SS-11 had ever been tested with independently targetable warheads or that the Soviets had secretly replaced any of the SS-11s with SS-19s, many American officials were concerned. The officials, particularly in the intelligence community and the Defense Department, argued that it undermined the treaty's verifiability for the Soviets to deploy MIRVed and unMIRVed missiles together and in fields made up of a single type of launcher. "We had no confidence," explained a high Pentagon official, "in our ability to verify in the future the Soviet claim that those 120 launchers weren't MIRVed. Therefore, it was our position that all 180 launchers at D-and-P must count as MIRVed."[85]

The Soviets did not appreciate the American viewpoint. Soviet negotiator Vladimir Semyonov argued that the United States knew that there were two types of launchers and two types of missiles at the two sites. Further, according to Semyonov, the SS-19 holes were distinguishable by virtue of the domed antennae on the associated command and control silos. In a series of meetings in Geneva, American negotiator Ralph Earle and Semyonov debated the issue back and forth. Earle would counter Semyonov's argument that the domed antennae was proof that there were "visible differences" between the launchers with the claim that the United States knew that a silo without such an antenna could still handle a MIRVed missile. After one too many predictable exchanges on the subject, Earle said, "Look, the existence of the antenna is irrelevant to the capability of launching a MIRV; we know you can launch one without it because we've seen you do it." In other words, a KEYHOLE satellite had "snapped" a picture of a launch complex without an antenna that U.S. intelligence was certain had been used to launch a MIRVed missile. Earle's comment effectively shut Semyonov up for a while, although it caused some concern that Earle had compromised "national technical means" by revealing to the Soviets more than they had previously known about the extent and precision of U.S. ability to monitor their strategic programs.[86]

The American position was in part a testament to the capabilities of the KEYHOLE satellites. The satellites had helped provide conclusive evidence that the domed antenna on SS-19 silos was not a functionally related observable difference that would allow the U.S. to distinguish SS-19 from SS-11 silos. The limitations of the satellites were also instrumental in determining the U.S. position, however. During the summer of 1977, the NSC's Special Coordination Committee debated what the U.S. position should be. The Pentagon and CIA, represented by Harold Brown and Stansfield Turner, maintained that the 120 unMIRVed launchers at Derazhnya and Pervomaisk were so similar to the 60 MIRVed launchers that the Soviets might someday secretly substitute SS-19s for the SS-11s when the missile fields were under cloud cover. Cyrus Vance and Paul Warnke saw the issue more in terms of getting the Soviets to accept an "essential verification rule", whereby MIRVed launchers must be visibly distinguishable from unMIRVed ones. It was more important, they argued, for the Soviets to accept such a rule for the future than to accept the application of the rule retroactively with respect to Derazhnya and Pervomaisk, since everyone agreed that there were 120 unMIRVed missiles there and that forcing the Soviets to count those 120 as MIRVed would ensure that they put 120 MIRVed missiles into those silos.[87]

Harold Brown was not convinced. In his mind, the inability to continuously monitor the 120 SS-11 silos plus his estimate that the Soviets were probably eventually going to put MIRVed SS-19s in the silos made it prudent for the U.S. to insist on counting all the Derazhnya and Pervomaisk silos as containing MIRVed missiles. Stansfield Turner agreed. CIA projections showed that the Soviets might build up to the MIRVed ICBM limit they had under SALT, and Turner considered it essential for the verification and ratification of the agreement that there be no ambiguities about exactly how many MIRVed ICBMs the Soviets had and exactly where they were located.[88]

Brown and Turner were joined in their position by Jimmy Carter and it became the U.S. position to insist that all ICBMs at Derazhnya and Pervomaisk be counted as MIRVed. That insistence was subsequently reflected in the provisions of SALT II.[89]

* * * *

On the morning of November 4, 1979, a crowd began to assemble around the American embassy in Teheran shouting anti-American slogans. At about 10:30 as many as 3000 demonstrators poured over the walls and forced their way into the basement and first floor of the chancellery.[90]

Less than a year before, the Shah, in the wake of increasing protests and riots, had fled the country he had ruled so autocratically for 25 years. It was an end that took the American intelligence community by surprise. A 60-page CIA study completed in August 1977, "Iran in the 1980s," asserted that "there will be no radical change in Iranian political behavior in the near future" and that "the Shah will be an active participant in Iranian life well into the 1980s." A year later a 23-page CIA Intelligence Assessment, "Iran After the Shah," proclaimed that "Iran is not in a revolutionary or even a 'prerevolutionary' situation." A month later the DIA issued an Intelligence Appraisal stating that the Shah "is expected to remain actively in power over the next ten years."[91] By August 1979, the Shah was long gone. In his place was an Islamic State ruled by the Ayatollah Ruhollah Khomeini.

From the time the Shah left, Iran became a priority U.S. intelligence target and among the resources used to gather information on it were the KEYHOLE satellites. Targets included Teheran, Iranian military facilities, and sensitive CIA signals intelligence facilities in the north of Iran, specifically at Behshar and Kabkan, which were used to monitor Soviet missile tests.[92] The KEYHOLE photography of Teheran would provide information on mass political and religious activity and on any conflicts that might arise between the military and Khomeini's fanatical followers. Photography of military facilities, along with communications intelligence, could give analysts clues about the state of the military and whether any military action against the new regime was imminent. Finally, photography of the intelligence facilities would tell whether they had been discovered by the new regime.*

But the most crucial use of the KEYHOLE satellites, and specifically of the KH-11, was yet to come. For although the Shah was gone from Iran, he was not absent from the minds of millions of Iranians, including the Ayatollah. When the Shah moved from his exile in Mexico to the

---

* The outposts were discovered several weeks after the Shah left. Ambassador William Sullivan engaged in negotiations with the authorities in Teheran, which apparently included the paying of ransom, to get the CIA employees out of the country safely.

United States for medical treatment, the resentment of past American support for the Shah spilled over onto the grounds of the U.S. embassy on November 4, and by nightfall the embassy and all the Americans inside had been captured by a group of student militants.

What followed was month after month of frustration for the Carter administration and for the relatives of the hostages. Economic sanctions such as blocking all Iranian assets in U.S. banks and their subsidiaries and embargoing oil from Iran failed to move the Ayatollah. Nor did diplomacy do the trick. With all other avenues failing, President Carter authorized a rescue mission on April 11, 1980.[93]

Planning for such a mission had begun almost from the moment the hostages were seized. The Army's Special Forces Operational Detachment — Delta, more commonly known as Delta Force — would be responsible for rescuing the hostages. Such a mission would require much advanced preparation, including precious intelligence information. A variety of questions needed to be answered in order to launch a rescue mission with any reasonable chance of success. Planners for the mission needed to know how the hostages were dispersed in the 27-acre compound, how many were in the embassy or in the chancellery and supply-residential area, what rooms they were held in, whether those rooms had windows, what types of locks were on the doors to the embassy, when the guards changed their shifts, and how many guards there were. In addition, information was needed on defenses, both on the embassy grounds and wherever U.S. forces would be moving. Finally, good information was needed on rendezvous locations for the helicopters and aircraft to be employed, sites that could hide Delta and the helicopters and possible pick-up points for the rescuers and the hostages.[94]

Many of the questions could not, of course, be answered from overhead photography. An action officer recalled that "we had a zillion shots of the roof of the embassy and they were each magnified a hundred times. We could tell you about the tiles; we could tell you about the grass and how many cars were parked there. Anything you wanted to know about the external aspects of the embassy we could tell you in infinite detail. We couldn't tell you shit about what was going on inside that building."[95] Thus, both human and signals intelligence were instrumental. Human sources included a deep-cover Iranian source who had gained access to the hostages, and communications intelligence coverage of Iran was stepped-up dramatically.[96]

The inability of the KH-11 to provide any useful information on what was happening in the embassy was also noted by a former Carter administration official. That official also noted, however, that the KH-11 photography was somewhat useful in providing information on Iranian defenses (although not as useful as might be expected) and was absolutely essential to the planning of the mission. Without its photographs no rescue could have been attempted.[97]

The value of KH-11 photography was also attested to by Charlie Beckwith, the on-ground commander of the rescue mission. Beckwith recalls that "the photos began to come in at once and it gave the intel people a chance to compare movement in the compound with photos taken the day before, or the week before. If there were changes the analysts could arrive at some deductions. One photograph showed 20 automobiles parked around the compound. If on the next day fifty were parked in the same area then something was happening and the message traffic might reveal what it was. If the analyst hadn't been alerted to the change by the photos they might not have spotted it in the cables."[98]

One thing the photos showed that was directly relevant to the planning of the rescue was the existence of poles distributed around the compound. Assuming that any rescue attempt would use helicopters, either the Iranian Revolutionary Guards or the militants had begun to place poles at all likely landing zones within the compound. Beckwith recalls that "without the photographs, the intel shop would not have picked up the construction."[99]

Also essential was photography of the Amjadieh soccer stadium. The rescue plan specified that three different Delta Force elements would be employed in the mission. A 40-member Red Element would secure the western sector of the compound, freeing any hostages found in the staff cottages and commissary and neutralizing the guards who were in the motor pool and power plant areas. A 40-member Blue Element would be responsible for the embassy's eastern sector and for freeing any hostages found in the residence of the Deputy Chief of Missions, the Ambassador's residence, and the chancellery. A 13-member White Element would secure the adjacent Roosevelt Avenue and eventually cover the withdrawal of the Red and Blue Elements to Amjadieh soccer stadium. Whether they would be picked up at the embassy compound or at the stadium would depend on whether poles spotted by the KH-11 could be removed.[100]

The overhead views of the soccer stadium provided basic information on obstacles such as the stadium wall and light towers, the layout of the stadium, roads in and out of the stadium area and facilities in the area between the stadium and the embassy. Similar information was obtained from KH-11 photography of additional stadiums, an athletic field, and race tracks that could be used as alternative landing zones in the Teheran area.

On April 24, eight RH-53D helicopters left the deck of the USS Nimitz in the Gulf of Oman. Meanwhile, three troop-carrying MC-130s and three fuel-carrying EC-130s departed from Masirah, Egypt where the Delta Force had made its initial stop. The planes would rendezvous with the helicopters at a site 200 miles southeast of Teheran that was designated Desert One.[101]

Once the helicopters had refueled and were boarded by Delta, they would proceed toward Teheran and the C-130s would return to Masirah. The helicopters would then fly for about two and a half to three hours, landing about an hour before sunrise at Delta's "hide site"—35°14′ N by 52°15′ E. After off-loading the Delta contingent, the helicopters would then be flown to their own hide site, 15 miles north of the Delta contingent, where they would spend the daylight hours hidden in the hills around Garmsar until it was time to pick up the hostages and Delta.[102]

The Delta contingent would be met by two American agents who had been infiltrated into Teheran several days earlier. The agents would lead the rescuers five miles overland to a remote wadi 65 miles southeast of Teheran, where Delta would remain throughout the daylight hours.[103]

Once it turned dark, the two agents would return to the wadi driving a Volkswagen bus and Datsun pickup truck. Six drivers and six translators who had arrived with Delta would be driven to a warehouse on the outskirts of Teheran where six Mercedes trucks were stored.[104]

Colonel Beckwith would then be taken to scout the route to the embassy. After reconnoitering the route, Beckwith would be returned to the hide site. The Mercedes trucks would already be there. The route to Teheran would take Delta north along the Damavand Road, with the exact route through Teheran to be determined at the time.[105]

All the potential routes had been studied in advance, using KH-11 photography. The photography allowed interpreters to get a perspective on the hide sites and routes, provide identification of obstacles such as check points, and develop alternative routing for the trucks, alternative landing areas for helicopters, and alternative departure points for the hostages and their liberators.

When the rescue team landed at Desert One they were equipped with a set of KH-11 photographs of primary and emergency landing and departure sites, but the rescue was to proceed no further. Problems with three of the helicopters, one of which already had turned back, reduced the helicopter force to five, one below the minimum considered necessary. Based on Charlie Beckwith's recommendation delivered via communications satellite, President Carter aborted the mission.[106]

Tragedy followed. In preparing to leave Desert One, one of the helicopter's rotor blades sliced into the left wing of a C-130, setting off an explosion of fuel. According to Lt. Colonel Ed Sieffert, the commander of the Sea Stallion helicopters who witnessed the event from less than 100 feet away, "It was instantaneous. Flame had engulfed the whole area by the time the rotor blade stopped turning." Eight men died. In addition, the crews of the three helicopters closest to the fire were unable to retrieve the classified material they were carrying, which included the KH-11 photographs.[107]

A West European magazine obtained the photos and planned to publish them but was dissuaded by an American emissary.[108] But what the magazine would not do, the Iranians did. In one of the multitudinous volumes of U.S. documents published by the Iranians appear assorted KH-11 photographs of the stadiums, race tracks, and athletic fields that represented the primary and emergency zones.

CHAPTER 8

Watching the Evil Empire

Among the casualties of the Iranian hostage crisis was the presidency of Jimmy Carter. Whether Carter might have triumphed over Ronald Reagan in the absence of the hostage crisis will never be known, but the hostage crisis firmly shut the door on his administration.

Ronald Reagan came to Washington with a solidly conservative agenda. In the national security policy area Reagan campaigned on a platform of a stronger defense and a foreign policy that would more aggressively confront the Soviets with their transgressions. A major buildup of U.S. military forces was one key aspect of the Reagan program. In addition, the Soviets would be challenged in any instance where they were believed to be violating the terms of an arms control treaty they had signed.

One way to build support for the planned U.S. buildup, which was to include deployment of 200 10-warhead MX missiles, the Trident II submarine, the B-1 and Stealth bombers, and improvements to the U.S. command, control, and communications system, was to provide evidence of Soviet military capabilities that would convince the public of the need for new U.S. forces. One vehicle for providing such information was the yearly issuance by the Defense Department of a slick volume entitled *Soviet Military Power*. Each volume provided a detailed breakdown, often using newly declassified data, of Soviet strategic land, sea, and air power. Maps indicated the locations of Soviet ICBM fields, airfields for strategic bombers, and key military production centers. Tables detailed the number of missiles, submarines, and aircraft of each type and their important characteristics. The text explained DOD's view of the threat from the weapons systems and the significance of new developments in Soviet capabilities. As well as serving as a propaganda tool for the Reagan administration, the various volumes of *Soviet Military Power* also served as an invaluable guide to Soviet weapons systems.

Each volume contained illustrations, including numerous artists' conceptions, of weapons systems and facilities such as ground-based lasers at Sary Shagan, the Nikolayev shipyard, or mobile air defense missiles that were clearly derived from satellite photos. However, in accordance with long-standing policy, no KEYHOLE photos were included. But at least one individual somewhere in the bureaucracy decided to provide visible evidence of a new Soviet intercontinental bomber that was under development in 1981.

When the many avid readers of *Aviation Week and Space Technology*, which includes the Soviet intelligence services, opened up their December 14, 1981 issue to page 17 they saw a rather peculiar, purposely-degraded photograph of a not-yet-operational Soviet bomber designated BLACKJACK by U.S. intelligence.

The BLACKJACK, which became operational in 1988, is similar in design and layout to the U.S. Air Force's B-1B although closer in size to a B-52. With an intercontinental mission the BLACKJACK has an unrefueled combat radius of about 7300 miles and a maximum speed of Mach 2 (1400 miles per hour). It has two large internal weapons bays, each of which can hold six AS-15/KENT air-launched cruise missiles for "standoff" missions or 12 short-range attack missiles to neutralize air defenses. The weapons bays could also be used to store gravity bombs to be used in a penetration mission.[1]

The picture that appeared in the December 14 issue of *Aviation Week and Space Technology* was one of the earliest views that the CIA and other interested intelligence agencies had of the new plane. The photograph, according to its caption, had been taken on November 25, 1981 as the plane sat on the tarmac at Ramenskoye Airfield, a strategic test center outside of Moscow. The photograph also showed TU-144 Charger transports, similar to the Concorde. Distinguishable features included passenger windows on the side of the aircraft.[2]

In the 1984 trial of a naval intelligence analyst for leaking other satellite photos, the prosecution confirmed that the BLACKJACK photo was taken by a KH-11. Analysis of the photo by an outside expert indicated that the photo was taken by 5504, launched on September 3, 1981, on a southbound pass across the Soviet Union during the late morning local time. It was also determined that the picture indicated a KH-11 capability to obtain a resolution between 5.46 at its apogee and 17.7

inches at its perigee.[3] Even to the casual observer the picture revealed the satellite's ability to take photos at quite a steep slant or oblique angle. Such a capability made it that much harder for Soviet operations security personnel to hide targets from KEYHOLE surveillance.

The BLACKJACK photo was only one of a vast number of photos produced by the KEYHOLE satellites in the early years of the Reagan administration. In 1982 there were two KH-11s in orbit every day of the year—5503, launched on February 7, 1980, and 5504. In addition, on May 11 a KH-9 blasted off on a 208-day mission.

Imagery returned shortly after that date included a photo of Saki Airfield in the southwestern Soviet Union, which showed the apparent construction of a ramp to assist heavily loaded vertical takeoff and landing (VTOL) aircraft in taking off. Imagery of the Shikhany Field Test Area, taken on May 19, showed a variety of vehicles—ARS-12/14 and AGV-3 chemical warfare decontamination vehicles, BRDM-2RKH chemical reconnaissance vehicles, and MT-LB tracked vehicles. To DIA analysts this combination of vehicles indicated Soviet experimentation with a unit to emplace chemical barriers or in some other way support chemical warfare operations in the field.[4]

Outside the Soviet Union, China and South Africa were among the targets of the KEYHOLE contingent. Imagery of the Xingtai Army Barracks, taken on May 18, showed tracked multiple rocket launchers, which had previously been noted with only three of China's 14 independent tank divisions. Two days later, imagery of the Harbin Tank Production and Rebuild Plant 674 revealed new self-propelled artillery with a rotating turret. Earlier, on May 11, South Africa's Simonstown Naval Base was photographed. The resulting imagery showed a clear view of the inner harbor, drydock, probable submarine maintenance buildings, submarine harbor, outer harbor, and ships in port.[5]

Imagery produced in 1983 provided the administration with the best evidence it would have of a Soviet violation of an arms control agreement.

At the beginning of June 1983, the NRO was operating three KEYHOLE satellites. Two KH-11s, 5504 (launched on September 3, 1981) and 5505 (launched on November 17, 1982), and a KH-8 were in orbit. The GAMBIT mission had begun on April 15 and would continue until August 21 with the fiftieth KH-8 being launched into 96 degree inclined, 89 by 184 mile orbit. On June 20, a KH-9 joined the reconnaissance fleet as a Titan 3D placed the spacecraft into a 96 degree, 105 by 142 mile orbit. Thus, from June 20 to August 21 the space photographic reconnaissance effort hit one of its peaks. Rarely was there ever to be such intense coverage again. Such coverage surely caused heartburn to those Soviet officials responsible for ensuring that vital military weapons systems were hidden from the KEYHOLE cameras.

And that coverage paid off, possibly with help from a human source in the Soviet Union. When analysts at NPIC examined the photos during the first part of August, alarm bells rang.[6] The photos showed a ballistic missile detection radar under construction near Abalakova, north of Krasnoyarsk in south-central Siberia, near a spur line running north from the Trans-Siberian Railroad. One official told the *New York Times* that "It is so big it seems to be about the size of a 50-story building."[7]

That the Soviets were building another large phased-array radar (LPAR) was interesting but hardly unusual. During the late 1970s, the Soviets began constructing a new class of LPARs to replace its older Hen House radars. The first of the new radars was located at Pechora in the north central Soviet Union, and the new class became known as the Pechora class. Four additional radars followed, all built in compliance with the ABM treaty, at Lyaki, Olenogrosk, Sary Shagan, and Misheleveka.[8]

What was so disturbing was that the radar was approximately 400 miles inside the nearest Soviet border. The ABM Treaty limited the deployment of early-warning radars to the periphery of each nation unless the radar was located at an ABM test center. The limitation was in keeping with the restriction of operational ABM sites to two (and subsequently one) each. Early-warning radars located deeper in either nation's interior would allow better "attack characterization"—that is, better estimates of the trajectory of incoming warheads. The better such information, the better ABM interceptors could be directed towards the incoming warheads. Thus, "battle management" would be enhanced and any ABM system would be more effective. Adding to the concern was the fact that the radar was located near three SS-18 missile fields, exactly the type of installation that an ABM system might be used to protect.[9]

Further, not only was the radar located 400 miles from the Soviets' nearest border (with Mongolia, to its south) but it was oriented to the northeast, across 3000 miles of Siberia. Its radar coverage would close a geographical gap (toward the northeast) in the Soviet early-warning coverage and provide warning of SLBMs fired by U.S. Trident submarines in the north Pacific.[10]

The discovery became a continuing issue in U.S.–Soviet arms control negotiations. The Soviet claim that the radar was intended for space tracking, which would make it acceptable, was universally dimissed by American experts, irrespective of their views of Soviet compliance with other aspects of the ABM treaty and other arms control agreements. Both the physical design of the radar and its orientation were considered clear indications that the radar was not intended for space tracking, and its northeast orientation precluded it from tracking launches from the two major Soviet launch sites at Plesetsk and Tyuratam. Disagreement centered around the Soviet motivation, with theories ranging from plans to break out of the ABM treaty and establish a nationwide defense to economic and environmental constraints.[11]

In addition to becoming a subject of frequent discussion in the arms control community and between the United States and Soviet Union, the Abalakova radar became a regular target of the KH-11. In late 1986 it was noted that the Soviets were in the process of finishing construction and that it might be in operation within a year. Analysts also noted, on the basis of the KEYHOLE photography, the similarities between the Abalakova radar and early-warning radars at Pechora in northern Russia and at Lyaki in Soviet Azerbaijan near Iran. The chief difference between these radars and the Abalakova radar was that the transmitter building at Abalakova is somewhat smaller. (Soviet radars consist of a building for transmitting signals and a building for receiving them.)[12]

Commercial satellite photography of Krasnoyarsk taken in 1987 by the French SPOT satellite gives some idea of what 1987 KH-11 imagery of the site showed. A direct overhead photo would have shown a fence around the front and sides and part of the back of the complex, a 27-story radar center to the west, and an 18-story radar center to the east. Behind the western radar center the photo would show the main buildings within the radar facility and four buildings between the eastern and western wings of the facility. Behind the main complex the photo would reveal an electrical power substation. Of course, the KH-11 images would reveal far more detail than the SPOT photos with their 33 foot resolution.[13]

The belated discovery of the radar at Abalakova, where construction may have begun in 1981, has different meanings for different observers. According to one official, "It took 18 months before the U.S. was able to detect the large-phased array radar at Pechora looking out towards the Arctic region, and a year after construction began to locate the one at Abalakova." Some have interpreted that delay as a dramatic indication of the limits of satellite photography. Others, such as *Aviation Week,* have argued that the belated discovery was the result of the decreased use of the KH-9 in the wake of the cost overruns associated with the KH-11. A third theory is that the failure was simply due to poor management— that failure to cover the Abalakova area for such a long period of time was due to a poor allocation of reconnaissance resources.*[14]

That Abalakova was only one of many Soviet arms control violations was a theme constantly sounded by many of the Reagan administration hawks like Richard Perle. A second complaint often mentioned by those on the Republican right, like former CIA analyst and Senatorial aide David Sullivan and former Staff Member of the Senate Select Committee on Intelligence Angelo Codevilla was that the Soviets had skillfully fed false data back through U.S. photo reconnaissance and signals intelligence satellites. Thus, they argued that estimates of Soviet activities and capabilities developed from the use of such resources were highly biased.

This problem was hardly unknown to the intelligence community, which referred to it as the cover, concealment, and deception problem prior to 1984 and as the denial and deception problem afterwards. To be sure that the United States was constantly aware of what the Soviet Union and other countries were doing to either deceive the United States or to simply deny it data, DCI William J. Casey established a National Intelligence Officer for Foreign Denial and Intelligence Activities.[15]

Essentially there are two parts to the problem. Deception involves actively misleading another nation's intelligence service. The United States and Britain placed great reliance on deception in protecting the secret of the location of the D-Day invasion, wrapping the truth, according to Winston Churchill's dictum, in a "bodyguard of lies." More recently, the United States fed false telemetry data to Soviet electronic intelligence satellites as they passed over the White Sands Proving Ground during a missile test.[16]

---

* In October 1987, Mikhail Gorbachev told Secretary of State Schultz that the Soviets were imposing a one-year construction moratorium on Krasnoyarsk. On October 27, 1988, the Soviet government stated that the moratorium continued to be in effect. External construction activities at Krasnoyarsk have not been noted since the moratorium began. It was reported in June 1989 that a KH-11 had photographed boxes being delivered to the site from a nearby rail station. In September 1989 the Soviets agreed to dismantle the facility.[17]

Deception of a photo reconnaissance satellite can involve putting up dummy versions of planes, missiles, or tanks to make it appear that a fixed number of targets are geographically distributed in a different manner than they really are, when some of the actual targets are actually hidden in hangars or underground. Alternatively, dummy targets can be employed to make it appear that one nation has more weapons of a given type than it really does.[18]

Denial is a somewhat simpler task. It involves preventing a satellite from photographing a particular target. That is where cover, concealment, and camouflage come in. Throwing a tarpaulin over a tank or missile during a satellite pass suffices to cover it from the camera's prying eyes. Cover does not prevent the photos from indicating that something is at a location, and it may not even prevent the determination of what type of weapons or equipment are at a location. But it certainly prevents determination of details about the objects under cover. Concealment completely hides the object, such as an airplane, from view by moving it into a hangar or some other structure. Concealment can also be achieved by making use of cloud cover and darkness to hide the target. Finally, camouflage hides the object from view by making it blend into the natural surroundings. Camouflaging a target may involve painting an object the same color as its background, use of vegetation as a screen, or a netting or covering that matches the surrounding terrain.[19]

The Soviet reputation for being skilled at denial and deception is an extension of the Russian reputation, specifically that of Grigory Potemkin. Potemkin became the lover of Russian Empress Catherine the Great and governor general of the southern provinces. In 1787 Catherine toured those provinces, which by then included Crimea, to find an apparently flourishing society. Subsequently, it has become legend that Potemkin skillfully and cynically built entire sham villages with cardboard houses and paste palaces to impress the Empress. In addition, Potemkin was said to have driven millions of unwilling slaves dressed up as farmers along with their cattle to the various places passed by Catherine, in order to create a false image of progress and prosperity. Although a biographer of Potemkin claims such charges are false and attributes them to a Saxon diplomat hostile to Potemkin, they have lived on as an example of Russian and Soviet duplicity.[20]

During the Second World War, deception played a key role in Soviet military operations. In the autumn of 1943, the Soviet army advanced along a broad front from Smolensk in the north to the Black Sea coast in the south. German intelligence correctly concluded that the Red Army's main effort would be in the Ukraine. As a result, the German forces successfully contained Soviet armies along the Dnepr River and in limited bridgeheads south of Kiev and south of Kremchung. A month of stalemate was ended as a result of a major Soviet deception operation. Through a variety of denial and deception measures—dummy radio networks, mock-up tanks, and night-movements—the Soviets were able to covertly shift an entire tank army from the Bukhrin area south of Kiev into a bridgehead across the Dnepr River to the north of Kiev. Once it arrived at its new location, the 3rd Tank Army joined the 38th Army in launching a surprise operation which drove German forces from the Kiev region.[21]

The beginning of a concerted Soviet denial and deception campaign came in 1966. In the early 1960s, according to *The Penkovskiy Papers,* the Soviet General Staff was deeply worried by U.S. reconnaissance satellites. An order was issued to improve the poor camouflage of Soviet nuclear weapons and facilities. In 1966, the Soviets began to build dummy SAM sites and submarines. As part of this deception effort, the Soviets attempted to hide their silos but not very successfully. Camouflage was brought in only as a last step. One set of satellite photos would show a silo under construction, and a month later, new photos would show only a railroad or oil storage depot.[22]

Two years later, in order to coordinate their denial and deception activities, the Soviets established a directorate of the General Staff to specifically handle denial and deception, or Maskirovka as it is known in Russian. The Chief Directorate for Strategic Deception (Glavnoye Upravlenie Strategiche Maskirovka—GUSM) was headed by the future Marshal Ogarkov, who served as Chief of the General Staff until dismissed by Mikhail Gorbachev.* Ogarkov was promoted ahead of schedule to general of the army and named first deputy chief of the general staff.[23]

According to Soviet defector and former GRU officer Vladimir Rezun, writing under the pseudonym Viktor Suvorov, the Chief Directorate is responsible for collecting and processing information about foreign satellites and determining when they pass over Soviet territory. In addition, it is in charge of camouflaging important targets, constructing decoy

---

* Being head of GUSM is certainly not a career impediment. Ogarkov's replacement as Chief of the General Staff, Marshal S.F. Akhromeyev, served as GUSM chief from 1974 to 1979. His replacement was probably Valentin Ivanovich Varenikov.

targets and implementing other measures to deny satellite coverage of Soviet targets.[24]

Today's Directorate, according to Suvorov, conducts its operations with the aid of American technology:

> ... A huge American computer, which has been installed at the Central Command Post of the Chief Directorate of Strategic Deception, maintains a constant record of all intelligence gathering satellites and orbiting space stations, and of their trajectories. Extremely precise short- and long-term forecasts are prepared of the times which the satellite will pass over various areas of the Soviet Union and over all the other territories and sea areas in which the Armed Services of the USSR are active. Each Chief Directorate unit serving with a military district, a group of armies or a fleet makes use of data provided by this same American computer to carry out similar work for its own force and area. Each army division and regiment receives constantly up-dated schedules showing the precise times at which enemy reconnaissance satellites will overfly their area, with details of the type of satellite concerned (photo-reconnaissance, signals intelligence, all-purpose etc.) and the track it will follow.[25]

The variety of techniques used by the Soviets in their denial and deception programs was spelled out in a 1979 NSC report. The report traces an evolution of techniques initiated between 1964 and 1974:

> *Disruptive painting* (1964): Used at ICBM bases and presumably designed to hide the missile complexes from air attack. Much of the paintwork was carried out between 1966 and 1968, but the paint brushes appear to have been retired.
>
> *Tonal blending* (1964): Later discarded, the technique sought to camouflage missiles by use of suitable colors.
>
> *Dummy roads and launch sites* (1966): Designed to confuse bomber pilots. Built mainly between 1966 and 1971. Dummy missile sites were built after 1968. Apparently five or ten dummy surface-to-air missile sites are still being built each year.
>
> *Satellite warning system* (1966): The Soviet counterpart of the U.S. SATRAN (Satellite Reconnaissance Advance Notice) system provides advance warning to military commanders that U.S.

reconnaissance satellites will be able to monitor their activities during certain time periods.

Missile covers (1967): Some of the covers for missiles may simply have been intended to conceal the missiles underneath. The use of covers was especially pronounced between 1972 and 1975.

Submarine tunnels (1967). The Soviets have built enough coastal tunnels to hide 20 or more nuclear-armed submarines from Western detection. According to the NSC report, in 1970 the Soviets began construction of three tunnels at naval bases "apparently for submarine berthing to provide protection against attack and to deny information on readiness status."

Submarine covers (1970): Used mainly between 1974 and 1975; the CIA was unable to determine whether canvas covers were being used for a valid environmental reason or to prevent the United States from determining whether the submarines underneath were ready for battle. According to one report the canvas covers were used to cover Delta-class submarines at Severomorsk and prevented not only an accurate count of the vessels under construction but also denied observation of the submarine-launched ballistic missiles they were being armed with.

Dummy submarines (1971): Dummies of nuclear missile submarines were observed in the period 1972–74 at bases in the Pacific and northern Soviet Union. In one case a severe storm tore through the Barents Sea. A subsequent batch of satellite photos showed one submarine bent in half, something which does not happen to real submarines. None has been seen since 1974. The CIA believed that neither the dummy submarines nor the canvas covers seriously hampered U.S. ability to maintain an accurate count of Soviet submarines, although it did make it tricky to know how many submarines were in port at a given moment.

Night tests (1973): A mobile missile, the SS-16 was tested at night in 1973 to reduce the information obtained by the United States.

Covered rail sidings (1973): used to conceal missiles and other equipment as they enter and leave production plants, to hinder U.S. estimates of the number of missiles being produced.

Covered submarine hulls (1974): Hulls of submarines were partly covered at the shipyard at Severodvinsk in 1974 and completely covered after 1974.[26]

According to the NSC report, the Soviet effort to hide details of their strategic programs "increased substantially" in 1974. During that year, the report stated that the United States detected "broad efforts" by Moscow to conceal its mobile missile program, the construction of strategic submarines, and the production of land-based rockets. After the United States raised the issue in a 1975 Standing Consultative Commission meeting, the Soviets apparently cut back on concealment activities.[27]

Since the report was written, additional examples of Soviet deception and denial have been reported. In one instance in 1980, a KH-11 may have spotted a Soviet exercise to determine the ability to make the SS-16 ICBM and SS-20 IRBM indistinguishable. The KH-11 spotted an SS-20 in its canister alongside an encapsulated SS-16 during 1980 war games. Officials speculated that the purpose was to allow Soviet spacecraft to compare the two so that steps could be taken to improve the similarity.[28]

In 1983 it was reported that the Soviets were using sliding roofs and tarpaulins to cover work on an upgraded enhanced range version of the SA-12 air defense missile, the SA-12B-Giant, on Kamchatka Peninsula and Sakhalin Island. The antennas and service modules were among the system components concealed from satellites.[29]

That same year *Aviation Week and Space Technology* reported that the Soviet Union was approaching its first launch attempt with a large Saturn 5–class vehicle capable of placing about 300,000 pounds in orbit. The first part of the 300-foot-tall Soviet booster was on a launch pad at Tyuratam shrouded in camouflage nets to prevent photography by the KEYHOLE satellites.[30]

In 1984 it became known that the Soviet Union was completing the construction of four large tunnels capable of hiding even the largest Soviet submarines. The tunnels, which are hollowed out of a coastline but have sea-level entrances, can accommodate even the huge Typhoon-class submarines, which are 557 feet long.[31]

The GUSM has also tried to deceive U.S. intelligence about the accuracy of its SS-19 ICBM by filling in impact craters from test warheads and digging false craters that would be photographed by KEYHOLE satellites. Photographs taken by a KH-11 during a morning pass showed Soviet troops concealing test craters from incoming SS-19 warheads launched the previous night from Tyuratam to the impact area on Kamchatka. Photographs also showed military personnel digging false craters that would indicate a less accurate warhead.[32]

The two most recent Soviet missiles have been the subject of denial practices. The SS-24, a 10-warhead missile that can be deployed either mobilely or in silos, is one. As of 1985, the United States did not have precise information on the nature of its rail-car launcher and whether it was distinguishable from other rail-cars. The missile was test-fired at night and efforts were made to conceal the launcher. One official told the *New York Times* that "We have never had a good look at it so that we can say with any degree of confidence what exactly the launcher looks like."[33] The SS-25 single-warhead missile has also been a problem. "They do not make the activities observable at a time when national technical means are capable of monitoring them," according to one official.[34]

Although the Soviet Union has the most resources to practice denial and deception, it is by no means the only country that tries to prevent the United States from obtaining satellite photos of military facilities. Just as the KEYHOLE satellites seek to pry into the activities of allies and adversaries alike, the targeted nations seek to protect their secrets from the satellites' high-powered optical systems. China has engaged in selective concealment of its space and missile capabilities in order to produce ambiguity in the perception of the outside world. According to Chong-Pin Lin, an expert on the Chinese space and nuclear programs, "Facilities, launch pads and control centers have been constructed in mountains with the cover of heavy vegetation or caves, protected from facile detection by foreign satellite reconnaissance."[35]

Israel seeks to hide its military forces from observation by the Arab countries, the Soviet Union, and the United States. In the weeks before the Israeli raid on the Iraqi Osirak nuclear reactor, Israel wanted to give no indication of the upcoming action. In a December 1983 New York talk to Jewish stockbrokers, Israeli Air Force Colonel Aviam Sella, who was subsequently involved in the Pollard spy case, revealed that Israeli fighter aircraft took off from a desert air base just outside of Eilat to conduct the mission. In the weeks leading up to the raid, the Air Force had made certain that the same number of planes in the same exact formation had regularly taken off from the base on what were meant to appear to be routine training exercises. A KH-11 monitoring the Israeli air base on the day of the raid would send back imagery no different from that on any other day.[36]

During the 1982 Falklands war, Argentine forces built mock shell craters on the airstrip at Port Stanley so that any KEYHOLE imagery would provide a misleading picture of the runway's status. The object was not to fool the United States so much as to mislead Britain, who the Argentines presumed were receiving KEYHOLE photographs from their American allies. They need not have bothered, because the heavy cloud cover that blanketed the Falklands neutralized the KH-9 and KH-11s in orbit at the time.[37]

The extent to which the value of KEYHOLE photography is limited by denial and deception, particularly by Soviet denial and deception, is a subject of dispute. Some Western observers believe that what the United States sees from KEYHOLE satellites is highly biased in a direction determined by the Soviets, and the Soviets apparently believe that their denial and deception activities are quite effective. According to a secret CIA study on the KGB and GRU,

> Every effort is made to construct decoys sufficiently convincing to pass inspection by Western air and space collectors. The KGB considers Soviet accomplishments in strategic deception so successful that decoy installations could become targets of nuclear missiles. Therefore, decoy installations are constructed only far from populated areas and other important installations, whether existing or planned.[38]

But many others believe that there is a significant limit to what the Soviets or any other country can do to hide from America's secret eyes. As CIA official Richard J. Kerr has noted:

> it is impossible for any country [to] stop all of its activities just because there is a satellite overhead. It cannot stop military training. It cannot stop the building of objects. It cannot stop movement of things. There are certain things that it can do. Those that are terribly sensitive, those that can be controlled precisely. Those things where there is a direct relationship between the knowledge that might be obtained and the possibility that a satellite was over. Those things you can do, but running a vast system is very difficult . . . It requires extraordinary communications. It requires extraordinary discipline. It requires agreement on objectives. A field commander in the field who wants to exercise his troops is not going to pay a lot of attention to an intelligence officer saying the Americans are looking at you.[39]

When deception is attempted photo-interpreters have ways to unmask the deception, according to Dino Brugioni. "If the Soviets put up dummy aircraft, you never see them being serviced. If they put up rubber dummies and decoys you see them smashed as the weather and seasons change." If patterns are painted on runways or other strategic locations to simulate streets, residential homes, or other non-threatening structures, a photo-interpreter would be clued in to the attempted deception by the absence of shadows or never-changing shadows. In addition, stereo interpretation of the photographs would reveal that the painted objects had no height.[40]

To former CIA official Admiral Daniel J. Murphy, the problem is grossly overstated: "It's a bunch of hogwash."[41] And the same CIA study that reported the KGB's optimistic view of its deception activities noted that "despite apparent Soviet pride in this program, little evidence of it is apparent in the West. Soviet decoy installations that have gone undetected are believed to be rare."[42]

Where the Soviets have done better, according to former NSA director Bobby Inman, is in taking advantage of the absence of satellites at given times, in the presence of cloud cover and night time, and in putting new activity outside normal tracks.[43]

The Soviet Union and Eastern Europe are under cloud cover for about 70 percent of the year. Thus, at the height of the 1980–1981 Polish crisis, when the United States feared a possible Soviet invasion, satellite photography was impeded by the heavy cloud cover that blanketed Eastern Europe at times. Analysts could not confirm reports in early January 1981 that Soviet landing craft were being loaded with troops at Riga, Latvia, and other ports on the Baltic Sea in preparation for a landing on Poland's northern coast.[44]

Cloud cover—along with Krasnoyarsk being outside of the normal area of KEYHOLE coverage——may have contributed to the delay in detecting the construction of the Abalakova radar. Detection of the radar was one of the last accomplishments of the KH-8/GAMBIT program. In 1984, 18 years after it began, the program concluded with the deorbiting of 1751 on August 13. The final KH-8 had been launched on April 17 and had operated for 118 days in its 79 by 193 mile orbit. The statistical summary of the program shows 50 successful launches in 51 attempts, a mean lifetime of 30.4 days, a mean inclination of 104.2 degrees, and mean perigees and apogees of 84 and 254 miles. The mean lifetime statistic hid the dramatic increase in the lifetime of the spacecraft. The launch of April 27, 1971 resulted in the first KH-8 with a lifetime of over 20 days. Lifetimes moved into the 30-day period in 1973, into the high 40s in 1974 and into the 70s in 1977. The final four KH-8 missions had lifetimes of 90, 112, 128 and 118 days. The ability of the KH-8 to return detailed photography was unsurpassed. By the later stages of the program, if not before, the two film capsules returned by each KH-8 contained imagery with six-inch resolution.[45]

While the final KH-8 mission was in progress, what would turn out to be the final KH-9 mission began. On June 25 a Titan 3D placed 1919 into a 105 by 163 mile orbit. The KH-9 joined 1750 and the seemingly everpresent KH-11 duo, 5504 and 5505. Until August 13, when the KH-8 deorbited, NRO again had four sets of space eyes in operation. From August 13 to October 17, when the unusually short-lived KH-9 was deorbited after only 115 days in orbit, NRO had three sets.

With the two KH-11s, the KH-8 and KH-9 launched in 1984, and the KH-9 launched on June 20, 1983 whose mission didn't conclude until March 21, 1984 after an astounding 261 days in orbit, there were substantial resources available covering a wide variety of targets.

Among the targets of the KEYHOLE contingent in 1984 was the Soviet space shuttle, more than four years before its first flight. The shuttle that was launched in November 1988 was not an exact duplicate of the U.S. system: its wing angle and nose were sharper, perhaps giving it greater maneuverability. Additionally, the U.S. orbiter is lifted by two recoverable solid rocket boosters and its own three main engines, a system designed exclusively for shuttle missions. The Soviet shuttle, named Buran (Snowbird in English), piggybacks on the 197 foot high Energia, the world's largest operational booster rocket, a multipurpose powerhouse designed to lift shuttles or unmanned spacecraft weighing up to 100 tons.[46]

Despite the differences in the two shuttles, there is a strong physical resemblance. They have been described by Nicholas Johnson, an expert on the Soviet space program at Teledyne-Brown Engineering, as "functional twins despite some technical differences." A comparison of basic orbiter dimensions indicates the degree of physical similarity: the length of the Buran is 36.4 meters while that of the U.S. orbiter is 37.2 meters; the Buran's height is 16.5 meters, the U.S. orbiter is 17.3 meters; the body width of the Buran is 5.6 meters while it is 5.5 meters for the shuttle; and the wing span for the Soviet shuttle is 24 meters while it is 23.8 meters for the U.S. shuttle.[47]

Indeed, it has been believed for years by some that much of the technology employed by the Soviets was acquired illicitly from the United States. In June 1984, an official of the DIA charged that the USSR had obtained a copy of the "blueprints" for the American orbiter. Whatever the actual validity of that view, it was enough for the Top Secret *National Intelligence Daily* to run a KH-11 photo of the Buran, next to a picture of the U.S. shuttle to highlight the similarities.[48]

The photography or the information derived from the KH-11 photos of the Soviet shuttle apparently found its way to *Aviation Week and Space Technology*. Following the public debut of the Buran, a reader, Edward L. Keith, wrote to the magazine and noted that "More than four years ago, *Aviation Week* published an article with a drawing of the Soviet shuttle. A close look shows some lettering that made no sense in 1984. The five Cyrillic letters BYPAH (pronounced "Buran") appear on the side, behind the mid-deck, in the drawing. Eventually the first Soviet shuttle to orbit the Earth was named the Buran. Maybe someday you will tell us readers how you knew the given name of the Soviet shuttle more than four years before the Soviets lifted the veil of secrecy from their shuttle program."[49]

The *Aviation Week* article where the drawing of the shuttle appeared noted that "U.S. Air Force imaging reconnaissance satellites are providing new data on the size and design of the Soviet Union's two large space boosters now undergoing advanced checkout on their launch pads at the Tyuratam launch site" but that the Soviet shuttle had not yet been mated on the launch pad. But while the Soviets had not put the shuttle on the launch pad for viewing by NRO's space eyes an accident did give the United States an opportunity to get a detailed look at the Buran. Sometime in 1983, coming back from a piggyback flight test the Bison bomber carrying the shuttle skidded off the runway at Ramenskoye. Before the Soviets could conceal the Soviet shuttle, a KEYHOLE had passed overhead and sent back one of its frequent pictures of the airfield.[50]

Another target of the KH-11 satellites in 1984 was the Nikolayev 444 shipyard in the Black Sea. In July of that year a KH-11 sent back imagery of the Soviets' first full-sized nuclear-powered aircraft carrier under construction. (July is among the best times of the year to obtain photography of Nikolayev, with the area being under cloud cover less than 35 percent of the time.) The carrier was designated Black Com II (for Black Sea Combatant II) by the United States. In one case the slant angle was such that the photo was taken from 504 miles away.[51] One computer-enhanced photograph taken at an oblique angle, but not as oblique an angle as the Blackjack photos, shows the general layout of the Nikolayev shipyard in the Black Sea, with what would appear to be a foundry in the foreground and assembly shops behind. Buildings housing the technical staff are shown to lie alongside the dry dock where the 75,000-ton nuclear carrier was shown under construction. The photograph also shows the stern section of the carrier Kharkov, the fourth of the Kiev class, in the process of being fitted out. Nearby, an amphibious landing ship, apparently of the 13,000-ton Ivan Rogov Class, is shown under construction. The photography is distinct enough to identify objects such as ladders and windows. A second photo gives a more detailed view of the carrier dry-dock and indicates the position of vertical silo-launched SAMs forward of the superstructure.[52]

The photos, to the great consternation of government officials, showed up in a trade magazine, *Jane's Defence Weekly*. The initial reaction of some individuals with an intelligence background was that the "leak" was authorized. Retired COMIREX chairman Roland Inlow subsequently testified "that my reaction was that somebody had decided to release these photographs . . . my reaction was much more ho hum than it was oh my God. I did not think of that as being unusual except that it was first." In fact, the pictures were provided to *Jane's* by Samuel Loring Morison, an analyst for the Naval Intelligence Support Center (NISC), which studies the technical capabilities of Soviet and other nations' naval forces. Morison, who also served as the Washington representative of *Jane's Fighting Ships*, cut off the NISC logo, the mid-July date and the notations Nikolayev Shipyard, SECRET/WNINTEL [Warning Notice Sensitive Intelligence Sources and Methods Involved], REL UK AND CAN [Releasable to the United Kingdom and Canada]. He then shipped the photos off to *Jane's* with an eye to enhance his future chances of a full-time position and, he claimed, to alert the public to the threat from Soviet naval developments.[53] The photos were reacquired from *Jane's* with the intercession of the director-general of British Defence Intelligence Staff. Fingerprints on the photographs were determined to be Morison's.

Caspar Weinberger was particularly annoyed. The photos showed that construction of the Soviet carrier was not as far along as had been claimed in the most recent issue of *Soviet Military Power*. Furthermore, Weinberger would have preferred to include some satellite photos rather than the sketches that were derided as "cartoons" by some who wanted more convincing proof of the claims in SMP. But as usual the intelligence community pressure not to publish was too great a stumbling block.[54]

When Weinberger saw *Jane's* publishing the satellite photos he could not, he ordered that Morison be prosecuted for espionage. The government's controversial thesis, one hotly disputed by the press as well as many liberal and conservative observers, that Morison could be tried for espionage without having passed information to any foreign government was upheld by the trial judge and appeals court. Morison was convicted and sentenced to two years in prison.

Morison had also provided *Jane's* with some issues of NISC's *Weekly Wire*. The copies received by the trade magazine concerned a massive 1984 explosion at Severomorsk, the aftermath of which was photographed by a KEYHOLE spacecraft.

Severomorsk, located on the icy Barents Sea about 900 miles north of Moscow on the eastern shore of a long bay leading to Murmansk, was the principal ammunition depot and home port for the Soviet Union's Northern Fleet. The installations at Severomorsk were said to be extensive, with warehouses for storage and plants for disassembling and maintaining ammunition. In mid-May the depot was rocked by a massive explosion, of such force that it was initially suspected of being a nuclear explosion.[55]

The death toll was estimated at between 200 and 300. Many of the dead were technicians sent into the fire caused by the explosion in a desperate but unsuccessful effort to defuse or disassemble the munitions before they exploded in a chain reaction over several hours.[56]

Initial intelligence concerning the explosion probably came from communications intelligence and seismic detection stations. Travelers and other human sources also contributed to the initial data set. Informed sources said survivors of the disaster evidently talked about it later and when their accounts reached Western Europe, CIA agents picked up the news. Norway and Sweden were said to have monitored the explosions with their own technical devices.[57]

To help assess the damage, a KH-11 was commanded to send back imagery on its next pass in the vicinity of the region. Ultimately, it was estimated that the blast destroyed:

- 580 of 900 SA-N-1 and SA-N-3 surface-to-air missiles
- About 320 of 400 SS-N-3 and SS-N-12 long-range ship-to-ship weapons
- All or nearly all of about 80 SS-N-22 missiles that are capable of carrying nuclear warheads
- An undetermined number of SS-N-19 antiship missiles
- Some SA-N-6 and SA-N-7 surface-to-air missiles[58]

KH-11 photos taken after the explosion as well as ones taken years earlier were crucial in producing the above estimates. Yet the estimates represent not only an example of the power of the KEYHOLE satellites, but also their limitations. KEYHOLE photography of the Severomorsk Weapons Storage Facility obtained on July 8, 1979 when KH-8, KH-9 and KH-11 satellites were in orbit, allowed interpreters to identify a new arched roof bunker, a weapons-handling and maintenance facility, storage and support facilities, and a weapons assembly facility. But determining exactly what was inside the facility was more difficult. A Defense Intelligence Agency document notes that

> this bunker, now almost completed, measures approximately 45 meters × 12 meters and will *probably* be used to store SS-N-14 missiles. A second, similar bunker is in the early stages of construction. . . .
>
> . . . Taking into account the measurements of the almost completed bunker and *assuming* only one-level stacking with a central aisle for maneuvering will be utilized, about 60 SS-N-14 missiles may be stored. Further, *assuming* the second bunker to be constructed will be the same size, a similar number of stored missiles would bring the total capacity to 120.[59][emphasis added]

CHAPTER 9

Disaster and Recovery

For such a complex task as deploying a reconnaissance satellite, the United States did extraordinarily well in the 25 years between the launch of the first camera-carrying Discoverer and August 1985. In particular, the three most recent programs, as of 1985, the KH-8(GAMBIT), KH-9(HEXAGON), and KH-11(KENNAN), had exceptional records. Seventy-four of 75 launches had been successful. By 1985, it appeared that if a mission were to fail it would not be the result of a launch failure, but the next two years were to bring an astonishing series of failures that crippled the entire U.S. space program.

The first eight months of 1985 involved only one launch of an intelligence satellite. In January the space shuttle DISCOVERY deployed a signals intelligence satellite, originally codenamed MAGNUM, into geosynchronous orbit. No new KEYHOLE satellites were deployed because two KH-11s were still operating, sending back their electro-optical images on a regular basis. The KH-11 launched on November 17, 1982, 5505, reached its 776th day in orbit on January 1. The more recently launched 5506, placed into orbit on December 4, 1984, provided late-morning coverage while 5505 provided early afternoon coverage.[1]

On August 13, after 987 days in orbit, 5505 was deorbited, which meant that a new KH-11 would be sent up in short order. Since August 23, 1981, when 5502 was deorbited, a pattern had been established in which two KH-11s were maintained in orbit, and when one was deorbited it was replaced shortly thereafter by a new KH-11.* The KH-11 deorbited on August 23, 1981 was replaced by the one put into orbit on September 3, 1981. When 5503, launched on February 7, 1980, was deorbited on October 30, 1982, it was replaced by the KH-11 launched on November 17, 1982. By the time one satellite was ready to be deorbited, both were circling the earth every ninety minutes—a faster speed than usual. After one satellite was removed from orbit, the other made a major maneuver several days later, which increased its period to 92.5 minutes. The satellite replacing the deorbited satellite was then launched and made a few maneuvers to get the same period and desired plane spacing with the other. After this, joint operations began, with one satellite providing late-morning coverage and the other giving afternoon coverage. The combination of viewing angles produced was a substantial aid in interpreting the images produced.[2]

On August 28, the replacement for 5505 was ready to go. A Titan 34D, carrying 5505's successor, lifted off from Vandenberg AFB. The Titan 34D is a massive structure consisting of a central core and two 90-foot long solid rocket motors, each of which is 10 feet in diameter and weighs more than 540,000 pounds. For polar or high-inclination orbits the boosters used on-board pre-programmed guidance and a ground-based radio guidance system.[3]

It was a particularly good day for a launch, with a clear sky, unlimited visibility, and negligible winds. Lift-off began normally, although the solid rocket boosters ignited a large brush fire in the Vandenberg pad area. According to standard procedure, only the solid rocket motors were ignited for lift-off and climbout. About two minutes into the flight, the Titan core vehicle's two liquid propellant engines were ignited, providing an additional one million pounds of thrust. The solid rocket motors burned for another several seconds after ignition of the core vehicle's two engines and then separated while the core vehicle engines continued to fire. The solids fell into the Pacific Ocean south of Vandenberg as the two-stage Titan continued to ascend and accelerate.[4]

---

* While the CIA and NRO desired to operate a two-satellite constellation, this was possible for only a short period during the first three years of the KH-11 operations, from June 14, 1978 to January 28, 1979. Between December 19, 1976 and June 13, 1978 and between January 28, 1979 and February 7, 1980 the United States operated only a single KH-11 because of limitations on funding imposed by the House Appropriations Committee.

But as the Titan continued to head towards outer space, telemetry showed what appeared to be problems developing in the No. 1 engine. Less than two minutes after the ignition of the core vehicle's liquid engines and about 45 seconds before their planned cutoff, the No. 1 engine shut down. Propulsion from the remaining core vehicle engine was not sufficient to maintain a proper altitude and velocity to achieve orbit. With no hope of rectifying the situation, the $500 million spacecraft was destroyed by triggering its self-destruct mechanism, sending its remains into the Pacific Ocean, 300 miles off the coast of Baja California.[5]

The disaster had immediate and far-reaching effects on the military space program. A number of major programs had relied on the Titan 34D launch vehicle, including the Defense Support Program early-warning satellites that could detect the infrared plumes of rocket and missile launches, the VORTEX signals intelligence satellite, and the Defense Satellite Communications System satellites. Although it was expected that all of these programs would eventually rely solely on the shuttle, that stage had not yet been reached. Naturally, the KEYHOLE program was affected. The Titan 34D was the launcher not only for the KH-11, but also for the KH-9. Until an investigation determined the cause of the failure and any necessary corrections were made, no further launches could be attempted. That investigative report was submitted on October 25, 1985, but the analysis, findings, recommendations, and related documents remain classified.[6] But whatever the findings, NRO and other relevant authorities were able to plan another launch for April of 1986.

In the interim another space program disaster had shaken not only the highly-classified reconnaissance program but the entire country, for this time the loss involved not only hardware, but seven astronauts. On January 28, 1986, the crew of the space shuttle Challenger met their tragic end minutes into the flight and just nine miles over the Atlantic. An explosion in one of the shuttle's solid rocket boosters, caused by deficient O-rings, separated the crew cabin from the orbiter and sent the astronauts to their deaths. In two minutes and forty-five seconds the Challenger's cabin and crew dropped 65,000 feet. The cabin slammed into 80 feet of water at 207 miles an hour, killing all seven astronauts inside.[7]

Although it may not have been apparent to most Americans, when the Challenger exploded so did NRO's plans for employing the shuttle in

1986 and 1987 to deliver a new generation of reconnaissance satellites into orbit. The very thing that so many had feared was to come true: giving humans a space role in space reconnaissance (as opposed to a ground role) would, in the event of a disaster, result in the prolonged grounding of the manned launch vehicles and consequently of their KEYHOLE passengers.

That the Challenger explosion had any impact on the military space program and its KEYHOLE component was the result of political decisions made many years earlier. Serious thinking about a space shuttle had begun in the late 1960s during the Johnson administration. Johnson's NASA administrator, Thomas Paine, remained in the job until 1971. Paine realized that the economics of any space shuttle would require that the Department of Defense and its large number of payloads use the shuttle rather than the expendable launch vehicles it had been using. And before he left he was able to get White House concurrence with his concept of the shuttle as a "national vehicle." In order to get such agreement he agreed to several Air Force demands, which Paine believes were made as deal-breakers.[8]

The Air Force requested a crossrange landing capability of 3000 miles so that a shuttle carrying one of the NRO fleet or other military satellites could not be forced to land in Communist territory. In addition, NASA was to ensure that there would be secure communications as well as avionics hardened against the effects of a nuclear explosion.[9]

The Air Force also developed plans to use the shuttle directly on reconnaissance missions. Two camera systems, codenamed DAMIEN and ZEUS, were to be employed on 20-day operational missions each year. The plan was ultimately eliminated when it was cut from the fiscal 1980 budget.[10]

Asked by OMB to produce a cost-benefit analysis of the proposed system, NASA produced a study indicating that the shuttle would be the most economical means of placing spacecraft in orbit if flights occurred frequently enough. Some took this to mean that if all military spacecraft flew the shuttle it would be cost-efficient. This view was questioned by the RAND Corporation in a 1970 study, which suggested that the shuttle would only produce savings if the nation had a more active space program.[11]

But the blue-suited Air Force was fighting a losing battle. Particularly devastating was the appointment of Hans Mark as Under Secretary of the Air Force and Director of NRO at the beginning of the Carter administration. Robert Seamans, a former high NASA official and Secretary of the Air Force, remembers Mark as being more committed to the shuttle than any other Pentagon official. "Hans Mark really bought the thing, absolutely hook, line, and sinker. And he felt very, very strongly that there should be a national commitment. That DOD and NASA should go right down the line. And this should be the launch vehicle of all time," Seamans recalls.[12]

Mark was instrumental in defeating attempts by the uniformed Air Force to remove funding for a shuttle launch pad at Vandenberg AFB.[13] Without such a facility, neither the KEYHOLE nor DMSP weather satellites could have been placed in their polar orbits. The ability to place other satellites like the JUMPSEAT electronic intelligence and SDS relay satellites in their proper orbit would have also been compromised. And without those payloads the economic rationale for the shuttle would have collapsed.

With Mark's enthusiastic support it was no surprise that the first of Jimmy Carter's presidential directives on National Space Policy, PD/NSC-37 of May 11, 1978, contained a strong endorsement for the shuttle program.[14] From that point, although there were to be problems and frustrations, it was really only a matter of time before the shuttle flew and only a matter of time before one exploded.

By 1980 Mark was stressing the necessity of having a backup to the shuttle. In a speech to the Electronic and Aerospace Systems Conference he continued to defend the space shuttle but also noted that "it is not clear whether a complete replacement of expendables by shuttle is such a good idea. Back in 1973 and 1974 we had to make those arguments or we would never have had a space shuttle." Mark went on to say that it was time to take a more realistic approach involving the development of new expendable launches while bringing the space shuttle on line. But by 1980 the die had been cast.[15]

Although the Challenger explosion meant certain postponement of the launch of the KH-11's successor and possibly of a new radar imaging satellite, the situation was not desperate. There was still one KH-9 remaining in the inventory. In addition, a decision was made almost immediately after the August 1985 loss of the Titan 34D to refurbish a KH-11 floor model, intended for use only in ground tests and demonstrations, for launch around May 1986.[16] As long as the upcoming Titan 34D launches proved successful, the temporary loss of the shuttle would not be a major problem.

But the worst happened. On April 18, a Titan 34D carrying a KH-9 sat on its launch pad at Vandenberg. The conditions for a launch were favorable, with scattered clouds at 25,000 feet, visibility of 20 miles and a temperature of 58 degrees. At 10:45 A.M. Pacific Standard Time, the launch vehicle lifted off the pad from Space Launch Complex 4. Within eight seconds, with the launch vehicle at 800 feet and just after the programmed roll maneuver to place itself in the proper flight azimuth, there were signs of trouble. Telemetry indicated anomalous launch conditions. One of the solid rocket boosters ruptured, sending out a 12-foot ball of fire and then totally exploding. Within a fraction of a second, the other solid rocket booster and the Titan core exploded because the initial explosion triggered the Titan's self-destruct system.[17]

Everything, including the KH-9 in the upper stage, was turned into an orange and black fireball that was described as "spectacular" by someone who saw it from 11 miles away. It would take seven months to repair the launch pad, but the result could have been much, much worse if the explosion had occurred closer to the ground. If the explosion had occurred before the Titan had cleared the launch pad the result would have been catastrophic. Most of the damage resulted from shrapnel impacts throughout the launch site facilities and an air-conditioning building next to the launch pad was destroyed when segments of the solid rocket motor crashed into it. But left relatively undamaged were the Critical Mobile Service Structure, its umbilical tower, and the launch services building.[18]

In addition to the damage caused to the pad area, the explosion sent 58 people to the hospital to be treated for skin and eye irritation. It also disrupted the activities of a variety of civilians. Schoolchildren were told to stay inside, and about 120 oil workers were evacuated from two drilling platforms in the Santa Barbara Channel. The National Park Service ordered the evacuation of rangers, campers, and tourists from the offshore Channel Islands National Park.[19]

Perry Solmonson, a freight pilot, was flying a small plane about 20 miles away when he spotted a column of smoke rising to 1000 feet. "We were just coming out of a turn when I saw this white plume of smoke with orange . . . at the top. . . . The size of the smoke made me realize it was awfully powerful. I wouldn't have wanted to be on the ground. It was awesome," said the 25-year-old pilot.[20]

To investigate the accident, a "mishap board" was formed with Colonel Nathan J. Lindsay, commander of the Eastern Space and Missile Center as Chairman and Aubrey McAlpine of the Office of Space Systems (i.e., NRO) as Vice Chairman. Ultimately, failure was attributed to the separation of thermal insulation in the solid rocket motor that ruptured, but sufficient evidence could not be extracted from the ruins to determine why the insulation separated from the metal casing.[21]

The explosion on April 18 was a sad end to a program that had represented a tremendous leap forward in U.S. photo-reconnaissance capabilities. According to one knowledgeable source, "a lot of people cried when the last one blew up."[22]

The statistical summary of the KH-9/HEXAGON program showed 19 successful launches in 20 attempts. For those 19 successful launches, the KH-9 spacecraft operated with a mean inclination of 96.4 degrees, mean perigee of 101 miles, and a mean apogee of 159 miles. The mean lifetime of the 19 missions was 138 days. The actual lifetime grew significantly within the first three years of the program, going from lifetimes between 50 and 100 days to well over 100 days. Over the next ten years, the lifetimes of the KH-9 missions grew even further, moving from 158 up to 275 before the short 115-day final mission of 1984.

The disaster at Vandenberg was followed within days by a disaster of far greater proportions in the Soviet Union, a disaster that the remaining KH-11 would be active in monitoring. On Saturday, April 26, a nuclear accident took place 80 miles from Kiev at the Chernobyl nuclear power plant, specifically in its reactor No. 4. As the result of a series of safety violations such as running the reactor without the emergency cooling system and removing too many control rods, a small part of the reactor went "prompt critical." The effect was the equivalent of half a ton of TNT exploding on the core. Four seconds later a second explosion blasted the 1000-ton lid off the reactor, destroyed part of the building, and brought the 200-ton refueling crane crashing down on the core. This was followed by a "fireworks" display of glowing particles and fragments escaping from the units, setting off 30 fires in the building. The huge blocks of graphite in the reactor core also caught fire, spewing out plumes of highly radioactive fission products.[23]

The first solid information that the United States received about the accident was from an official Soviet statement on Monday, April 28th. There had been indications of unusual activity around Kiev on Sunday, probably from communications intelligence, but it was not clear what was happening. Only on the following day was the situation clarified.[24] Once alerted to the disaster, the intelligence community responded by turning its full set of resources on the Kiev area. A VORTEX signals intelligence satellite sucked up all military and relevant civilian communications within several hundred miles of Chernobyl, and the lone KH-11 was programmed to obtain photography as soon as possible. Its last visit had been almost two weeks before. The first opportunity for 5506 to provide imagery came on Monday afternoon. However, given its orbital path, the photo had to be taken from a considerable distance and even with computer enhancement didn't show much. Even had 5506 been closer to its target, the smoke hovering over the reactor area would probably have obscured the site. The following morning the distance was still too great to produce a good photo, but by evening the KH-11 had approached close enough to return the first good imagery of the accident site. The picture was reported to be "good and overhead."[25]

With the photos in hand, analysts at NPIC began assessing the situation. The photos revealed that the roof of the reactor had been blown off and the walls were pushed out, "like a barn collapsing in a high wind," said one source. Inside what was left of the building was an incandescent mass of graphite. Some tendrils of smoke and the blackened roof of the adjoining building indicated that at some point the fire had been more active. The graphite settled down into a glowing mass while radioactive material from a pile that had contained 100 tons of uranium was still being vented through the open roof and into the atmosphere.[26]

The photos revealed activity in the surrounding areas, activity that was quite remarkable given the perilous situation at Chernobyl. The KH-11 photos showed a barge peacefully sailing down the Pripyat River and men playing soccer inside the plant fence less than a mile from the burned-out reactor. The photos of the town of Pripyat showed that there had been no evacuation.[27]

To supplement such data collected after the event was announced to the world, the intelligence community looked back through information collected by other forms of surveillance. Tapes from the Defense Support Program satellite responsible for monitoring the Soviet Union that were recorded the day of the explosions showed infrared images of a sudden flash in the vicinity of Kiev, apparently the explosion that shattered the reactor. Communications traffic from the day before the explosion indicated that emergency measures were being undertaken even then and that reports of those measures were being sent to Moscow.[28]

Among those briefed with the satellite photos was the House Permanent Select Committee on Intelligence. "We were shown satellite pictures of the reactor building from before and immediately after the explosion," said committee member George D. Brown, Jr. (D-California) after a closed door hearing on Thursday, May 1. "They were dramatic, with the roof beams collapsed and debris scattered around the plant. No bodies were visible," he added.[29]

KH-11 photos taken on Thursday morning May 2 showed no smoke at all, leading an interagency panel to believe that it was possible that the fire had been put out as the Soviets contended, although only the day before, the panel had predicted that the fire might burn for weeks. Some analysts said they detected shimmering over the reactor, suggesting that the graphite was still burning. It appeared that the Soviets were dumping dirt or sand on the fire from helicopters, and one KH-11 photo showed a helicopter hovering directly in the plume of radioactivity.[30]

The satellite photos provided data that allowed analysts to determine the fallaciousness of reports that Unit 3, adjacent to the damaged Unit 4, was affected by a meltdown or fire. Additional satellite photos allowed a federal task force to come to the conclusion on May 3 that there was no problem with the other Chernobyl reactors. Lee M. Thomas, head of the task force, announced on the basis of the photos that "we see no problems with the other units."[31]

The KH-11 undoubtedly continued to monitor the cleanup, sending back pictures of the Mi-8 helicopters (with lead shields on their floors). The helicopters flew hundreds of missions, day after day, dropping sacks into the broken roof of the reactor from heights of more than 650 feet. They then sealed the roof shut with tons of lead pellets, which rolled into whatever cracks remained between the sand bags.[32]

Although the immediate crisis was over before the end of the year, the KH-11 mission to monitor the impact of the disaster will continue for years to come. Not only will the final KH-11s and their successors monitor Chernobyl and other nuclear facilities but they will watch the crops in the surrounding areas. It was feared the widespread dusting of farmlands with fallout could prove catastrophic, because vast quantities of radioactive top soil would have to be scraped up and trucked away for safe disposal. In addition, spring rains could carry contaminated soil into local streams and even into underground water tables where it would be virtually impossible to eliminate. The damage would not be evident for months.[33]

In addition to monitoring the Chernobyl situation and its aftermath, the lone KH-11 was also busy with its traditional mission of monitoring Soviet strategic developments. Both offensive and defensive developments caught the attention of the KH-11's optical system in 1986 and 1987.

New SAM deployments in the western Soviet Union were spotted under heavy concealment near three large tracking radars located along the Baltic coast. It was also reported that several hundred SA-12B "Giant" surface-to-air missiles were spotted in the western Soviet Union, spread out near Muckachevo, Baranovichi, and Skrunda, the sites of three large phased-array radars.[34]

Also "dozens" of mobile Flat Twin and Pawn Shop anti-missile radars were spotted throughout the Soviet Union. The Flat Twin is a tracking radar consisting of two large flat rectangular phased-array antennas and the Pawn Shop radars are missile guidance radars with three antennas covered by spherical domes. Such antennas would be moved by vans and would take about two months to deploy.[35]

5506 was also busy monitoring the deployment of the new SS-25. By early 1987, KH-11 photography and other intelligence sources indicated that about 100 of the new 10-warhead ICBMs had been deployed in remote areas of the Soviet Union. At the same time, 5506 was watching for the initial deployments of the rail-mobile SS-24 missile.[36]

The NRO entered October 1987 with only one set of eyes in space, a KH-11 that was almost three years old. October was also the conclusion of an 18-month program that ultimately established new standards and procedures for solid rocket motor construction, inspection, repair, and in-flight monitoring. The program included an exhaustive study of possible failure modes, processing standards, and inspection and repair techniques. Earlier in 1987, test firings of the two 5.5 segment solid rocket motors validated new booster inspection methods and cleared the way for the resumption of Titan 34D launches.[37]

On October 25, a launch attempt ended in a scrubbed countdown when Air Force meteorologists determined that the winds at Vandenberg were too strong to permit a launch. Four days earlier, two valves in the thrust vector control system of the vehicle's solid rocket boosters were identified as possibly defective. Finally, on October 26 at 1:32 P.M. Pacific Standard Time, the last KH-11, the floor model never intended to be placed into orbit, lifted off into a 97.8 degree inclination orbit with initial perigee and apogee of 95 and 653 miles.[38]

The Titan was packed with extra sensors on the first and second stage engines, providing real-time temperature and pressure measurements from inside the engine compartments. Other sensors indicated that the propulsion, guidance, and altitude control systems operated normally from lift off to attainment of orbit.[39]

An extra two months had been taken for testing and checkout, partially due to the functioning of 5506. Edward Aldridge, Under Secretary of the Air Force and Director of NRO, noted that "we were ultraconservative. Our in-orbit [spacecraft] capabilities were doing very well. We did not want to pressure the system. We did not need to, [and] we wanted to make sure when this one flew that it was going to work."[40]

Aldridge's comments notwithstanding, it was certainly a relief for the consumers of overhead imagery, COMIREX and NRO, to have the KH-11 constellation back at full strength. It was not just a question of restoring full coverage. For over 18 months, they had had to face the prospect of a failure in the one KH-11 in orbit eliminating all overhead coverage of the Soviet Union and China. Now a failure in 5506 would only reduce coverage, not eliminate it.

The two satellites would be busy in the following years monitoring a variety of developments. As the Soviets continued to deploy SS-25 SICKLE missiles and began to deploy SS-24 SCALPEL missiles, the KH-11s continued to send back imagery of new deployments. KH-11 photography revealed that the Soviet Union deployed 10 SS-24 missile systems at bases near three Soviet cities between January and October 1988, doubling their number. Each SS-24 rail car contains four launchers with four of the ten-warhead missiles and four reloads. The SS-24s were photographed on rail cars loaded in long garages at Plesetsk and Kastroma in the northwestern Soviet Union and at a newly discovered facility at Gladkaya about 60 miles south of Krasnoyarsk. Single SS-24

trains were detected at Plesetsk and Gladkaya and three trains were photographed at the Kastroma base. KEYHOLE photography also indicated
a 50 percent increase in the number of road-mobile SS-25s, bringing
their number to 150. In addition, in 1988 the KH-11s looked for the
initial deployment of the BLACKJACK bomber. The 1988 version of
Soviet Military Power revealed that 11 of the new bombers had been
produced and that the initial BLACKJACK regiment could be expected
to be deployed before the year was out.[41]

Another target of the KH-11 was the new Soviet aircraft carrier,
Tbilisi, first photographed at Nikolayev in July 1984. The 1988 satellite
photographs showed that the completed ship has a ski-jump above the
bow, no deck-edge lifts to port, and only two to starboard. Previously
it had been assumed that there was a third lift. The photograph was
significant to naval intelligence analysts because the number of lifts
is generally related to the size of the carrier's air group. The two-lift
limit seemed to indicate that the Tbilisi air group would be significantly
smaller than that of United States super-carriers.[42]

On the defensive side, in December 1988 the KH-11s spotted concrete
foundations for what appeared to be large, phased-array radars. One
radar was detected near Sevastopol in Soviet Crimea and the second at a
remote location in far eastern Siberia. In accord with the ABM Treaty,
both were located on the Soviet periphery. Based on the KEYHOLE
imagery of the sites, photo-interpreters identified the future radars as
"Pechora-class" large phased-array radars that would probably be completed in 1990 or 1991. The detection of the radars was apparently the
result of a three-year KH-11 effort to detect new Soviet early-warning
radars that would fill in a network providing 360 degree coverage.[43]

Soviet strategic developments were not the only targets of the KH-11
cameras. A massive earthquake struck Yerevan in Armenia on December
9, 1987. A KH-11 was employed to help the United States assess the
damage and the photography it returned indicated that a nuclear power
station in the area of the quake escaped damage.[44]

Some of the activities of Colonel Qadaffi were also observed. In about
December 1987, the U.S. intelligence community spotted a construction
site at Rabta, 40 miles south of Tripoli. Analysts were immediately suspicious, explained General William F. Burns, director of the Arms Control
and Disarmament Agency, because "a facility quite isolated in the desert
with some capability to do manufacturing makes you wonder why."[45]

Additional suspicion was generated by Libya's early-1988 purchase of materials required for construction of a chemical weapons facility.

Unlike common pharmaceutical plants, chemical weapons plants need special storage areas, handling equipment, and security measures.[46] Not until July 1988 had enough information been gathered from KH-11 and SR-71 photography, communications intelligence, and human sources (Arab and western workers at the site and sources in Hong Kong and Singapore helped trace the dummy corporations through which the plant's components were bought), to convince analysts at the CIA that Moammar Qadaffi was definitely building a chemical warfare facility at Rabta.[47]

Although the CIA was convinced, some of America's European allies were not so easily persuaded. Apparently, the Europeans did not consider it conclusively demonstrated that the facility shown in the KH-11 photos was to be used for making chemical weapons. A senior West German official was quoted as complaining that "the American intelligence consisted of two satellite photos, which can be interpreted in different ways, and one gigantic inference: the fact that the plant is situated far from a city and is heavily protected means something is wrong."[48]

But another foreign official who saw all the evidence disagreed: "The Americans have done a superb bit of detective work on this." The official told *Newsweek* magazine that in addition to the satellite photos, the evidence included blueprints for the plant and details of the clandestine commercial network through which it was assembled. The news weekly further reported that according to U.S. intelligence sources, 100 yards from the chemical plant was a machine-tool shop that had already started to manufacture stainless-steel bomb casings.[49]

An operational Titan 34D program represented one essential element of recovery for the KEYHOLE program. A second necessary element was to have an operational space shuttle. Several satellites had been constructed in such a way that the shuttle was the only launch vehicle that could safely place them into orbit. Modifications were made in several of those satellites to allow them to be deployed by expendable launch vehicles, but the first two versions of a satellite system known as LACROSSE still required shuttle launch.[50] LACROSSE was an essential part of America's space imaging program because it was a partial solution to the cloud cover problem.

Cloud cover is a serious impediment to monitoring the USSR. The mean monthly cloud cover over Krasnoyarsk, for example, never falls below 49 percent and rises to 76 percent in December. Likewise, the mean monthly cloud cover for Novaya Zemlya, a nuclear test facility, never falls below 60 percent and reaches as high as 83 percent in October. As one defense official put it, "the weather in the Soviet Union is crappy all the time." Producing a photograph of a particular area or target may take several years. The cloud cover over Tallin, where CORONA satellites first detected possible ABM deployments in 1961, was so extensive that it was another eight years before additional photography could be obtained. Possibly a complete picture will be formed only by constructing a photographic montage of the area made up of pictures taken on a number of orbital passes.[51]

It is impossible to take a standard photograph through cloud cover. Photography requires a satellite to gather up the visible light reflected off an object, but since visible light cannot penetrate cloud cover, a standard photographic camera is useless when significant cloud cover blankets a target. However, imagery can still be obtained. Imagery consists of any photographic-type image of an object and may be produced by sensing devices that rely on different parts of the electro-magnetic spectrum. Radio waves constitute part of that spectrum and can be used to produce imagery.

In the years before World War II, the British pioneered the development of radar in which radio waves were emitted in the direction of possible targets. An object that collided with those waves would send them back toward the emitter. Receipt of the waves would allow detection of an object and possibly identification, depending on the quality of the radar system. Subsequently, it was determined that it was possible to produce an image of the target using radar. The virtue of radar lies in its ability to produce imagery through most forms of cloud cover since radio waves can pass through the clouds on their way to and from the target and also during darkness because the radio waves operate in darkness as well as light. United States experiments with imaging radar have been conducted using satellites and the space shuttle.

SEASAT-A, an experimental ocean survey satellite launched in 1978, carried a synthetic aperture radar with a resolution of 82 feet that provided all-weather photographs of ocean waves and ice fields. Synthetic aperture radar increases resolution of the radar system by using the motion of the satellite to make the relatively short antenna that can be carried by satellites behave as if it were a much longer antenna.* In the

---

* The 82-foot resolution could have been substantially improved by changing the frequency and power of the transmitted signals.

absence of that type of radar, the antenna required by the SEASAT-A to produce images of equal resolution would have had to be enormous. For example, a conventional radar imaging system at an altitude of 150 miles and operating at a wavelength of 20 centimeters would require an antenna of 1.2 miles.[52]

In November 1981, the orbiter Columbia carried Shuttle Imaging Radar-A (SIR-A). SIR-A unexpectedly penetrated up to 16 feet into the dry sands of the eastern Sahara desert in Egypt, revealing to analysts at the U.S. Geological Survey and Jet Propulsion Laboratory traces of ancient sub-saharan rivers that had carved out valleys as broad as those of the present Nile.[53]

Almost three years later, in October 1984, the ill-fated Challenger carried SIR-B to orbit on mission 41-G. Although the antenna had dimensions of 35 by 7 feet, the console was, due to miniaturization, about the size of a shoebox. Targets of the radar antenna included Montreal, Quebec, parts of New Hampshire and southwest Maine, and Mount Shasta, California. Given the cloud cover that blanketed many of the targets, the mission was a dramatic demonstration of the additional capability provided by radar imaging. The mission produced excellent photographs that gave a clear indication of what radar imaging could do over Eastern Europe and the USSR in autumn and winter during cloud cover or night.[54]

The radar images can be transmitted from space in either analog or digital form. Either way, the images can be turned into standard photographs to reveal activity during the night or under cloud cover, but digital transmission provides the interpreter with more flexibility because it can be manipulated by computer. That manipulation may bring out relevant features that were obscured in the original image.[55]

"You can digitally manipulate data in such a fashion that you create very strong contrasts," says John Ford, supervisor of the Imaging Radar Geology Group at the Jet Propulsion Laboratory where SIR-A and SIR-B were built. As he told William E. Burrows, "that allows you to see things which were completely obscured before this processing technique has been applied." In addition, radar images obtained from directly above the target can be digitally rearranged to create the perspective of seeing the target from all sides. Such a technique can be of immense use in the analysis of foreign weapons systems and military installations.[56]

LACROSSE began as a CIA project codenamed INDIGO. A synthetic aperture radar rather than cameras or charged-couple devices would provide imagery. George Bush, when he was Director of Central Intelligence, approved $250 million for research and development of two satellite systems that would employ radar imaging to produce images. In addition to INDIGO, funds would also be devoted to development of a radar ocean reconnaissance satellite to detect and image Soviet ships. That program, CLIPPER BOW, was subsequently cancelled.[57]

The CIA planned to put the imaging radar on the advanced models of the KH-11 rather than to devote a separate platform to the radar imaging mission, but the plan was vetoed by Defense Secretary Harold Brown on the grounds that the plan placed too many surveillance assets on a single platform.[58]

As was the case with the KH-11, LACROSSE was the subject of intense political battles. DCI William J. Casey was instrumental in keeping the program alive. By 1983 it was apparent that LACROSSE would eventually cost more than $1 billion. Vast cost overruns and numerous problems had appeared in the development stage, and funding of about $200 million was needed to keep LACROSSE alive for 1983. The Martin Marietta Corporation was the principal contractor, and General Electric was doing the ground processing—handling of the signal after it arrived at the ground stations. Martin Marietta needed the money then, and the "drop dead date" was approaching. Hundreds of millions of dollars had to be provided or the project would die. Among those opposing continued funding was Edward Boland, Chairman of the House Permanent Select Committee on Intelligence. Part of Boland's opposition came from his belief that the developmental problems were insurmountable and part from his feeling that NRO had lied about the cost.[59]

However, Barry Goldwater's Senate Select Committee on Intelligence retained $200 million for LACROSSE, leading to a meeting between Goldwater, Patrick Moynihan, the Vice Chairman of the Senate Committee, Boland, and the House Committee's Vice Chairman, Ken Robinson. Goldwater was passionately in favor of the project. He argued that the U-2 and SR-71 spy planes had cost overruns and problems but had ultimately added a vital new dimension to U.S. intelligence-gathering capabilities. In addition, he pointed to the 26th Tactical Fighter Wing in Germany that monitored the border between East and West, sending real-time radar imagery to ground stations. LACROSSE's contribution to the nation's security could be immense, was worth any cost, and "it could work to prevent war."[60]

Boland was not completely convinced by Goldwater's argument, but Goldwater was able to raise the prospective cost of disagreement beyond what Boland was willing to pay. He threatened to take the issue to the Chairman of the Senate Armed Services Committee, John Tower. Boland knew that he would wind up facing the entire Senate on the issue and decided he would rather settle the matter between the two intelligence committees. As Goldwater slowly hobbled towards Tower's office, Boland backed down, and dispatched an aide to intercept Goldwater. Boland agreed to one year's funding, and Goldwater agreed that further cost overruns would result in the project being terminated.[61]

Boland had not been LACROSSE's only critic. Donald Latham, Assistant Secretary of Defense for Command, Control, Communications, and Intelligence, opposed LACROSSE on a variety of grounds. He argued that LACROSSE was not cost efficient, was of limited utility, and was very expensive. In addition, it was suggested that it was too much of a war-fighting satellite because of its ability to track mobile missiles with its all-weather capabilities.[62]

On the other hand, some have seen LACROSSE's ability to monitor mobile missiles as making it essential to verification of the December 1987 Intermediate Nuclear Forces Treaty and any future START treaty. David Boren, who had become chairman of the Senate intelligence committee in 1987, pushed the Reagan administration to approve $12 billion for six of the satellites and threatened to hold the INF treaty hostage until the White House agreed. In February 1988, administration officials assured Boren that Reagan would ask Congress for supplemental appropriations for the satellites. When Boren subsequently learned that President Reagan had not even been briefed on the matter, he furiously announced an INF filibuster; his committee also voted to slash the CIA and the National Security Agency budgets. This time the White House took Boren seriously, promising to secure partial funding for the program.[63]

The first LACROSSE was completed in October 1987, but the shuttle program was still trying to recover from the explosion of the Challenger. It did not resume operations until September 29, 1988 when the orbiter Discovery was successfully launched on a mission to deploy a Tracking and Data Relay Satellite (TDRS) into geosynchronous orbit 22,300 miles above earth. The 5000 pound TDRS, one of which was lost in the Challenger explosion, was the second of a three-satellite constellation to allow for continuous communications between shuttle and satellites in low-earth orbit and their controllers on earth. Before TDRS, voice and data transmissions were only received during 15 percent of NASA spacecraft orbits, when a satellite was passing over or near one of several ground stations. TDRS would allow space shuttles and other spacecraft to communicate during 90 percent of their orbits.[64] Part of TDRS's power comes from two large solar arrays rotating on booms aligned on the satellite's north-south axis. On the east-west axis are two 16-foot antennas to receive radio signals from satellites. The spacecraft's main body contains 30 antennas that track and return data from as many as 20 satellites. TDRS can relay the equivalent of a 20-volume encyclopedia in one second. Controlled from Goddard Space Center in Greenbelt, Maryland, the TDRS network would downlink its data to its single ground station at the White Sands Proving Ground.*[65]

Given the success of STS Mission 26, NASA set 7:00 A.M., December 1 for launch of the Atlantis, with the countdown to begin at one minute past midnight November 28. The countdown would last 78 hours, with 35 hours of built-in hold time.[66]

The launch site was not Vandenberg AFB but Cape Canaveral. There was no choice in the matter. Originally, it had been intended to launch several shuttle missions into polar orbit from Vandenberg, missions that would deploy the advanced version of the KH-11 and LACROSSE into the traditional polar orbit. Shuttles launched from Vandenberg into polar orbit could also retrieve reconnaissance satellites for refurbishment and repair or be used to refuel satellites. The launch pad at Vandenberg designated for shuttle missions was Space Launch Complex-6 (SLC-6), better known as "Slick-Six," which had been the intended launch pad for MOL missions. But problems plagued SLC-6. Its main engine duct, the cavity under the shuttle where the flame from its engines would be deflected and drawn off, was believed to trap the flame's hydrogen,

---

* The Soviet Union may have sought to establish a ground station near the White Sands facility to intercept the communications from LACROSSE and other satellites downlinked to White Sands. An Anaheim, California firm that was linked to an attempt to smuggle equipment to the Soviet Bloc in early 1982 had already begun building a satellite ground station at Las Vegas, New Mexico, about 200 miles north of White Sands, that could have eavesdropped on the communications being received at White Sands. The company, Land Resources Management Inc., had an agreement with NASA to receive LANDSAT signals. The operation could have been used as a cover for the interception of far more sensitive signals.[67]

creating the possibility of an explosion before the launch vehicle was able to get off the pad. In addition to this problem, there was a high incidence of dense fog.[68]

Given the problems associated with SLC-6 and the cost of correcting those that were susceptible to correction, it was decided in 1987 that the facilities should be transitioned to minimum caretaker status. As a result, the launch processing system was deactivated, the propellant system was shut down, and portable support facilities were placed in storage. It would require four years to reactivate the site following a decision to launch the space shuttle from the west coast.[69]

So Cape Canaveral was to host the first LACROSSE launch. Launching a reconnaissance satellite from the Cape imposed one substantial constraint. Given the limitations on the shuttle flight path, to avoid overflying populated areas, the maximum inclination that could be attained was 57 degrees. With that inclination, coverage would stop 100 miles north of Moscow. Excluded from LACROSSE coverage would be the naval facilities around the Kola peninsula, the SS-11, SS-17, and SS-19 ICBM fields at Yedrovo, Kozelsk, Kostroma, Yurya, Pervomaisk, and Derazhnya as well as the ballistic missile test range and satellite launch facility at Plesetsk.[70]

As expected, the launch approached under a blanket of official secrecy. Even the landing day and time would not be divulged until 24 hours before touchdown at Edwards AFB in California, and not even the wives of the four married astronauts were told when their husbands would return. In a letter to Kennedy Space Center employees, center director Robert S. McCartney wrote, "It is of vital importance to our national security to protect the status and capabilities of operational space systems, the deployment of those systems and the products and services they provide."[71] An Air Force spokesman, Lt. Col. Joe Purka, was to explain after the launch that "We have a consistent policy for protecting those missions that are sensitive to national security. Our objective is to ensure that any potential adversary has some doubt as to what the payloads are and what its purpose is."[72]

The five astronauts selected to carry out the mission represented a blend of talents. Mission commander Robert L. Gibson, a 42-year-old Navy pilot, had already logged more than 300 hours in space, having commanded both the Challenger and Columbia orbiters. The pilot, Guy S. Gardner, was making his first space flight. Gardner, who flew 177 combat missions in Southeast Asia, has a master's degree in astronautics from Purdue. The key to the mission's success were the three mission specialists, Colonel Richard M. Mullane, Lieutenant Colonel Jerry L. Ross, and Commander William M. Shepherd. Colonel Mullane was a mission specialist on the first flight of Discovery in August 1984, and Lt. Colonel Ross had supervised payload operations for all space shuttle flights. He flew aboard the Atlantis in November 1985 when he made two six-hour walks in space to demonstrate construction techniques for a space station. William Shepherd, making his first shuttle flight, served on the Navy SEAL underwater demolition team.[73]

The day for the launch was set for December 1, but that day came and went without a launch. A high-altitude wind caused NASA officials to postpone the launch. The mission was scrubbed before 9 A.M., although clouds had vanished and gusty winds at ground level had dissipated, because of a "rather large shear," or abrupt changes in direction in ascending layers of winds blowing at nearly 70 mph at 45,000 to 55,000 feet. The shuttle would have been accelerating through its period of maximum stress at nearly twice the speed of sound as it encountered those high-speed shears. This "could have exceeded the load capability of the wings," said Lawrence G. Williams, an engineering official at Johnson Space Center in Houston.[74]

That NASA would be extraordinarily cautious was understandable. Each LACROSSE was a several hundred million dollar investment. Also, given the political need to launch a satellite that could add substantially to U.S. verification capabilities, and given its ability to peer under the clouds, a failure would have been quite upsetting. One Pentagon official noted before the launch that "this may be the most important job that NASA has ever done for the Air Force and NASA better not screw it up."[75]

The next day was also a close call. The window for launching Atlantis ran from 6:32 A.M. to 9:32 A.M., with the intended lift off reported to be 7:00 A.M. The countdown clocks at Kennedy Space Center, blacked out for most of the count by Pentagon officials, suddenly flashed on at about 9:20 A.M., nine minutes before the planned ignition. Weather problems on two fronts had emerged. After hours of disheartening weather reports and with only minutes to spare before the launching would have had to be postponed again, winds above 25,000 feet grew calmer and space officials decided to resume the delayed countdown. The official clock then stopped at 31 seconds before ignition because two transoceanic sites designated for emergency landings were reporting bad weather.

Astronaut John Casper, flying a reconnaissance plane over Zarraguza, Spain, one of the landing sites, reported that an unacceptable cloud ceiling at 6000 feet would drift safely out of the landing area before the shuttle might need it. After several minutes delay the countdown resumed, producing a cheer from the launch teams.[76]

At 9:30 A.M., the shuttle's engines fired and the Atlantis lifted off on its journey. Nine seconds into its flight, the vehicle began a dramatic 102-degree roll and then veered sharply northward as opposed to the usual eastward route required to launch the geosynchronous satellites normally launched from the cape. The maneuver indicated that the shuttle, rather than going into an equatorial orbit, was going into a 57-degree orbit. That maneuver confirmed the reports that it was LACROSSE that was in the cargo bay.[77]

Four minutes into the launch there was a scare when a sensor indicated a temperature spike in main engine No. 3, but the computer instantly recognized this as a problem caused by a faulty sensor, not the engine, and kept the engine going. Eight and a half minutes later the three main engines were shut off as the Atlantis passed about 140 miles east of Atlantic City, New Jersey.[78]

The crew did not wait long to deploy their payload. On the seventh orbit, about seven hours after launch, the payload was hoisted from the bay and released into space by the Canadian-built remote manipulator arm. After Atlantis was maneuvered a safe distance away, the two solar panels attached to the central body of the satellite were instructed to unfold. In what was undoubtedly the beginning of some anxious moments, they failed to respond. Ross and Shepherd were prepared to conduct an extravehicular activity to repair the satellite had the solar arrays not deployed, but a second set of radio commands freed the solar arrays.[79]

Once deployed, the Atlantis remained nearby to monitor the spacecraft visually and to ensure that its systems were functioning. Gibson first maneuvered the orbiter to point about three miles from LACROSSE. He then maneuvered Atlantis about five miles away and increased that separation at about five-mile increments. If performance was unsatisfactory, the crew had the option of blowing the arms off the spacecraft, rendezvousing with the instrument core, retrieving the core with the robot arm, and restowing it in Atlantis' cargo bay for a return flight to Earth. Pyrotechnic packages on the arms could separate them without harming the spacecraft.[80]

After the astronauts deployed the newest reconnaissance satellite they conducted some reconnaissance of their own. The crew used the orbiter's 50-foot arm for a television inspection of the belly of the Atlantis to check for possible damage to the vehicle's thermal protection tiles. They also conducted visual surveillance of earth and ocean targets. In one experiment they employed the USAF Spaceborne Direct View System, involving hand-held optics used in connection with specific ground test targets. The system is designed to test an astronaut's ability to monitor targets or surface targets of military interest. The astronauts took detailed photos of Greenland, the Grand Canyon, and the Himalayas.[81]

Despite great secrecy, the flight of Atlantis and its deployment of LACROSSE was followed with great interest by a variety of observers. One was undoubtedly the Soviet Union. According to media reports there was one Soviet intelligence-collection ship off the coast of Florida during the launch period. If so, it was not within 200 miles. The data base of the Navy Operational Intelligence Center, responsible for monitoring the movements of Soviet ships, shows only three Soviet ships within 200 miles of Cape Canaveral during the period November 28 to December 4: the bulk carrier *Chkavlovsk* on December 1 and 2, the cargo ship *Nikolay Gogol*, and the cargo ship *Novomoskovsk* on December 4.[82] The Soviets were probably monitoring the launch from either their Lourdes, Cuba communications intelligence complex or from what appears to be a geosynchronous signals intelligence satellite, Cosmos 1961, stationed over the Western Hemisphere.

The flight was also monitored by some dedicated space-watchers who followed the mission with their home computers and high-powered binoculars. One was Ted Molczan, a Toronto engineering technologist who specializes in energy conservation. Molczan had been in contact with about 150 satellite watchers from as far away as Australia about the Atlantis' probable orbit. Toronto skies were cloudy the day of the Atlantis launch and Molczan was unable to see it pass, but others did and in late afternoon with the craft on its 6th orbit, he started to receive calls. Two sightings were from the Toronto region and one was from Ashtabula, Ohio. "To get three confirmations like that in the space of half an hour was very exciting. We had more than a dozen people looking for it." Knowing the times of the sightings allowed him to calculate the height of the orbit with the computer, and Molczan was able to quickly determine where the shuttle could be seen for the rest of the mission. According to Molczan, watchers sighted only the Atlantis on the sixth revolution but the satellite was sighted by the seventh revolution, three miles ahead of the shuttle. Sightings for the next two days indicated that the shuttle stayed close to the satellite. Molczan's assessment was that "if we could figure it out, you could be sure that the Russians figured it out too."[83]

Time lapse pictures taken near Denver during Atlantis' seventh orbit showed two streaks of reflected light representing the Atlantis and the satellite it deployed. By December 6, just before Atlantis' re-entry, members of the group in Australia observed LACROSSE about 180 miles from Atlantis.[84]

After spending most of its orbit flying close to LACROSSE, Atlantis withdrew to a position 40 miles away. At 2:29 P.M. on December 6, the shuttle fired its re-entry rockets. The shuttle dropped quickly from the north, producing two sonic booms, made a 315-degree turn, and landed at 3:36 P.M. in light wind conditions of one to two knots. Thus, the Atlantis mission came to a halt 4 days, 9 hours, 6 minutes, and 19 seconds after blast-off. It had completed 69 orbits, totaling 1.7 million miles. The crew exited at 4:29 P.M. It appeared that the nose tiles, which protect the shuttle when it re-enters the atmosphere, were deeply pocked. "It looked like they were hit with a baseball bat," said Ralph Jackson, a NASA spokesman. "They're the worst I've ever seen." Indeed, Atlantis had about twice as many tiles damaged as Discovery on Mission 26.[85]

The fully-deployed satellite the astronauts left behind has a round core element from which the rectangular radar antenna hangs parallel to the core, and the two solar panels extend horizontally for 150 feet from the core while a communications antenna extends vertically upwards. The solar panels help provide power to the 10,000-watt radar.[86]

Within five weeks after its initial deployment, LACROSSE used its on-board rocket engines to reach a 437 by 415 mile orbit, about 150 miles higher than its original orbit. Since then it has been transmitting its digital imagery, with resolution in the 5- to 10-foot range, via the Tracking and Data Relay Satellite System (TDRSS) to its ground station at White Sands, New Mexico. In March 1989 the shuttle Discovery deployed another TDRS to complete the constellation.[87]

The successful launch of LACROSSE was not only another step in the recovery process but a step forward in broadening America's ability to monitor world events—as it sharply expanded the conditions under which America's secret eyes in space could see what was taking place below. It was also a step toward achieving greater redundancy in satellite imaging—as LACROSSE joined the final two spacecraft of the KH-11 program, whose codename had been changed from KENNAN to CRYSTAL, in returning imagery. Additionally, the capabilities of the final KH-11s had grown dramatically over the years. At their peaks, the final KH-11s could operate whenever needed rather than being restricted to operating a few hours each day. Further, stations other than Ft. Belvoir were now able to receive KH-11 imagery, including a site in West Germany and probably Buckley Air National Guard Base in Colorado, America's primary downlink for early warning information. As yet unattained was a constellation of three photographic and two radar imaging satellites, which will be able to deliver an image every four to five seconds. A step toward achieving that goal was not far away.[88]

CHAPTER 10

New Systems, New Targets

George Bush began his Presidency seeking to avoid a confrontation with the Democrat-controlled Congress. In his first months in office he managed to reach agreements with the legislative branch on several potentially contentious issues, including the new budget and aid to the Contras. But on another issue the new President seemed on a collision course with the Chairman and ranking Republican member of the Senate Select Committee on Intelligence, Senators David Boren (D-Oklahoma) and William Cohen (R-Maine).

As noted earlier, Ronald Reagan, as part of the price for gaining the Senate intelligence committee's assessment that the United States could verify the Intermediate Nuclear Forces treaty, agreed to a five-year, $15 billion satellite modernization program. The program involved funds for development and deployment of what the Senators considered sufficient, if not optimal, numbers of LACROSSE, KH-11 follow-on, laser-monitoring, and signals intelligence satellites. Without such a program, the Senators believed it would be difficult to verify the INF treaty and impossible to verify any treaty produced by the Strategic Arms Reduction Talks.[1]

But Bush, faced with the need to cut the defense budget by $6.3 billion, eliminated most of the funding for satellite modernization. Under the Bush plan the deployment of some satellites would have been delayed from 1992 to 1994, and the $15 billion authority for the systems would be stretched out over 10 years rather than the original five.[2]

The Bush plan cut $500 million from LACROSSE funding through the end of fiscal 1991, reducing the funding to $200 million. Boren and Cohen were informed of the cuts at an intelligence committee hearing

229

by DCI William Webster, who apparently presented them as a fait accompli. At that or a subsequent meeting Webster apparently indicated that the cuts would force the intelligence community to reconsider its entire upgrade program for spy satellites.[3]

The two senators reacted in the same manner that they did when the Reagan administration seemed reluctant to approve funding for the modernization program. Boren declared that the cuts "would seriously jeopardize our own near-term national security and could slow completion of a strategic arms reduction treaty. . . . If they don't keep their word, we will keep it for them by cutting other areas of the budget," Boren threatened, suggesting personnel cuts as a possibility. He indicated that he would have the support of Senator Robert Byrd, chairman of the Appropriations Committee, in making his threat stick.[4]

Boren and Cohen summoned Bush's national security adviser, Brent Scowcroft, to a closed committee hearing in early April. Boren, Cohen, and the other advocates of the original modernization plan apparently got their views across quite effectively by repeating that the proposed budget cuts and resulting delays in deployment were an "unacceptable" violation of the 1988 Reagan pledge and would jeopardize START ratification.[5]

The message Scowcroft brought back to the President led to an April 13 meeting between Bush and Senator Boren. Boren repeated his position that failure to adequately fund the satellite modernization program could cause ratification problems for any START agreement. Before the week was out, Bush had restored the money. On Monday April 17, Boren issued a statement in which he declared that "I am very pleased with the personal attention the president is giving to national security issues" and said that Bush had "made it clear that he intended to honor the commitment of the previous administration to modernize our national technical means."[6]

The first step in the modernization process occurred only a few months later. At 8:37A.M. on August 8 the orbiter Columbia and its five man crew blasted off on STS mission 28. Before the mission was over, five days later, the crew activated, and launched, a secondary payload to perform SDI-related experiments. They also conducted a variety of other experiments and engaged in the direct observation of several targets, including ship wakes and preplanned ground activities such as flares, hand-held optical reflections, and smoke. But the most important aspect of their mission took place on their first day in orbit. Approximately seven and a half hours after reaching orbit, mission specialist Commander David Leestma, Lt. Col. James C. Adamson, and Major Mark Brown deployed a satellite into an initial 188 × 196 mile orbit at a 57 degree inclination.[7]

The satellite launched on August 8 is the first of the successors to the KH-11 and has been referred to as the "KH-12". In fact, no satellite or optical system with such a designation exists. After years of repeated press disclosures concerning the KH-8, KH-9, and KH-11, NRO decided to abandon such designations and refer to the satellites by a purely random numbering scheme. However, since the new satellite is an improvement to the KH-11 it is sometimes referred to as the Advanced KENNAN or Improved CRYSTAL.[8]

The Advanced KENNAN is identical to the KH-11 in one very important way. It will continue to rely on an electro-optical system and transmit its data in real time. At the same time there will be some noticeable differences between the satellites. The Advanced KENNAN has an infrared imaging capability. In addition to being able to produce imagery based on near infrared emission the new satellite's optical system's sensitivity extends into the thermal infrared portion of the electromagnetic spectrum. Thermal infrared imagery is produced by sensing the heat produced by objects. Since detection of such heat can be conducted equally well during nighttime as daytime, nighttime imaging will be feasible with such a system.[9]

In addition, the Advanced KENNAN carries the Improved CRYSTAL Metric System (ICMS) to allow the imagery that is returned to be used for terrain mapping purposes. Specifically, the ICMS makes it possible to instruct the satellite to send back specific images with the reseau crosses desired by mappers, but to exclude the crosses from all other images.* Plans to equip the Advanced KENNAN with the ICMS were made to allow the Advanced KENNAN data to be used by the Defense Mapping Agency in developing terrain contour maps for U.S. cruise missiles. Placing such a capability on the Advanced KENNAN was also a means of circumventing DMA's request for its own mapping satellite.[10]

---

* Reseau crosses, also known as fiduciary marks, are markings placed on a frame of film or other form of image prior to a target's being photographed. The distances between the markings correspond to known distances between points on the ground. They therefore make the geometric reproduction of the image onto a map easier by allowing precision measurement of the distances between different points in the image.

A third difference between the KH-11 and its successor may be their respective weights. As noted earlier the KH-11 has been estimated to weigh approximately 28,000 pounds. Originally, the weight of the Advanced KENNAN was estimated at 32,000 pounds at liftoff.[11]

The deployment of the new spacecraft was originally envisioned as being performed by the space shuttle. Thus, NASA's Baseline Reference Mission Description included a description of "DOD Reference Mission 4," which involved

> a payload delivery mission of a modular spacecraft weighing 32,000 lb. at lift-off. The mission will deploy a spacecraft weighing 29,000 pounds in a 150 n. mile circular orbit at 98 degrees inclination within two revolutions after lift-off. A passively cooperative stabilized spacecraft weighing 22,500 pounds will be returned from a 150 n. mi. circular orbit and returned to VAFB. The mission length [of the shuttle], including contingencies will be 7 days . . . For mission performance and consumable analysis, a cradle weight of 2500 lbs. will be assumed to be included in the ascent payload but must be added to the retrieved payload weight. . . . Contingency EVA [Extra-Vehicular Activity] capability will be provided.[12]

The original plan for a 32,000-pound spacecraft was modified on several occasions. At first only 2000 pounds was added, but eventually the total planned increase constituted 25% of the Advanced KENNAN's total weight. Eventually, plans for the new spy satellite called for it to weigh 40,000 pounds. The size of the Advanced KENNAN required the Air Force to increase the volume of two of its huge C-5A cargo planes to carry the 62-foot long and 17-foot square Space Container Transportation System (SCTS). The SCTS is an environmentally clean container in which the new satellite can be transported from Lockheed to its launch site.[13]

The planned increase in spacecraft weight was due almost exclusively to increases in the hydrazine that the new satellite would carry. For it is the hydrazine that allows spacecraft maneuvering, including the crucial maneuvering required to keep the satellite in its proper orbit. Increasing the fuel supply from 6500 to almost 15,000 pounds would enhance the satellite's capability in several ways. It would allow more frequent changes in its altitude, so that it might be boosted to higher altitudes when broad area pictures are desired and lowered in altitude when a closer look is the objective. Additionally, an enhanced maneuvering capability increases the chance, by permitting more frequent orbital maneuvering, that the NRO could defeat attempts by the Soviet Union and other nations to camouflage or conceal targets. The enhanced maneuvering capability could also be employed to dodge Soviet ASAT weapons in the event of war.*[14]

The actual weight of the satellite launched on August 7 was reported to be much less than 40,000 pounds.[15] At least part of the disparity may have resulted from a need or decision to launch the new satellite with much less additional hydrazine than originally planned. Given its launch from Cape Canaveral into a 57 degree orbit, the satellite could be plucked out of orbit or refueled in orbit during a future mission from Cape Canaveral.

In addition, the new satellite is reported to be hardened against nuclear effects and to contain a variety of protection devices to prevent damage from Soviet laser weapons. Attempts to protect against such threats may involve giving a satellite the ability to operate without ground control, encasing satellite components in sheet metal, using gallium arsenide in place of silicon to manufacture the CCDs that go into the sensors and make them more resistant to nuclear radiation, and coating lenses to make them laser resistant. Optical systems can be shielded by special shutters or filters that are triggered by warning of laser illumination. Future versions of the Advanced KENNAN or its successors may also carry the Satellite On-board Attack Warning System (SOARS), presently under development by TRW. The initial version of SOARS will include sensors to detect attack by radio frequency microwaves, lasers, and projectiles, and will have a transmitter to provide real-time warning to allow satellite ground controllers to enact countermeasures.[16]

---

* Soviet ASAT weapons are considered to fall into four classes. One is their Galosh ABM system outside of Moscow. The second is their co-orbital ASAT. Normally, after two revolutions in space it comes close enough to its target to destroy it by exploding a conventional warhead. The system has not been involved in an in-space test since June 1982. It had a less than inspiring success rate when testing stopped, with 11 of 20 tests being estimated as failures, although the bulk of the failures could be attributed to use of a new sensor system. Ground-based lasers at Sary Shagan and Dushanbe constitute the third class. The extent to which they truly threaten U.S. satellites has been the subject of debate, among Western defense experts and between the U.S. and USSR. Finally, there is electronic warfare, or radio-electric combat in Soviet terminology, which includes jamming uplink and downlink transmissions or even taking control of a satellite.[17]

Another means of avoiding Soviet countermeasures and preventing other nations from evading U.S. coverage would be to give an imaging satellite a stealth capability, which was essentially the objective of Leslie Dirks in 1962. The same methods used in producing stealth bombers, fighters, and cruise missiles might be applied to satellites. Two possible methods involve replacing metals with lightweight composite materials that absorb radar signals and smoothing body parts so they deflect radar signals rather than reflect them.[18]

The Defense Department is reportedly hard at work on applying stealth techniques to satellites. In April 1984, NASA launched a four-ton cylinder carrying experiments to develop new space-age materials, including secret ones for making stealth satellites.[19]

When the plans for the Advanced KENNAN were finalized it was assumed that the space shuttle would be employed for placing the spacecraft into orbit and retrieving it for refurbishing and refueling. This would substantially extend its lifetime, possibly to eight years. As NASA's "DOD Reference Mission 4" profile indicated, in addition to an Advanced KENNAN being placed into orbit another could be retrieved from its orbit by the shuttle and returned to Vandenberg. Or as "DOD Reference Mission 4Y" postulated, a one-day mission might be undertaken simply to remove an ailing satellite from orbit and return it to Vandenberg.[20] There the particular Advanced KENNAN could be repaired or refurbished and redeployed on a future shuttle mission. Alternatively, in some cases the shuttle's astronauts could repair or refuel satellites without even returning them to earth. Such procedures would help to considerably extend the life of the ultra-expensive reconnaissance satellite.

The viability of such operations was demonstrated over several shuttle missions, examples of where the shuttle's civilian operations had military implications. During the seven-day Mission 41-C of the Challenger in April 1984, astronauts George D. "Pinky" Nelson and James D. van Hoften devoted seven hours to repairing a Solar Maximum Mission observation satellite. Solar Max's attitude control system had malfunctioned, making it impossible to point the telescope toward the sun.[21]

As the Challenger and the Solar Max sped around the earth, the satellite was plucked out of its orbit by the shuttle's manipulator arm and deposited in the shuttle's open cargo bay. Nelson and Hoften proceeded to remove the malfunctioning spacecraft's 550-pound attitude control module and replace it with a new one. After checkout by the Goddard Space Flight Center at Greenbelt, Maryland, the repaired satellite was then relaunched so that it could resume its mission.[22]

The Challenger's commander, Robert L. Crippen, observed that "our impression from this particular mission is that satellite servicing is something that's here to stay." That point seemed to be underlined when the next Challenger mission, 41-G of October 1984, also involved a servicing exercise. Dr. Kathryn D. Sullivan and Lieutenant Commander David C. Leestma devoted three hours at the far end of the shuttle's cargo bay practicing satellite refueling. The astronauts pumped hydrazine back and forth between two tanks in several practice maneuvers.[23]

A third demonstration of satellite retrieval capability came on shuttle mission 51-A of November 1984. Again a satellite was retrieved from orbit, but rather than being retrieved by the shuttle's manipulator arm, two astronauts propelled by their manned maneuvering units (MMUs) grabbed the Palapa-B and Westar 6 communications satellites and brought them to the shuttle. The two satellites had been accidentally fired into useless orbits earlier in the year. When they landed with the shuttle at Cape Canaveral on November 16 they were taken away to be prepared for a second launching.[24]

But the carefully-developed scenario for shuttle launch and retrieval of the KH-11 follow-on spacecraft appeared to be shattered by the Challenger accident as well as the mothballing of the SLC-6 at Vandenberg. The disaster, as previously noted, served to demonstrate how vulnerable the overhead reconnaissance program was to disruption when manned vehicles were the only means of placing spacecraft in orbit. A year after the loss of the Challenger, Colonel Donald C. DePree, the Space Division's Deputy for Space Transportation Systems [i.e. the shuttle] observed that the shuttle is "an excellent vehicle, but why not use it only for those missions where there is a payoff to having [a] man in the loop? Why use man just to accompany satellites into orbit, throw them out the bay, and come back home?" And former Defense Secretary Robert McNamara remarked that "The whole shuttle program is an illustration of how not to do it for most missions. We didn't need a man. . . . By substituting manned vehicles for unmanned vehicles we screwed up our whole collection program."[25]

The screwing up of the whole collection program was exactly what Edward Aldridge, in his dual roles as Under Secretary of the Air Force and NRO Director, worried about when he appealed to the House Armed Services Committee in 1984 for funding for a new generation of Complementary Expendable Launch Vehicles to launch "national security payloads." Aldridge, answering the questions of one skeptical congressman, noted that "if we got into a situation where we grounded the flights for some reason because men are on board, we would not have any access to space, and that seems to me unacceptable."[26]

Aldridge's pleas produced an appropriation of slightly over $2 billion, with which the Air Force ordered 10 of the new launch vehicles. First designated the Titan 34D7, they were subsequently renamed Titan IV. In the wake of the Challenger explosion, the order was increased from 10 to 23, to allow a greater number of satellites to be launched on the unmanned rockets.[27]

Original plans called for the first Titan IV to be launched in October 1988. The rocket had been placed on the pad in May of that year in preparation for the October launch date. But there was to be an eight-month delay before the new rocket would get off the launch pad. Causes for the delay were hardware problems (including the payload shroud), refurbishment of the launch pad, and testing of new ground systems. Additional delay resulted when lightning struck close to the launch pad on May 22, 1989.[28]

The Titan IV is 204 feet long when it carries a payload shroud 86 feet long and 16 feet in diameter (a smaller version of the payload shroud is 56 feet long). Its two 7-segment solid rocket boosters are able to produce a total of 2,725,000 pounds of thrust. Such a rocket, with the aid of the appropriate upper stage, could place a 10,500-pound satellite, such as an advanced signals intelligence satellite, in geosynchronous orbit.[29]

But the new rocket's first mission—placing a two-and-a-half-ton Defense Support Program early-warning satellite in geosynchronous orbit—appeared to proceed without a hitch in June 1989. Lt. Col. Ron Rand, public affairs director for Air Force launchings in Florida, described it as "a beautiful launch. You couldn't ask for a better mission."[30]

With one aging and one relatively new KH-11, a brand new LACROSSE in orbit, and another scheduled for launch within a year, the NRO and CIA were in a far better position than they had been in years. With the Titan IV appearing to have easily passed its first test it would not have been surprising if the first Advanced KENNAN was not launched until the Vandenberg Titan IV facility became available in 1991, allowing for a polar orbit. Nor would it have been surprising if one of the two Titan IVs scheduled to be available to launch satellites into low earth orbit from Cape Canaveral in 1989 had been used to place the first Advanced KENNAN into orbit.

But it was surprising when *Aviation Week and Space Technology* reported that the Columbia would carry the follow-on to the KH-11. The public did not know of the plan to establish the five-satellite constellation or that the Titan IV launch did not go as smoothly as reported. An engine problem during the June flight was severe—severe enough that it appeared that it would be seven and a half to ten and a half months before another Titan IV could be launched.[31]

But despite the problems with the Titan IV and the Titan 34D the Columbia launch marked the beginning of the end for Department of Defense and NRO use of the shuttle. In the three years preceding the August 7 mission the Air Force had mothballed the $3.3 billion shuttle launch facility at Vandenberg as well as the shuttle operations control center at the new Consolidated Space Operations Center in Colorado. In 1988 it had terminated the Los Angeles-based Manned Space Flight Engineer Program, with its cadre of 32 astronauts. Established in 1979 by Air Force Under Secretary Hans Mark, the program provided and trained the mission specialists to launch secret military and intelligence payloads. On June 30, 1989 the Air Force Manned Spaceflight Control Squadron in Houston was disbanded. It had 134 personnel at its peak. At Cape Canaveral the Air Force's Firing Room 4, which had been specially protected against Soviet electronic eavesdropping, was also closed down. The launch of the Columbia thus represented the first time there was no military role in control of a military shuttle flight.[32]

Presumably, by 1991 any problems plaguing the Titan IV will be corrected and all low earth orbit missions employing Titan IVs will be launched from Vandenberg into the standard polar orbit, as scheduled. Whether it will be possible to refuel or retrieve the satellites launched from that location is uncertain. It was estimated by NRO Director Aldridge that it would take four to five years before the mothballed Vandenberg shuttle launch facility could be used once a decision to fly from the west coast had been taken.[33]

However, planned upgrades on the Titan IV solid rocket motors may allow the KH-11's successors to at least be launched with a full load of fuel. It is projected that the solid rocket motor upgrade program will allow a 40,000-pound satellite to be launched into a 100 nautical mile circular orbit from Vandenberg (and a 48,300-pound satellite to be placed into an 80 by 95 nautical mile orbit from Cape Canaveral).[34]

However they get into space, and at whatever inclination they operate, the new imaging satellites will continue to transmit their electro-optical data through relay satellites. However, the present relay system, the Satellite Data System spacecraft, is scheduled to be replaced by the mid-1990s by the MILSTAR (Military Strategic and Tactical Relay) satellite network, eventually consisting of seven to ten satellites in both geosynchronous and elliptical orbits. As of late 1989, plans called for launch of the first three Block A satellites into geosynchronous orbit from Cape Canaveral by the end of the 1994 fiscal year, with two Block B satellites being placed in similar orbits the following fiscal year. The first of five launches from Vandenberg to place MILSTARs into elliptical orbits would not take place before 1996. The relay function would be only one small part of MILSTAR's responsibilities, which would include providing worldwide jam-resistant communications between the "National Command Authority" (i.e., the President and Secretary of Defense) and military forces in the field. To give the satellites a maximum chance of survival from hostile action their design has incorporated almost all means of hardening against nuclear blast. To ensure secure communications they would transmit on extremely high frequencies, with the messages often being encrypted, and would have the capability to frequency-hop, antenna-hop and burst transmit.[35]

If a time lapse occurs between the termination of the SDS relay system and the full operations of the MILSTAR network, or the program is cancelled or reduced, the Advanced KENNAN satellites may make use of NASA's Tracking and Data Relay Satellite System.*

Despite TDRSS's ability to transmit huge volumes of data, that capacity may not be sufficient to deal with the extremely high data rates of the spy satellites, especially LACROSSE.[36]

A special system dedicated to controlling and relaying the data for the entire reconnaissance fleet was proposed within the intelligence com-

---

* In August 1989, just as TRW delivered the first MILSTAR, the House Appropriations Committee voted to kill the program. Its ultimate fate is still uncertain.

munity in the mid-1980s. The proposed system, codenamed KODIAK, would have consisted of four geosynchronous satellites, with one satellite in view of the Washington-area downlinks at Fort Belvoir and Fort Meade. That satellite would have been capable of downlinking the information in such a narrow beam that it would be virtually immune to interception. The other three satellites, in addition to having control functions, would be able to receive data from reconnaissance satellites and then transmit it either to the downlink satellite or to another satellite that would then transmit it to the downlink satellite. Limitations in funding led to the proposal being killed in 1987. Funding for KODIAK would have required cuts in the SDI budget that the administration was not willing to make.*[37]

Without a system such as KODIAK the Consolidated Satellite Test Center HQ will continue to play a crucial role in the control of U.S. intelligence satellites. Joining it in exercising control over such satellites will be the Consolidated Space Operations Center (CSOC) in Colorado Springs, Colorado. CSOC was established to share the burden of controlling most military satellites with the Sunnyvale facility.[38] By late 1989 it was exercising primary control over the Global Positioning System satellites and was also involved in control of the Defense Meteorological Satellite Program and Defense Support Program satellites. Involvement in reconnaissance satellite programs will probably begin in 1990 or 1991.[39]

Establishing a second control facility was considered advisable because the Sunnyvale facility is vulnerable to earthquake damage and is within bazooka range of the nearby public highway—a situation which, it was feared, might be exploited by Soviet Spetznaz (special forces) in the event of a limited conflict with the Soviet Union. (Obviously, in a nuclear exchange Sunnyvale would be a major target.) Also, Soviet AGIs routinely patrolled off the coast of nearby San Francisco, intercepting a variety of military and commercial communications. In addition, one Air Force officer told the House Appropriations Committee that "we have experience in the California area . . . in which kooks or terrorists periodically take actions against the public power companies.

---

* Another proposed system that never received approval was a massive imaging/signals intelligence platform, apparently proposed by the CIA's Directorate of Science and Technology. Referred to as "Battlestar Galactica" by its detractors, it was thought to place too many reconnaissance eggs in one basket.

There is a history of anti-nuclear protests and so on in the region and we are concerned about any of those kinds of acts. Whether they were deliberately directed against the DOD or the government in general, it could impact our operation."[40]

The imaging satellites operated by NRO, along with their SIGINT counterparts, will have a variety of targets to keep them busy throughout their lifetimes. Some of those targets will be vestiges of the 40-year nuclear confrontation between the superpowers, and others will be reflections of newly emerged threats. Still others will be the result of changing conditions within countries such as China and the Soviet Union.

The Soviet military establishment will continue to be the top priority of the KH-11, Advanced KENNAN, and LACROSSE spy satellites for the foreseeable future—both because the satellites are particularly suited to monitoring objects like missiles, aircraft, and radars and because the military capabilities of the Soviets will still be of concern to U.S. national security officials.

The Soviet ICBM fields spread out over the Soviet Union will continue to be a prime objective of the satellites' imaging sensors. As of 1988, 20 such fields, located in all areas of the Soviet Union but the forbiddingly cold northeast, held over 1400 missiles. That total includes 380 SS-11, 60 SS-13, 110 SS-17, 308 SS-18, 320 SS-19, 170 SS-25, and about 60 SS-24 missiles.[41]

The images sent back to the analysts at NPIC and other intelligence community interpretation centers will allow analysts to estimate whether the Soviets are constructing new silos (and if so, for what type of missiles), whether they are destroying old missiles, and if new missiles are being deployed (and if so, their dimensions.)

When in 1985 the Soviet Union continued construction work on seven sites for SS-20 medium-range missiles, interpreters were able to determine that they were SS-20 sites because the design patterns were similar to those at known SS-20 bases. A year earlier, interpreters determined that the Strategic Rocket Forces were installing the new SS-25 missiles in Soviet western military districts. Observation of special equipment required by the SS-25, including its silo and very large wheeled transport, was the key.[42]

KH-11 photographs of the SS-25 also helped convince West European defense ministers that the SS-25 was, as the United States had charged, a violation of the SALT II agreement. The provisions of the agreement limited both the United States and USSR. to only one new type of missile. The Soviets had already declared the SS-24 to be their new missile and claimed that the SS-25 was simply a modified version of the SS-13. However, according to West Germany's Defense Minister Manfred Woerner, the photographs showed the SS-25 to be 10 percent longer, 11 percent wider and possessing 92 percent more throw-weight or explosive power than the SS-13. Norway's Anders Sjaastad said he had been dubious before the meeting about the allegations "but the material presented today was very convincing and left no doubts in my mind."[43]

An important aspect of monitoring Soviet ICBM strength will be monitoring of the rail-mobile SS-24 and road-mobile SS-25 missiles, unless those missiles are limited or eliminated as a result of a START agreement. Rather than being able to leisurely revisit fixed sites the KH-11, Advanced KENNAN and LACROSSE will have to locate the missiles and then keep track of them as they are moved from one location to the next.

A secondary component of the Soviet strategic force is its bomber fleet. But however secondary they are, the bombers have a substantial capability to inflict damage on the United States. Spread out across the Soviet Union, the bomber bases constitute targets for a variety of reasons. Imagery provides data on the number of bombers, type of bombers, and the capabilities of the aircraft. Additionally, real-time monitoring of the sites provides indications and warning data, such as indications of a heightened state of alert. Thus, one can confidently assume that Dolon Airfield, a main operating base for Soviet Bear H intercontinental bombers (and probably Blackjack bombers in the future) has frequently found itself a target of the KH-11. It has also been a target of France's commercial SPOT satellite. Rather than reveal what its KH-11 photos showed of Dolon, the Department of Defense purchased a SPOT photo for incorporation into the 1988 version of *Soviet Military Power*. The photo showed jet engine blast scars, aircraft parking aprons, a new parking area, and a main runway.[44]

In addition to watching ICBM fields, monitoring intermediate- and shorter-range missile deployments has also been a major responsibility of the KH-11 and its predecessors. Given the mobile nature of the missiles, the location and monitoring of those missiles has represented a significant effort. On December 8, 1987, Ronald Reagan and Mikhail Gorbachev signed the "Treaty Between the United States of America and the Union of Soviet Socialist Republics on the Elimination of Their Intermediate-Range and Shorter-Range Missiles." The treaty required the United States to eliminate its European-deployed Pershing IB, Pershing II and ground-launched cruise missiles (GLCMs). The Soviet Union was obligated to eliminate all SS-20 (designated RSD-10 and Pioneer by the Soviets), SS-4 (R-12), SS-5 (R-14), SS-12 (OTR-22) and SS-23 (OTR-23) missiles. As of the signing the United States had deployed 429 of its missiles while the Soviet Union had deployed 857 of the missiles covered by the treaty.[45]

The treaty, in a dramatic break with Soviet precedent, allowed for on-site inspection to monitor the destruction of the missiles and associated equipment and compliance with the obligations to produce no new missiles. But satellite surveillance, or "national technical means" as it is called in the treaty text, will remain a prominent part of verifying compliance. The KH-11s, LACROSSE, and Advanced KENNAN will all be involved in ensuring that the missiles are removed, for example, from the SS-20 missile-operating bases at Postavy, Petrikov, and Slutsk and the SS-4 missile operating bases at Vyru and Ostrov. Altogether there are a total of 68 missile-operating, launcher storage, testing, elimination, and launcher-production sites for the spacecraft to glance at as they fly by.[46]

Along with the ICBM fields, the space launch and strategic test centers represent prominent targets for America's secret eyes in space. At Plesetsk, 500 miles northeast of Moscow, is the launch site for most military satellites and the test site for Soviet solid-fuel ICBMs. The Soviet equivalent of Vandenberg AFB, Plesetsk began operating in 1966 and consists of four major launch complexes. The other major launch/test complex is located at Tyuratam, 188 miles southeast of Baikonaur in the central Soviet Union. As the center for all manned launches and many COSMOS satellite launches, and a principal test center for liquid-fueled ICBMs, it is an area of considerable interest to the analysts at CIA, DIA, and the Air Force's Foreign Technology Division who study Soviet space and missile programs. Among the items of interest to those analysts is the Energiya booster, which carried the Soviet space shuttle to orbit in 1988. In the fall of 1987, those analysts were able to observe, courtesy of 5506, the second Energiya heavy-lift booster being erected on its launch pad at Tyuratam, left on the pad for tests, and then removed.[47]

Test facilities that will remain under the watchful eyes of NRO's space assets are the cruise missile and bomber testing center at Vladimirovka in the North Caucasus Military District, and the strategic aircraft test center at Ramenskoye.[48]

Other components of the Soviet strategic forces establishment are its missile development and production, its nuclear weapons production, and its testing facilities. As the KH-11 did, the Advanced KENNAN will periodically send back pictures of the Uranium Enrichment Plant at Troitsk in the Urals Military District, a possible plutonium production facility at Beloyarsk in the Siberian Military District, a reported main production, final assembly, and disassembly facility for nuclear warheads at Cholyabinsk in the Urals Military District, and other similar facilities throughout the Soviet Union. And as made clear by an unclassified CIA study, *The Soviet Weapons Industry: An Overview*, the Dnepropetrovsk Missile Development and Production Center is also a target. The center, according to the CIA study, is "by far the largest missile-producing plant in the world, with more than 2 million square feet of floor space devoted to the major fabrication and final assembly of strategic missiles and space launch vehicles." Above that caption in the study is a drawing of the facility and its large number of buildings and roads from an overhead vantage point—clearly the result of an artist referring to a high-resolution photo.[49]

Nuclear weapons testing facilities are located at six sites, the most important of which is Semipalatinsk in the Central Asian Military District. Semipalatinsk is the primary center for Soviet nuclear weapons tests. Although the main means for evaluating tests are seismic detection and communications intelligence, the KH-11 and its predecessors have played a significant role in alerting the intelligence community to upcoming tests by noting preparations such as the moving of equipment or the drilling of holes where explosives would be placed.

Ronald Reagan's promotion of the Strategic Defense Initiative, popularly known as Star Wars, was for many the first exposure to the world of beam weapons—for example, particle beams. The military laboratories of the superpowers, however, had been exploring the military uses of lasers, particle beams, and other similar exotic weapons for years before Reagan's promotion of SDI. Although the imaging satellites are useless for direct observation of activities inside a building, they are useful for monitoring equipment placed outside research facilities. In 1980, a KH-11 provided imagery of a directed-energy device being constructed at Sary Shagan in Kazakhstan and codenamed TORO. "The argument that once existed between Air Force, Defense and Central Intelligence agencies over whether the Soviets are involved in developing charged-particle beam weapons and their rate of progress is no longer valid," one Pentagon official told *Aviation Week and Space Technology* at the time. Further, "If we were looking for a smoking pistol two years ago, we've got one now . . . we can see the Russian machine taking shape from the overhead stuff" one beam weapons expert told the trade magazine. According to Defense Department officials, the photos showed rows of magneto-explosive generators all lined up behind required shielding with wires leading to an intermediate location, to an electron injector, and then to accelerating modules of a betatron.[50]

Soviet strategic and tactical defensive installations will also continue to be prominent targets for the KEYHOLE satellites. Images of the Moscow Galosh ABM-1B facilities at Novopetrovskoye and Klin, with their interceptors and radars, periodically land on the desks of the photo-interpreters and analysts who monitor Soviet strategic defenses.

Another part of the strategic defense network that will be the subject of considerable attention from the KH-11 and other imaging satellites consists of the Pechora-class phased-array radars on the periphery (and in the case of Krasnoyarsk, the interior) of the Soviet Union. Other radars that elicit interest from the intelligence community and their space eyes are the phased-array radar at Sary Shagan, used for tracking incoming ICBMs during ABM tests, and a long-range OTH radar system near Kiev that faces southeast toward Soviet ICBM fields. Of particular interest are sites such as Skrunda, Baranovichi, and Mukachevo where it is expected that Pechora-class LPARs will replace the older Hen-House radar systems.[51]

In addition to the radars on the periphery of the Soviet Union there is also a collection of naval facilities that are a continuing object of space imaging. The naval facilities include the shipyards at Severodvinsk (Delta and Typhoon submarines), Kerch (surface ships), Nikolayev (surface ships), Gorky (submarines), Zelenodsk (surface ships), Leningrad (submarines), Kaliningrad (surface ships), Komsomolsk (submarines) and Novolitovsk (surface ships).[52]

A 1986 KH-11 photograph of a new submarine at the Komsomolsk shipyard on the Soviet east coast led to a dramatic revision of U.S. estimates of Soviet submarine capabilities. Alerted to the existence of the sub, subsequently named the *Akula*, the Navy had an attack submarine that was waiting to slip into the Sea of Japan, shadow the new sub, and monitor its sea trials. The acoustic signals and other signals collected by the sub were shipped to the Naval Intelligence Support Center for analysis. NISC's analysts concluded that the Soviets were 10 years more advanced than U.S. analysts had believed they were. Unlike the standard noisy Soviet submarine, the Akula was astoundingly quiet.[53]

Of more immediate concern than the activities that will produce future ships are the ports and naval bases from which the present fleet operates. Among the KEYHOLE targets are the Petropavlovsk base of the Far Eastern Military District, where 75 percent of the submarines in the Pacific Fleet berth; Sevastapol, the headquarters of the Black Sea Fleet and its main naval and submarine base; and Vladivostok, the Pacific Fleet headquarters, one of three major naval bases in the Pacific and home port for surface ships, submarines, and two VTOL (vertical takeoff and landing) aircraft carriers. Also closely watched is Iokanga, a northern fleet submarine base and reportedly home for the Typhoon submarines.

Soviet facilities for the collection of signals intelligence, either through the interception of signals or the receipt of data obtained by Soviet signals intelligence satellites, will also come under scrutiny. Satellite photographs will reveal the type of equipment stationed at such facilities as well as any expansion of collection activities. In particular, the KH-11 and its successors will continue to survey the Soviet Union and Eastern Europe in search of satellite dishes devoted to the interception of U.S. and other nations' satellite communications. In a 10-year period beginning in 1979 the KH-11 and other KEYHOLE satellites have returned images of 12 GRU LOW EAR satellite dishes in the Soviet Union, Cuba and Eastern Europe.[54]

Aside from the military targets that constitute the vast majority of KEYHOLE targets in the Soviet Union, there are a variety of additional targets that are worth photographing. Ethnic conflict attracts the attention of the KEYHOLE imaging sensors. Such photography allows CIA analysts to estimate the size of mass demonstrations, levels of destruction from rioting, and the extent to which refugee camps have been established. Thus, when ethnic conflict in Armenia produced large crowds in front of a cathedral in that region's provincial capital, that event was photographed by a KH-11. Such ethnic conflict gives every sign of continuing. In June 1989, ethnic violence in Uzbekistan killed about 100 people and drove thousands into refugee camps.[55]

The KEYHOLE satellites watch developments in Soviet crop production on a more consistent basis. In a speech to a Rotary Club meeting in

December 1988, Congressman Dan Glickman, a member of the House Permanent Select Committee on Intelligence, acknowledged that the KH-11 was used to monitor Soviet crops. He noted, "I frequently receive briefings on grain forecasts which are derived from satellite photographs over the Ukraine."[56]

Similarly, military and other activities in China will remain a major focus of U.S. overhead reconnaissance activities. Lop Nur, China's only nuclear test site and the end-point of missile flight tests from Shuangchengzi, will be one prominent target. Other targets related to China's nuclear weapons, missiles, and space programs include the Bohai Gulf SLBM test launch site south of Huludao, the Changxing missile test center and launch site for IRBM and SLBM tests, the Jinxi Missile Test and Development Center for the test, development, and launch of SLBMs and ICBMs, the new space launch center at Chongqing, and the Lanzhou Gaseous Diffusion Plant. Other prominent targets will include Chinese army, navy, and air force bases, weapons production facilities, and military test centers such as the headquarters of the North Sea Fleet and 1st Submarine flotilla at Quingdao, the Harbin bomber production plant, and the Changcun laser research and development center.[57]

Future political turmoil will also trigger interest from NRO's space eyes and ears. Even before the situation in Tiananmen Square reached its climax on June 4, the situation there was attracting the attention of KH-11s in orbit. Overhead photography could produce some idea of the extent of the violence, the number of people involved in the protest, whether the military or security forces were attempting or preparing to suppress the protests, and what obstacles had been established to block movement by the military and security forces.

At the same time limitations were evident, particularly in attempting to locate the Chinese leadership. As one intelligence official explained, "All of the top leaders have gone underground to their bunkers. Satellites can take pictures of tanks but not individuals who are hiding." It can also be difficult to sort out people or different military factions, according to an intelligence specialist: "It's hard to figure out from overhead pictures which are the people and which are the troops and which tanks belong to whom."[58]

At one time the Sino-Soviet Bloc constituted the great preponderance of KEYHOLE targets. Prior to the operation of the KH-9 in 1971, 90 percent of all KEYHOLE photos were of Sino-Soviet Bloc targets. Plans to photograph targets outside the bloc required approval from the White House. The tremendous area that could be covered by KH-9 photos allowed marginally more attention to be devoted to non-Communist nations, and non-Communist targets jumped to 20 percent. But the KH-11 and its ability to return real-time electro-optical photos brought about a dramatic change in the distribution of targets between the Communist and non-Communist world. By the mid-1980s, fully half of the KH-11 targets were outside the Soviet Union, Eastern Europe, and China for two reasons. For one thing, the real-time capability of the KH-11s increased the value of photos taken during crisis situations or potential hot spots. Such photos could guide policy and action. Second, without the constraint of film capacity, a far greater number of images could be produced, allowing an expansion of the target set from 20,000 potential targets to 42,000 potential targets.[59]

The non-Communist targets are similar in many ways to those in the Communist world. The airfields, naval bases, and troop concentrations of a wide range of nations are of great interest to CIA and DIA analysts responsible for tracking and predicting events in the non-Communist world. The airbases of the highly volatile Middle East represent one set of targets, and the seven major and four minor Israeli airbases are not overlooked by the KH-11s passing overhead. Nor are the airfields of Egypt, Saudi Arabia, Iraq or the other Arab nations of the Middle East. The air defense system, naval bases, and other military facilities of such nations also make prominent KEYHOLE targets.[60] Such photographs establish the order of battle for each nation, identify new equipment whose acquisition has been reported (with varying degrees of confidence through other intelligence sources) and provide a damage assessment in the event of actual conflict.

But beyond providing standard military intelligence, the observation of non-Communist targets is also required by the growing and disturbing trend toward nuclear weapons proliferation, ballistic missile production and deployment, and chemical weapons production. Nations with nuclear programs that have developed, may have developed, or may be on the verge of developing nuclear weapons include Israel, India, Pakistan, and South Africa. Nations that have either expressed interest in developing nuclear weapons or are judged by experts to be candidates to develop the weapons include Argentina, Brazil, Libya, Iraq, Iran, Taiwan and North Korea. The KH-11 and its successors will photograph the power reactors, uranium mining sites, uranium purification, conversion and enrichment sites, reprocessing facilities, and research reactors that make up each nation's nuclear network. Such photography can help analysts trace the build-up of nuclear related facilities and assess how sophisticated a nation's nuclear program may be. Since the early 1980s KEYHOLE imagery has shown the presence of a large research reactor at Yongbon, north of the North Korean capital of Pyongyang. The appearance, in 1989 photos of a second facility, which appears to be a nuclear fuel reprocessing plant near the Yongbon reactor, has increased concerns. Such a facility could extract plutonium from nuclear fuel rods that are reprocessed in the nearby research reactor. Likewise, the Pakistani enrichment facility at Kahuta and the Israeli facility at Dimona as well as Argentinian nuclear facilities have been the subject of repeated attention. Although Pakistan has pledged that it will not develop an atomic bomb, Bush administration officials remain skeptical. "It's a question of whether the glass is half full or half empty," an Administration official said. "There's been some progress. . . . The jury is still out."[61]

The KEYHOLE satellites will also peer down in attempts to locate possible construction of nuclear test sites and to monitor the activity at known test sites. If a U.S. military attaché reports on rumors of construction of a test site in India or Pakistan, or if the Brazilian press alleges that a nuclear test site is being built on a Brazilian military reservation in Cachimbo, that location will be subject to the closest scrutiny as soon as a KEYHOLE satellite can be programmed to photograph the alleged test facility.[62]

Of mounting concern among U.S., Western, and Soviet officials is the development or procurement of ballistic missiles by a growing number of Third World nations. In a March 1989 speech in Los Angeles, DCI William Webster predicted that by the year 2000, 15 developing nations would be producing their own ballistic missiles.[63]

Such developments are troubling because the development of such missiles, combined with nuclear weapon programs, makes it easier for nations to inflict nuclear devastation not only on their immediate neighbors but on nations a considerable distance away. Thus, revelations that Iraq has a strategic missile program along with a nuclear weapons program gives national security and intelligence officials one more headache.[64]

Other nations heavily involved in ballistic missile development have been Israel, South Africa, India, and Argentina. Israel flight-tested the

Jericho IIB to a distance of 500 miles in May 1987. One report, relying on U.S. government sources, stated that the missile would eventually have a 900-mile range—sufficient to reach Riyadh, Baghdad, Benghazi, and the southern Soviet Union. A subsequent report indicated that South Africa tested a new intermediate-range ballistic missile developed with Israeli aid. The IRBM, a modified version of Israel's Jericho II, was detected at Arniston, a test range near Capetown.[65]

On May 22, 1989, a group of Indian officials watched as India successfully tested the medium-range ballistic missile AGNI, launching it from the newly-developed Balasore test range. One top Indian official noted that "being able to hit a target 1500 miles away with precision has certain advantages." Such a missile would allow India to threaten Pakistan and many key Chinese industrial and population centers with nuclear devastation. Though the Indian official claimed that the missile, if developed further, would be a conventional weapons system, DCI William Webster told a Senate committee a week earlier that the CIA had found "indicators" that India was interested in developing a thermonuclear weapons capability.[66]

The Argentinian program has also been of great concern to U.S. intelligence analysts, and KEYHOLE photography has provided early warning of plans to test the missile. A KH-11 photo showed a CONDOR missile on its launch pad, but by the time two more passes were made, the missile had been removed.[67]

It has become clear that more attention would need to be devoted to spying from above on nations that might desire to deploy such missiles. In 1988, the United States discovered that Saudi Arabia had purchased and deployed Chinese DFA-3 ballistic missiles (designated CSS-2 by the U.S. intelligence community), missiles that have a range of over 1000 miles. The purchase was arranged by Prince Bandar bin Sultan, the Saudi ambassador to the United States. Prince Bandar visited Beijing in 1986, although Saudi Arabia does not have diplomatic relations with the People's Republic of China, under the pretext that he was trying to persuade it to stop selling arms to Iran. Carrying the deception further, the Saudis told U.S. officials that they had "offered" to compensate the Chinese for the loss of arms sales to Iran by purchasing the same weapons, including Silkworm missiles, for Iraq. In addition, the Saudis claimed they offered to have the Iraq-bound missiles shipped through Saudi territory to protect them from Iranian attack.[68]

The Chinese shipped the CSS-2 missiles along with those destined for Iraq. After both sets of missiles were counted leaving Saudi ports, those destined for the kingdom were trucked south into the great Saudi desert, where, the Saudis told the United States, they were building an ammunition depot that they wanted to keep far from their populated areas for security reasons. Actually, the depot was a training and storage area for the Chinese missiles. The Saudi cover was blown in January 1988 when the U.S. discovered that trucks carrying some of the presumed Iraq-bound missiles were traveling south, rather than north, from Saudi ports.[69] If the U.S. reconnaissance arsenal had been at full strength and more closely monitoring developments in Saudi Arabia the Saudi ruse might have been discovered earlier.

The production of chemical weapons is a third area of concern, leading the KH-11, and undoubtedly its successors, to target suspected facilities. As already noted, KH-11 imagery was instrumental in the identification of and diplomatic action taken concerning the Libyan Rabta facility. Also, according to one press account, in late 1982 a probable chemical weapons production and storage facility was identified (probably via satellite photography) at the Dimona Sensitive Storage Area in the Negev Desert. The Iraqi use of chemical weapons during its prolonged war with Iran has raised even more concerns.[70]

Third World internal and external conflicts have also become prominent targets for the KEYHOLE satellites. As expected, in 1986 America's lone KH-11 did not ignore the Iran–Iraq battlefield when it passed overhead. Its photographs showed Iranian forces moving across the Shatt al Arab by boat and over pontoon bridges into the bridgehead in the Fau Peninsula. The observations supported Iranian claims that their troops had pushed west from Fau and had reached the eastern shore of the Khauh Abdallah waterway opposite Kuwaiti territory. The photos also showed that the Iranians were still a significant distance away from the Iraqi naval base at Umn Qasr on the Gulf.[71]

The KH-11 was also busy examining the conflict in Cambodia. In 1985, KH-11 photographs showed that the Kampuchean resistance forces were pushing deeper into the Cambodian interior and were successful in fighting the Vietnamese. The photos also confirmed reports of the resistance radio concerning attacks on the bases and main transportation lines of the Vietnamese troops.[72]

Internal Burmese events in 1988 were also considered sufficiently important to make that country a KEYHOLE target. Dissatisfaction with the autocratic, xenophobic, and socialist government of long-time strongman Ne Win led to mass demonstrations in Mandalay. The KH-11s were programmed to provide photos of those demonstrations as they passed silently overhead.[73]

The KH-11 and other imaging satellites will also continue to periodically photograph targets within the continental United States. In a 1980 court case, CIA Deputy Director for Science and Technology Leslie Dirks specified the reasons for such photography.

> First, reconnaissance photography is taken of targets in the United States the foreign counterparts of which are of substantive intelligence interest. By comparing overhead photographs of known U.S. targets with their foreign counterparts, [the CIA] is able to make sophisticated and knowledgeable determinations regarding foreign targets. Second, reconnaissance photography is taken of domestic locations in order to determine the performance parameters of collection systems. By analyzing such photography [the CIA] is able to calculate the resolution of photographs generated by a system and to assess performance parameters.[74]

So in addition to sending back photographs of Chinese, Soviet, and other naval bases, radar sites, and missile fields, the KEYHOLE satellites will periodically be programmed to send back pictures of a naval base in Hawaii, a radar site in Texas, or a missile field in Missouri. Photo-interpreters can then determine what certain components of a military facility look like from an overhead view and be better able to recognize them in a photo of a foreign facility. This is particularly important when new types of structures, such as fiber-optics factories, are introduced.

Dirks also indicated that U.S. targets can be used to assess performance parameters such as resolution. Such calibration missions are flown against a variety of U.S. targets, including some unlikely targets. Richard Bissell recalls a 1960s satellite photograph of a football game in Butte, Montana in which the players, but not the ball, were visible, as well as an early 1970s photograph of the U.S. capital. More recently, KEYHOLE satellites were programmed to take photos of a night football game at RFK stadium in Washington to test the satellite's ability to return imagery under low light level conditions. In another instance Bissell recalls an early 1970s photograph of the U.S. capital, in which cars (but not their make) could be distinguished.[75] The precise dimensions of the targets can be independently determined, which is obviously not the case with foreign military installations, and the atmospheric, climatic, and meteorological conditions that affect system performance can also be more precisely determined.

Some KEYHOLE targets in the United States have become targets at the request of civilian, non-national security-related agencies. As soon as the U-2 operation began in 1956, some of those involved realized the value of overhead photography for civilian purposes. At that time Arthur Lundahl went to the president and the National Security Adviser and informed them of the potential. In 1967 after a formal DCI study, the DCI approved agreements with civilian agencies to share information produced by military aircraft and satellites.[76]

The CIA has routinely turned over satellite photos to officials of agencies such as the Environmental Protection Agency, provided that the civilians who are granted access to the pictures have Top Secret security clearances and vow not to discuss the pictures with anyone who is not cleared. These photos can be instrumental in assessing damage from oil spills, hurricanes, and tornados, conducting national forest inventories, forecasting snow runoff, and early detection of crop disease. Photos are handed over by bilateral arrangements at meetings of the obscure Committee for Civil Applications of Classified Overhead Photography of the U.S. The Committee, formed in 1976 at the suggestion of the Rockefeller Commission to prevent the abuse of classified military photography, consists of middle-level civil servants from seven agencies, and some military people.[77]

Since 1967 the Committee on Imagery Requirements and Exploitation (COMIREX) has sorted through the varied requests of the intelligence community, military, and civilian agencies for satellite photography. Even with the advent of electro-optical imaging, priorities still must be set, for an imaging satellite can only look at one target at a time, unlike the signals intelligence satellites that can gather varied signals simultaneously. The major function of COMIREX, operating from Intelligence Community Staff headquarters in downtown Washington, is to determine which targets get imaged, how often, by which systems (e.g. the KH-11 or LACROSSE), and when.

The varied functions of COMIREX have been summarized by former COMIREX chairman Roland Inlow:

> COMIREX performs the interagency coordination and management functions needed to direct photographic satellite reconnaissance, including the process of deciding what targets should be photographed and what agencies should get which photos to analyze. It also evaluates the needs for, and results from, photographic reconnaissance, and oversees security controls that are designed to protect photography and information derived from photography from unauthorized disclosure.[78]

At the trial of Samuel Loring Morison, Inlow elaborated on the need for COMIREX:

> COMIREX is essentially a coordinating activity. The way to think about it is that the intelligence community consists of a number of organizations—the Central Intelligence Agency, the Defense Intelligence Agency, the intelligence components of the Army, Navy, Air Force, State Department and so on. As you would appreciate, these organizations have a different view of needs at any particular time; and they each have their own priorities for the kind of information that they would like to have collected. It is necessary to have some mechanism to coordinate all of those needs and assign them to, in this case, satellite collection. A satellite can only do one thing at a time essentially. COMIREX activities . . . funnel all of these requirements for information together, coordinate them and then set the priorities for collection.[79]

With regard to the distribution of the data Inlow went on to explain that

> once the imagery is collected then the way to think about it is that information goes out in kind of a fan back out to all of these organizations. There is also a coordinating aspect to that in the sense that it is, you know, simply inefficient to have every organization in the intelligence community attempting

to analyze and evaluate and extract information from every image that is collected. So, there is a coordination process there which essentially attempts to divide up that activity in a rational way. COMIREX is at the middle of the coordinating process where the needs for information comes in. The guidance on the ordering of exploitation goes out the other way. In the course of that it requires a continual review of the needs for information which the United States government has in the intelligence field. Knowledge of the technology that is available or emerging in a collection sense that could be applied to that, and then the formulation of recommendations to the government concerning new activities or new programs that seem to make sense in that connection[sic].[80]

If the Air Force desires a photograph of a new Soviet bomber, such as the BLACKJACK, it instructs its COMIREX representative, who is a member of the Air Force Office of the Assistant Chief of Staff for Intelligence, to submit a request for satellite coverage. The Air Force representative will specify not only what the Air Force wants photographed, but where the target is likely to be found (in the case of the BLACK-JACK, Ramenskoye test center) and why the information is needed.

Such requests may be debated, along with requests by the other agencies, at COMIREX meetings. Although disagreements may at times be serious, according to several of those who have been involved in such meetings, they never become acrimonious. Disagreements are settled amicably, without table pounding. Former NPIC director Arthur Lundahl did remember some whining: "the Navy would whine in the wings, 'We've got to know all about this submarine at this or that particular place.' And I said, 'Listen, you have to make your case, you have to set it all down on paper as to why it's important and they would have lay side by side with other competition for coverage, and if indeed your case is right, it will get covered first. If it isn't right, it might get covered later or it might never get covered at all.'"[81]

The main conflict over KEYHOLE targeting and priorities is between those consumers with national intelligence responsibilities (e.g. the CIA) and the parts of the military more concerned with tactical intelligence. The military, with its mission of being able to fight a war, wants as much warning as possible plus up-to-date and detailed coverage on Warsaw Pact capabilities. Day-to-day coverage is seen as valuable on a tactical level for revealing movements of troops or weapons and small changes in capabilities; at the national level, in the absence of a crisis it is of little interest. COMIREX serves as a means of prioritizing claims of the CIA, DIA, military services, and other consumers, attempting to distribute a strictly limited resource in such a way as to at least minimally satisfy the legitimate requests of several competitive bureaucracies.

The conflict over strategic versus tactical intelligence has become greater in recent years, partly due to the deployment of the KH-11. Its real-time capability has meant that military commanders have a potential means of receiving information about military developments by opposing forces in short periods of time, which enormously increases the potential for early warning. Former Marine Corps Director of Intelligence Lt. Gen. Harry T. Hagaman noted that "We opened that magic door . . . and many eyes were opened to what was actually out there."[82] The demand for such information has led to a host of programs to see that such "national assets" as the KH-11 are exploited for tactical purposes. Most prominent is the Tactical Exploitation of National Space Capabilities (TENCAP) program. Other such programs include National Intelligence Systems to Support Tactical Requirements (NISSTR), the Defense Reconnaissance Support Program (DRSP), the Joint Tactical Fusion Program, and Fleet Imagery Support Terminals.

Although the principals on COMIREX may debate these issues from time to time along with other basic issues it is the COMIREX subcommittees and working groups that do the bulk of the work. Not only does the volume of work require a significant supporting staff, but the real-time capabilities of the KH-11 no longer allow for leisurely discussions of targeting priorities. Opportunities may come and go in a brief period of time and only a full-time staff will allow the United States to seize those opportunities.

COMIREX components include the Imagery Collection Requirements Subcommittee, the Operations Subcommittee (OPSCOM) that is responsible for day-to-day operations, the Exploitation Research and Development Subcommittee (EXRAND), the Imagery Interpretation Keys Subcommittee (KEYSCOM), the Current and Standing Requirements Working Group, the TK Modification Working Group, the Imagery Planning Subcommittee (with its Exploitation Program Working Group), the Mapping, Charting and Geodesy Subcommittee, and the COMIREX Automated Management System (CAMS) Task Force.[83]

The COMIREX Automated Management System was the creation of Roland Inlow. CAMS was developed in the belief that statistical and operations research techniques could be used to aid the target selection process. Since the decision to photograph a target at a given moment limits the targets that can be photographed afterwards, a system was needed that considers the various requirements and constraints in developing a reconnaissance schedule. CAMS was developed to take into account the location of requested targets, the slant angle desired, the altitude necessary to produce the required level of detail, the general priority assigned to the target, and the weather conditions. With such data the CAMS algorithm will produce an attainable schedule of photography.[84]

The general priority assigned to a target may vary substantially. A crisis may move a target like Tiananmen Square up the list substantially. Other less dramatic reasons may also explain why particular targets are photographed at a particular time. As Inlow testified, "there are a host of reasons for taking a photograph on any particular day. It may be that, as simple as it's time to take some more photographs of an installation that is important and you wish to have additional information. It can be for some important special reason. There's an analyst somewhere who is doing some analytical work and that person wishes to have updated current information."[85]

Estimates of weather conditions fed into CAMS are only derived from raw Air Force weather satellite data if that data is received within approximately ten minutes of the KEYHOLE satellite's scheduled pass. Otherwise, they are derived from a sophisticated weather model produced by Air Force Global Weather Central at Offut Air Force Base and are based on over 20 years of data. DMSP and other meteorological systems feed data into the model to further enhance its capabilities.[86]

Once the list of targets is worked out and priorities developed, COMIREX consults with NRO, whose staff calculates the desired orbital parameters and determines when and where to turn the optical system on and off.[87] They are likely to be quite busy in the future.

CHAPTER 11

Still Secret
After All
These Years

By 1987, Congressman George Brown Jr. had been a member of the House of Representatives for more than 20 years. Unlike so many of his fellow colleagues, Brown was not a lawyer: he holds an undergraduate degree in Industrial Physics and has completed graduate work in physics, political science, and industrial relations. Much of his committee work was oriented toward space, technology, and agriculture. Among other assignments, he served as ranking Democratic member of both the House Science, Space, and Technology Committee and its Subcommittee on Space, Science, and Applications.

Congressman Brown was also a member of the House Permanent Select Committee on Intelligence and was particularly interested in the space reconnaissance activities of the intelligence community. It was not just that he considered such activities important to national security; he also believed that satellite photography from the KH-11 and future systems should be used more widely in support of activities such as crop estimation, urban and regional planning, and environmental change monitoring.[1]

To effectively use the KH-11 and its successors for such purposes would require a substantial loosening of the secrecy attached to such systems and to the photography they produce. Congressman Brown pursued that idea, not only in the Intelligence Committee's usual executive sessions but also in public speeches on the floor of the House. That is when he began to upset the intelligence community, or at least those responsible for the security of the space reconnaissance program. The program's security guardians did not like it when members of the executive or legislative branches stepped outside the narrow limits of acceptable speech on the subject of overhead reconnaissance. It did not matter if something was widely reported or even if there were clear indications that the Soviet Union understood that aspect of U.S. reconnaissance capabilities. In 1987 it was only permissible to acknowledge the existence of photographic satellites, but the Congressman's speeches went beyond that. Relying on numerous public sources, he discussed the KH-11 and the NRO. In a February speech on the floor of Congress, Brown noted that the United States was currently operating only one KH-11 and that the "KH-12" would be launched on the second space shuttle mission once such missions resumed. Later that spring he compared the KH-8, KH-9 and KH-11 spacecraft and noted that the name NRO was officially classified although "anyone can read about [it] in various unclassified articles, reports and books."[2]

Before long a letter arrived for the Chairman of the House Permanent Select Committee on Intelligence, Louis Stokes, with copies also sent to the House leadership. The letter was classified Secret. Indeed, the paper on which it was written was itself classified Secret, because it bore the NRO letterhead.[3] Signed by the then Director of NRO, Edward Aldridge, the letter expressed concern over the statements of Congressman Brown, reminded the chairman of the sensitivity of information concerning space reconnaissance and the responsibility of committee members to protect such information, and asked that Congressman Brown be instructed to restrict such comments to closed sessions of the Intelligence Committee.

Brown understood the intelligence community's point of view: "They were trying to maintain this culture which says that . . . this is an intelligence matter and the intelligence community has briefed the committee about [it so] even if it is available in the open literature you're not supposed to talk about it." He just didn't agree, believing that "It's one of the most irrational things I've ever seen but it doesn't keep them from getting very emotional about it."[4]

Faced with strong intelligence community opposition to his statements and a lack of desire by the House Intelligence Committee to get involved in "pissing matches," Brown decided to resign. His resignation allowed him to speak freely and he began doing so in the statement he issued to explain his resignation. Brown claimed that "in the civilian sector, we are letting our capabilities completely collapse, while in the classified realm, we are operating the best systems money can buy. The intelligence community doesn't want me bringing attention to this contradiction, yet I feel it is of national importance because the U.S. is losing its competitive position in a technology which it created." Referring to the Soviet acquisition of the KH-11 manual in 1978, Brown said, "I would guess they have read the manual by now, but as a member of the Intelligence Committee, I'm not supposed to mention the KH-11's existence. This is ridiculous and I'm tired of it."[5]

In addition to his public statement, Brown left the committee with a long classified memorandum in which he spelled out the reasons he thought significant declassification was necessary. Among the reasons were the need to preserve the integrity of the classification system and the benefits to commercial and civilian satellite interests.[6]

The pervasive secrecy that has surrounded the KEYHOLE program is caused by a variety of factors.[7] One is the belief that such pervasive secrecy is needed to preserve the "essential integrity" of the system. The KEYHOLE satellites, it is argued, provide crucial intelligence on the Soviet Union, China, and other areas of the world and make it possible for the United States to monitor arms control compliance. Revealing details of that capability would make it easier for the targets of U.S. surveillance, whether adversary or ally, to engage in denial and deception activities.

This view was exemplified by testimony during a Freedom of Information Act case in 1980 when then CIA Deputy Director for Science and Technology Leslie Dirks claimed that "damage to the national defense or the foreign relations of the United States" could result from:

(a) the disclosure of performance capabilities of overhead intelligence collection systems; or

(b) the disclosure of the fact a particular photo-reconnaissance satellite system was operational at a particular point in time; or

(c) the disclosure of the types of targets against which overhead photo-reconnaissance was [operating], or

(d) the disclosure of photo-reconnaissance satellite coverage of foreign nations.[8]

Much of this information, Dirks claimed in court testimony, could be determined from the release of even a single photograph. In addition, release of a single photograph, according to Dirks, would allow a foreign nation to determine that a particular satellite was employed for photo-reconnaissance purposes.[9]

A second long-standing argument for tight secrecy has been the sensibilities of other nations to U.S. reconnaissance activities. As explained in Chapter Three, the atmosphere that existed when the first reconnaissance satellites were launched was extremely threatening. The Soviets had just shot down Francis Gary Powers and his U-2 and were threatening to do the same to DISCOVERER/CORONA and SAMOS. One way to reduce the chances of this happening was to not humiliate the Soviets by announcing to the world that the U.S. was regularly photographing secret Soviet military installations.

Since that time the potential reaction of other nations to the revelation that U.S. satellites were photographing their secret installations has been added to the rationale. In his memoirs, former CIA Director William Colby noted that a prime obstacle to declassification was the "diplomatic objection that other nations would create difficulties if they were compelled to admit that many of their tightly-protected secrets were in fact not secret at all."[10]

National security concerns are not all that keeps the guardians of the security of the space reconnaissance program from loosening their grip. There are a variety of bureaucratic interests. As discussed earlier, Pentagon lawyers have repeatedly dissuaded senior officials from releasing any photographs on the grounds that releasing even one satellite photograph would make the entire holdings of NPIC vulnerable to Freedom of Information requests. Although numerous reasons to deny specific requests for such photography could be offered, often legitimate reasons, it would still be necessary to search for the photos requested, evaluate their sensitivity, and then deny access. Such a process is clearly far more time-consuming and costly than a policy which considers all KEY-HOLE photography to be classified and requires no search and review. Since NRO and the CIA see no benefit in having to devote resources to answering FOIA requests, a policy of total non-release suits their bureaucratic interests just fine. And since it is still the institutional opinion that the release of any satellite photos could only compromise "sources and methods," this reinforces the bureaucratic imperative.

Thus, despite President Carter's initial inclination to release KEY-HOLE photographs and the support of key advisers for such an option, he was dissuaded from doing so. President Ronald Reagan's use of aerial photographs to illustrate Soviet military activities in the Caribbean represented a compromise among his national security advisers concerning the disclosure of overhead photography. Reagan and national security adviser William P. Clark considered declassifying photographs taken by KEYHOLE satellites that showed new military installations and armament factories inside the Soviet Union. They too were dissuaded from using the photos on the grounds that it would end the long-standing policy of not declassifying such photos.[11]

To maintain such a policy, facilities that had been photographed by a KEYHOLE satellite would be rephotographed by U-2 or SR-71 reconnaissance planes if it were desired to make photographs of such facilities public. As noted earlier, this was done during the Carter administration, according to Bobby Inman, to allow photographic evidence of Soviet naval installations in Somalia to be made public. In the Reagan administration it was the dreaded Nicaragua that occasioned duplicative aerial overflights.[12]

Another rationale for tight secrecy that combines both bureaucratic and national security concerns is the general argument expressed by a Defense Department official that "once we start answering questions and opening doors, where do we stop?"[13] For simple bureaucratic reasons, the prospect of more detailed questions and probes into the affairs of NRO and the reconnaissance program is considered undesirable by many national security officials. Few government officials like to have the public closely scrutinizing their activities. If they have managed to avoid it for almost 30 years, they see no reason to change the situation now.

But there is also a national security component to the opposition to more openness. Even the most die-hard people in NRO realize that there are numerous details about reconnaissance activities that are well-known and would be quite harmless to officially declassify, such as the existence of NRO or some details of the CORONA program. However, it is feared that making such information public would make it easier for others to ferret out the truly sensitive information. The theory is that to protect the important information, it is best to wrap it in a buffer of trivial information. When it is harder to find out even trivial details, it becomes that much harder to get to the family jewels.

Indeed, it is really this last rationale that allows the pervasive secrecy to continue even as other arguments lose their validity with respect to all or parts of the KEYHOLE program. There may often be reasons for keeping secret some aspects of current reconnaissance programs. According to Bobby Inman, there is good evidence that the Soviets do not always fully understand the capabilities of new systems. "It's in that period that you get your most valuable data," says the former Deputy Director of Central Intelligence. And while Soviet ignorance does not last long, Inman points out that there are other countries not as sophisticated as the Soviets. Disclosures of advance capabilities may risk depriving the United States of information that could be critical in times of crisis.[14]

But the concept of protecting present advanced capabilities does not justify the continued secrecy about basic capabilities of programs that ended 15 to 25 years ago. It is difficult to see how national security is protected by continued secrecy concerning the SAMOS, ARGON, or LANYARD programs, which were terminated over 25 years ago. Likewise, the revelation of details concerning the CORONA program, which concluded 17 years ago, is hardly likely to damage national security. It seems particularly strange to treat as Top Secret information concerning satellites (e.g., SAMOS, early CORONA) that were no more capable than the present SPOT commercial observation satellites. In 1987 the Soviet Union began selling satellite photography of locations outside of the Soviet Union, Eastern Europe, China, and Cuba to anyone willing to pay the price. The best of the photographs had five-meter resolution and were produced by a camera system known as the KFA-1000, carried on board some Soviet reconnaissance satellites.[15] In light of such developments, "protecting" information about early reconnaissance programs serves no purpose. What is being protected is the concept that reconnaissance is not a subject for official discussion and the public should remain in ignorance.

Leslie Dirks' claims for the necessity of extreme secrecy do not withstand close scrutiny either. To claim that the disclosure of the types of targets against which KEYHOLE satellites are used would damage national defense is taking a valid point and stretching it to the extreme. Such revelations may be damaging, for example, if the nation believes the U.S. is unaware of a facility's existence, but one need only turn to appendix A of a volume called *Nuclear Battlefields* to find a list of well over 100 Soviet nuclear-related installations that have clearly been the subject of KEYHOLE photography. Likewise, the installations in the Soviet Union, of which SPOT photographs appeared in the 1988 version of *Soviet Military Power*, were certainly first viewed by the KH-11 or its predecessors.[16]

Outside the Soviet Union, there are also a variety of targets that quite clearly are the targets of the KH-11. To "reveal" that the Iran–Iraq battlefield, Kahuta, Dimona or known missile test sites are subject to periodic surveillance is to reveal nothing at all. If such sites were not subject to surveillance, that would be news.

Whether the release of any satellite photos would automatically provide significant information to nations that might wish to avoid KEYHOLE coverage was a major issue in the trial of Samuel Loring Morison. Prosecution witnesses claimed that it would reveal if the satellite was operational and if it had picked up real or deceptive objects. The CIA's Deputy Director for Science and Technology, Richard Evans Hineman, testified that the photos provided by Morison would indicate the slant range at which the KH-11 could photograph a target and that it could see into the shadows. The former head of NPIC and NIO for Foreign Intelligence Activities and Denial, General Parker "Hap" Hazzard, asserted that the photographs would possibly allow the Soviets to determine if they had successfully hidden some object they attempted to hide or the spectral response of the satellite to water.[17]

That view was disputed by former COMIREX chairman Roland Inlow. Under questioning by defense attorney Mark Lynch, Inlow explained that in his view, damage could come in three ways: by revealing to the Soviets that satellites were watching them; by revealing the technological capabilities of the satellite; and from the information content of the photographs. On all three counts Inlow concluded the damage was zero. The Soviets clearly knew of the satellite program [and, as was pointed out by other defense witnesses, would have to assume that any satellite was operational] and that they were building an aircraft carrier.[18]

As for the satellite technology, Inlow noted that Kampiles had provided the manual that explained the KH-11's basic technology. Any changes that might have improved the KH-11's resolution were not disclosed by the photos, for the photos that the Soviets had were those in *Jane's* — six or seven generations removed from the originals.[19]

The argument that the United States cannot officially release information concerning the KEYHOLE program, especially not photographs, because of foreign sensitivities had a great deal of merit in 1960 in the aftermath of the U-2 incident and in the presence of continued Soviet threats against Discoverer and SAMOS. But the Soviets long ago dropped their objection to satellite surveillance. Once they fully developed their own systems and saw that their diatribes against U.S. satellite surveillance were producing no results, they ceased complaining.

President Carter's acknowledgment of U.S. photo-reconnaissance satellites produced no paroxysms of protest. According to one Yugoslav reporter, "the world press and other so-called mass media have published the recent news from Cape Canaveral . . . inconspicuously and obscurely and almost without commentary—if they published it at all. In some places it was even accidentally thrown into the wastepaper basket."[20]

Further, it is clear that the United States, Soviet Union, and China are not going to be the only participants in space reconnaissance, particularly photo-reconnaissance. France has already produced the SPOT commercial satellite, which has been used to acquire photographs of a variety of military installations, and these photos have been published. In addition, France, along with Italy and Spain, is developing the HELIOS military reconnaissance satellite. Israel has placed an experimental satellite in orbit and is clearly on its way to deploying a photo-reconnaissance satellite. India will probably be next.[21] Nobody is complaining.

It should also be noted that the United States has employed SR-71 and U-2 aircraft to overfly numerous foreign countries for intelligence purposes, an act which is clearly illegal under international law. And while some nations complain or try to shoot the planes down, the United States has not stopped.

The third argument, that the release of a single photo would open up all satellite imagery to the Freedom of Information Act is one that is never made in public. It would be impolitic for government officials to suggest that all photographs were being classified not to protect the nation's security but to protect the government from having to comply with the law. Whether it would in fact be burdensome cannot be known until some imagery is declassified. It does not appear that release of U-2 or SR-71 photography has resulted in a torrent of requests for other photos or a burdensome situation for the CIA. Of course, even if it would be burdensome that is hardly a justification for circumventing the law.

Even the seemingly invulnerable argument that keeping all information classified is the best means of protecting the truly important secrets is subject to challenge. For it is not clear that the best means of protecting secrets is to classify, classify, classify. Numerous observers, including government panels, have noted that overclassification serves to devalue secrecy. Prolonged exposure to information that is clearly overclassified only serves to cast doubt on the need to protect any particular item of information. Further, the wider the net of secrecy, the more difficult it is to physically protect and keep track of truly important secret documents such as KH-11 manuals.

The associated justification that any revelations would be followed by more probing questions and demands for information is hardly complimentary to those who presently control the information. Apparently they believe themselves to be so weak-willed that once they divulge any information they would only have to be asked before they would divulge the deepest, blackest secrets of the NRO.

The rationale for declassification of much information concerning the KEYHOLE program goes far beyond the fact that the justifications for absolute secrecy are flawed. One rationale for greater openness is based on a concern for history and the informed discussion of public policy. History may seem like a trivial concern to some, but it is the basis for understanding the past and acting in the future. The history of the KEYHOLE program is a history of one of the most significant military technological developments of this century and perhaps in all history. Indeed, its impact on post-war international affairs is probably second only to that of the atom bomb. The photo-reconnaissance satellite, by dampening fears of what weapons the other superpower had available and whether military action was imminent, has played an enormous role in stabilizing the superpower relationship.

The inability of scholars to explore in detail the technological developments and role of the KEYHOLE satellites in a variety of situations limits our understanding of past and present international events. And though documents concerning such events may someday be declassified, present-day secrecy can severely limit access to the individuals involved in such events.* And those individuals are not immortal. Many of those

---

* In 1988, Merton Davies and William R. Harris were able to publish an account of RAND's early involvement in the development of reconnaissance satellites: *RAND's Role in the Evolution of Balloon and Satellite Observation Systems and Related U.S. Space Technology.* The volume reveals much new information concerning the 1946–1958 period, but does not include any new information on CORONA or other operational programs.

involved in the early programs have passed away. By the year 2000 several more pioneers will also be gone.

An informed public debate concerning important public policy issues also suffers from the present secrecy. Some understanding of the capabilities of the KEYHOLE (as well as the SIGINT) satellites is necessary to reasonably debate issues like arms control verification or to competently argue whether adequate steps have been taken to procure a sufficiently robust strategic reconnaissance capability.

Also limited by the secrecy surrounding KEYHOLE and other reconnaissance systems is the discussion of international events and foreign government activities. Some of those who advocate declassification of KEYHOLE photography have suggested that it would allow the public to make a more informed judgment about aspects of U.S. defense policy. In 1985, former NRO Director Hans Mark wrote that "I believe that the American people should be informed about reconnaissance systems. It would make it much easier for our political leaders to justify a number of important military and foreign policy initiatives if people really knew what our adversaries around the world are doing."[22]

Similarly, when General Bernard W. Rogers, Supreme Allied Commander for Europe, was asked during a 1984 Congressional hearing "What more can we do to get the true message of the [Warsaw Pact] threat across to our people, both in Europe and here?" he called for the immediate release "of some overhead photography that shows the threat." Not surprisingly, Rogers' efforts were frustrated. In a 1984 interview he noted that he had been trying to win release of photography since 1980 but that "the damn intelligence community won every time."[23]

The atmosphere between the U.S. and Soviet Union has changed dramatically since 1984 and 1985, but the notion of supplying more information about certain international events remains valid. Future KEYHOLE photography can serve to reassure the public that the Soviet Union is complying with terms of arms control agreements or can make it clear that it is not. They can show Soviet Bison bombers with tails cut off in compliance with SALT II (as in 1985) or they can show the presence of the Krasnoyarsk radar. New threats may be more easily combatted with public and international support gained by the release of satellite photography. The perilous situation that is emerging with regard to proliferation in nuclear weapons, chemical weapons, and ballistic missiles certainly cannot be solved via the use of KEYHOLE imagery. It can, however, be more effectively addressed in public and international forums. Such programs flourish in secrecy and are retarded by publicity. As was demonstrated during the Cuban missile crisis, photographs cannot be easily ignored. U.S. Ambassador to the U.N. Adlai Stevenson not only claimed that the Soviets had placed offensive missiles in Cuba, but proved it with aerial photos of the missile sites that presented dramatic evidence of the Soviet actions. Governments inclined to turn a blind eye find it harder to do so when confronted with dramatic evidence.

In addition, the official secrecy concerning the existence of the National Reconnaissance Office has no justification. Its existence has been widely reported: it has been mentioned in a Senate report, in the memoirs of a former Director of Central Intelligence, and in an oral history interview of a former NRO official. Given the admission that the U.S. conducts overhead reconnaissance, there is no rationale for maintaining the existence of NRO as a secret. Thus, Bobby Inman, carefully avoiding use of the term NRO, has "never understood or concurred in [the] rationale that [the management structure] needs to be classified.[24]

The practice of establishing agencies that do secret work and whose very existence is secret is one that should be limited to very special circumstances. That practice is hardly consistent with the notion of open and democratic government. Special circumstances do exist when the activity that an organization engages in must be unacknowledged, either because it is imperative that an adversary not know that such work is being undertaken or because the activity cannot be acknowledged in even the most general sense. The circumstances under which the Manhattan Project was established constitutes one example of such special circumstances, and the existence of NRO in 1960 may arguably be another. The existence of NRO in 1990 is not.

Beyond allowing for a more informed public, release of some photography or information derived from the photography would allow use of the satellites for broader purposes. Secrecy has decreased somewhat within the national security bureaucracy. The first progress in this regard was made in the early 1970s when some of the photography was downgraded from TOP SECRET RUFF to SECRET. This change was initiated by DCI William Colby after an early 1970s tip by General Daniel Graham, then Colby's deputy for intelligence community affairs. Graham discovered that target folders shown to bomber pilots had sketches of the targets rather than the satellite photos from which those sketches had been derived. The pilots were not cleared at the SCI/TOP SECRET RUFF level and therefore were not permitted to see photos of the targets they were to risk their lives trying to destroy. That discovery, Colby recalls, sent Graham and Colby "up the wall."[25]

Another loosening of distribution restrictions within the national security agencies occurred toward the end of the Carter years. A plan to radically revise and simplify the SCI system had been developed under the supervision of DCI Stansfield Turner. The overall system, known as APEX, would have five subcompartments that include one for overhead photography. Beyond simplifying the SCI system, APEX also vested control of all APEX information in the DCI. As a result it was strongly opposed by the NSA, the CIA's Directorate of Operations, and probably by the Navy, because all of those organizations had important sources of information that they wished to continue to control. Despite opposition, President Carter signed PD 55, "Intelligence Special Access Programs: Establishment of the APEX Program," on January 10, 1980. However, before the program could be implemented a new administration and a new DCI were in office and APEX was permitted to die stillborn. But as Turner writes in his memoirs, "Anticipating APEX, the agencies in charge of photo-reconnaissance shifted a good deal of their product into top secret, secret, and confidential. Wider distribution has meant that more analysts benefit from our expensive overhead reconnaissance systems."[26]

Some feel that even wider distribution is appropriate. Thus, Army Brigadier General Robert L. Stewart, deputy commander of the Army's Strategic Defense Command, told *Aviation Week and Space Technology* that "the world of space reconnaissance is too closed. This is an area that doesn't need to be guarded that closely any more—maybe the details do, but not the basic principles." According to Stewart, "if there were less secrecy in U.S. reconnaissance satellite programs, the Army could make much better use of these advanced spacecraft."[27]

In addition, many believe that the data produced by KEYHOLE satellites could be useful in a variety of non-national security applications. In some instances, the far greater resolution of KEYHOLE satellites could provide valuable information not available from the SPOT and LANDSAT systems. According to Fred Doyle, the science adviser for cartography of the National Mapping Service of the US Geological Survey, there have been a number of studies on the civilian application of KEYHOLE photography during which the members of the study group "wax enthusiastic" about the potential use of such data and produce "glowing reports." But inevitably, says Doyle, three days before the reports come out the CIA or DIA proclaims "not only no, but hell no!, we ain't going to declassify, even the historical stuff."[28]

As a result the use of such imagery has remained limited because the capacity for and utility of dealing with classified satellite imagery is limited for many of the civilian agencies. Handling highly classified material is expensive, in terms of security clearances that must be obtained and security procedures that must be followed. In addition, the photography itself cannot be disclosed nor can information that would indicate the resolving power of the optical system. So, for example, often details that could be transferred from the photograph to a map must be omitted because of what the intelligence community considers those details would reveal about KEYHOLE capabilities. According to Fred Doyle, "Everybody likes to look at gee-whiz pictures. When we get access to some of the DOD [i.e., KEYHOLE] material everyone gets enthusiastic about it but the realities come down to what's a civil agency going to do with it and how are you going to fit it into your program."[29]

One particular arrangement that fell apart over security concerns involved the U.S. Geological Survey (USGS). During construction of the Alaskan pipeline, USGS wanted support from the CIA/NRO in obtaining overhead photographs that would allow the people on the ground to have a detailed understanding of the terrain. USGS and COMIREX worked out detailed arrangements to protect the photography, but reached an impasse when it came to the requirement for the map makers to eliminate the reseau marks and other markings that would normally be transferred from the imagery to the topographic maps that were to be created. The reseau markings would indicate the amount of territory photographed and would show that a satellite had produced the original photograph. Other markings would indicate the inclination of the satellite and would lead, it was believed, to the conclusion that it was a KEYHOLE satellite. By the time an agreement could be worked out, the photos would no longer be useful and so the project was canceled.[30]

Other potential applications are evident from hearings held by the House Committee on Science and Technology. In the hearings it has been noted that DMA has initiated a "universal rectifier" project that will allow imaging satellites to beam pictures of a fire or a flood or earthquake area, run the photos through the "universal rectifier," and superimpose them on a map. This would give virtually instantaneous data.[31]

The value of historical overhead data was testified to by then Representative Albert Gore, Jr., who recalled that he had made use of the National Photographic Interpretation Center: "there was tremendous controversy in west Tennessee over the location of abandoned hazardous dump sites, and a series of photographs going back over a period of 30 years showing activity in that selected geographic area was extremely helpful in answering many questions."[32]

Satellite photography can also be employed to prevent flooding from the melting of snowpacks. According to former CIA photo-interpreter Dino Brugioni, satellite photos of snowpacks can be measured by photogrammetrists so that hydrologists can compute the runoff. Based on such computations, downstream dams can be "drawn down" to prevent flooding.[33]

Greater availability of high-resolution imagery or the information that could be extracted from it for agriculture would help both government analysts and the private sector. A commodity analyst with a major grain-exporting company cited several advantages of high-resolution imagery. Such imagery allows more accurate estimates of acreage devoted to agricultural crops, makes it easier to distinguish between some crops (e.g., hard wheat vs. soft wheat), makes it possible to determine at an earlier date what the crop is (as it pokes through the soil), and makes it easier to identify the extent and nature of any diseases that have afflicted crops. A measure of the utility of the high-resolution imagery lies in the fact that CIA estimates of Soviet crop yields have been consistently more accurate than those of the Department of Agriculture.[34]

Making the data from such imagery available to commercial firms could, according to the commodities analyst, prevent a strong escalation in prices that would cost the American consumer billions of dollars. Soviet secrecy concerning its import plans means that greater-than-normal demands caused by poorer-than-normal harvests have surprised the market and produced significant increases in prices. Companies that use grain, such as General Mills and Pillsbury, are caught in a situation where they are not aware of events in the Soviet Union as early as they might be and therefore do not amass inventories necessary to cope with the increased demand, resulting in a strong escalation in prices.[35]

George Brown believes that "about 90% of what is now classified should be declassified across the board." Included would be most of the technologies (at least in parts), information about most of the programs and, as a beginning, NRO itself. In addition, he "would like to see completely declassified the backlog of imaging data that's been collected by NRO over the first umpteen years of its existence."[36]

Among the costs of maintaining present secrecy, Brown believes, is the potential loss of a $10 billion dollar market to the Japanese. The Japanese have been developing small antenna dishes that will allow every home in Japan to receive direct broadcast from satellites. While the same technology has been developed by the United States, it has been developed for the NRO and CIA to allow reception of data from imaging satellites using a small hand-held pointable dish. The technology remains classified.[37]

U.S. satellite reconnaissance has played a crucial role in preventing Cold War from turning into nuclear war. As capabilities have advanced, the ability to monitor and perhaps ameliorate crisis situations has increased dramatically. It will continue to play a vital role in a changing world—with respect to national security concerns, economic affairs, environmental concerns and disaster prevention and relief. But it could play an even more vital role if it were less secret.

Abbreviations
and Acronyms

ABM	Anti-Ballistic Missile
ACDA	Arms Control and Disarmament Agency
ARDC	Air Research and Development Command
ARPA	Advanced Research Projects Agency
ASAT	Anti-Satellite
CAMS	COMIREX Automated Management System
CCD	Charged Couple Device
CIA	Central Intelligence Agency
COMINT	Communications Intelligence
COMIREX	Committee on Imagery Requirements and Exploitation
COMOR	Committee on Overhead Reconnaissance
COPOUS	Committee on the Peaceful Uses of Outer Space (UN)
CSOC	Consolidated Space Operations Center
DCI	Director of Central Intelligence
DCID	Director of Central Intelligence Directive
DIA	Defense Intelligence Agency
DDRS	Declassified Documents Reference System
DDS&T	Deputy Director for Science and Technology (CIA)
DMSP	Defense Meteorological Satellite Program
DRSP	Defense Reconnaissance Support Program
DSP	Defense Support Program
EXRAND	Exploitation Research and Development Subcommittee (of COMIREX)
FBIS	Foreign Broadcast Information Service
FROG	Film Readout GAMBIT
GMAIC	Guided Missiles and Astronautics Intelligence Committee (of USIB)
GRU	Glavnoye Razvedyvatelnoye Upravleniye (Chief Intelligence Directorate, Soviet General Staff)

GUSM	Glavnoye Upravleniye Strategiche Maskirovka (Chief Directorate for Strategic Deception, Soviet General Staff)
IAC	Intelligence Advisory Committee
ICBM	Intercontinental Ballistic Missile
ICMS	Improved CRYSTAL Metric System
IPS	Imagery Planning Subcommittee (of COMIREX)
INF	Intermediate Nuclear Forces
IRBM	Intermediate-Range Ballistic Missile
JPL	Jet Propulsion Laboratory
KEYSCOM	Imagery Interpretation Keys Subcommittee (of COMIREX)
KGB	Komitet Gosudarstvennoy Bezopasnosti (Committee for State Security)
KH	KEYHOLE
LPAR	Large Phased-Array Radar
MIDAS	Missile Defense Alarm System
MILSTAR	Military, Strategic, Tactical Relay
MOL	Manned Orbiting Laboratory
NASA	National Aeronautics and Space Administration
NFIB	National Foreign Intelligence Board
NID	National Intelligence Daily
NIE	National Intelligence Estimate
NISSTR	National Intelligence Systems to Support Tactical Requirements
NISC	Naval Intelligence Support Center
NPIC	National Photographic Interpretation Center
NREC	National Reconnaissance Executive Committee
NRO	National Reconnaissance Office
NSA	National Security Agency
NSAM	National Security Action Memorandum
NSC	National Security Council
NSCID	National Security Council Intelligence Directive
OMB	Office of Management and Budget
OPSCOM	Operations Subcommittee (of COMIREX)
PFIAB	President's Foreign Intelligence Advisory Board
SAC	Strategic Air Command
SAINT	Satellite Interceptor
SALT	Strategic Arms Limitation Talks
SATRAN	Satellite Reconnaissance Advance Notice
SCF	Satellite Control Facility
SCI	Sensitive Compartmented Information
SCTS	Space Container Transportation System
SDI	Strategic Defense Initiative
SDIO	Strategic Defense Initiative Organization

SDS	Satellite Data System
SIGINT	Signals Intelligence
SIR	Shuttle Imaging Radar
START	Strategic Arms Reduction Talks
TCP	Technological Capabilities Panel
TDRS	Tracking and Data Relay Satellite
TENCAP	Tactical Exploitation of National Space Capabilities
TK	TALENT-KEYHOLE
USGS	United States Geological Survey
USIB	United States Intelligence Board
WDD	Western Development Division
WNINTEL	Warning Notice: Sensitive Intelligence Sources and Methods Involved
WS	Weapon System

Glossary

Apogee the farthest point from the earth reached by a satellite

ASAT Anti-Satellite Weapon. ASAT's can be of several types—co-orbital, direct ascent (missiles), or ground-based laser or electronic equipment

Circular orbit an orbit in which the perigee and apogee are approximately equal

Elliptical orbit an orbit shaped like an ellipse, where the apogee and perigee are not equal

Electro-Optical Imagery imagery created from electronic signals, in which the signals convey the different light levels in the scene under observation

Geosynchronous orbit a satellite 22,300 miles above the equator that revolves around the earth at the same speed as the earth turns and appears to hover over the same spot

Ground track the territory on earth over which a satellite passes

Hardening methods used to enhance a satellite's ability to withstand attack or radiation bombardment

Inclination the angle made by a satellite's path and the equator. It determines which parts of the earth the satellite will pass over. (For example, a satellite with a 57 degree inclination will overfly all territory between 57 north latitude and 57 degrees south latitude. Inclinations under 90 correspond to eastward launches. Inclinations over 90 degrees correspond to westward launches.)

Infrared Film film sensitive to the near infrared portion of the electromagnetic spectrum. It is dependent on the reflective properties

277

of objects and can be black and white or color. (Color infrared film is also known as "false color" film because objects are recorded in different colors than they appear in nature.)

Oblique   imagery obtained from an angle, rather than from directly overhead

Outgassing   the release of gases from a spacecraft

Perigee   the closest point to earth reached by a satellite

Period   the time a satellite needs to make one complete revolution around the earth

Polar orbit   an orbit in which a satellite travels over or almost over the earth's two poles so that the satellite eventually passes over the entire surface of the earth as it rotates below

Radar Imagery   imagery produced by bouncing radio waves off an object or area and creating an image using the returning pulse

Real-time imagery   imagery that is transmitted as the event being viewed is occurring

Resolution   the minimum size required of at least one dimension of an object for it to be identifiable by photo analysts

Semisynchronous orbit   an orbit in which the satellite makes two revolutions around the earth during a day

Sun-synchronous orbit   a near polar orbit, in which a satellite passes over points on the earth's surface at the same local time each day so that the shadows cast by the sun are approximately the same

Thermal Infrared Imagery   imagery produced by sensing the heat emitted by the target. It can be produced under conditions of darkness

Visible-Light Imagery   imagery produced by sensing the reflection of light off a target (standard photography)

Notes

Notes For Chapter One: Pioneer Reconnaissance

1. Interview with Merton Davies, Santa Monica, CA, December 1, 1988.
2. Ibid.; telephone interview with Merton Davies, February 22, 1989.
3. Davies interview; Davies telephone interview; Merton E. Davies and William R. Harris, *RAND's Role in the Evolution of Balloon and Satellite Observation Systems and Related U.S. Space Technology* (Santa Monica, CA: RAND, 1988), pp. 6–7.
4. Davies and Harris, *RAND's Role in the Evolution*, p. 7.
5. Ibid., p. 7.
6. Davies interview.
7. Robert L. Perry, *Origins of the USAF Space Program, 1945–1956* (Washington DC: Air Force Systems Command, June 1962), p. 30.
8. Ibid.; Philip J. Klass, *Secret Sentries in Space* (New York: Random House, 1971), p. 74; *RAND 25th Anniversary Volume* (Santa Monica, CA: RAND, 1973), p. 3.
9. Thomas White, Memorandum for Deputy Chief of Staff, Development, Subject: [deleted] Satellite Vehicles, 18 December 1952, File 2-36300 through 2-36399, RG 341, Entry 214, Modern Military Branch, National Archives. The studies were

   Flight Mechanics of a Satellite Rocket (RA-15021)

   Aerodynamic, Gas Dynamics and Heat Transfer Problems of a Satellite Rocket (RA-15022)

   Analysis of Temperature, Pressure and Density of the Atmosphere Extending to Extreme Altitudes (RA-15023)

   Theoretical Characteristics of Several Liquid Propellant Systems (RA-15024)

   Stability and Control of a Satellite Rocket (RA-15025)

   Structural and Weight Studies of a Satellite Rocket (RA-15026)

   Satellite Rocket Powerplant (RA-15027)

   Communication and Observation Problems of a Satellite (RA-15028)

279

Study of Launching Sites for a Satellite Projectile (RA-15029)

Cost Estimate of an Experimental Satellite Program (RA-15030)

Reference Papers Relating to a Satellite Study (RA-15032)

Proposed Type Specification for an Experimental Satellite (RA-15013) (Summary)

10. Davies and Harris, *RAND's Role in the Evolution*, p. 14.
11. Ibid.
12. Ibid., p. 16.
13. Ibid., p. 17; Lee Bowen, *The Threshold of Space: The Air Force in the National Space Program 1945-1959* (Washington DC: USAF Historical Division Liaison Office, September 1960), p. 5.
14. White, Memorandum for the Deputy Chief of Staff, Development.
15. Ibid.
16. Davies and Harris, *RAND's Role in the Evolution*, p. 23.
17. Ibid.; Directorate of Intelligence, DCS/O to Assistant for Evaluation, DCS/D, subj: Research and Development on Proposed RAND Satellite Reconnaissance Vehicle, March 17, 1951, RG 341, Entry 214, File 2-19900 to 2-19999, Modern Military Branch, National Archives.
18. Directorate of Intelligence, DCS/O to Assistant for Evaluation, DCS/D, subj: Research and Development on Proposed RAND Satellite Reconnaissance Vehicle.
19. Ibid.
20. Davies and Harris, *RAND's Role in the Evolution*, pp. 23–25; S.M. Greenfield and W. W. Kellogg, *Inquiry into the Feasibility of Weather Reconnaissance from a Satellite Vehicle* (Santa Monica, CA: RAND, 1951), p.v.
21. Perry, *Origins of the USAF Space Program*, p. 31.
22. See Jeffrey Richelson, *American Espionage and the Soviet Target* (New York: Quill, 1988), pp. 100–126, 248–253.
23. Perry, *Origins of the USAF Space Program*, p. 32.
24. Davies and Harris, *RAND's Role in the Evolution*, p. 26.
25. J. E. Lipp, R. M. Salter, R. S. Wehner, R. R. Carhart, C. R. Culp, S. L. Gendler, W. J. Howard, and J. S. Thompson, *Utility of a Satellite Vehicle for Reconnaissance* (Santa Monica, CA: RAND, 1951), pp. 31, 38.
26. Perry, *Origins of the USAF Space Program*, p. 32.
27. Davies and Harris, *RAND's Role in the Evolution*, pp. 26–27.
28. Ibid., pp. 27–28.
29. Ibid., p. 28.
30. Ibid., p. 28.
31. Ibid., p. 30.
32. Davies interview.

33. Perry, *Origins of the USAF Space Program*, p. 38 n. 12.
34. Davies and Harris, *RAND's Role in the Evolution*, p. 47, citing Lt. Col. V. M. Genez, Director/Intelligence, Deputy for Development, ARDC, Memo for the Record, "Conference with RAND Corporation re: FEED-BACK Program," August 13, 1953, cited in Perry, p. 39.
35. Davies and Harris, *RAND's Role in the Evolution*, p. 47.
36. Perry, *Origins of the USAF Space Program*, pp. 35–36.
37. Davies and Harris, *RAND's Role in the Evolution*, p. 48.
38. Ibid., p. 49.
39. Ibid.
40. Ibid., p. 52.
41. J. E. Lipp and R. M. Salter (eds.), *Project Feedback Summary Report, Volume II* (Santa Monica, CA: RAND, 1954), pp. iii–iv; Davies telephone interview.
42. Lipp and Salter (eds.), *Project Feedback Summary Report, Volume II*, pp. 105–108.
43. Davies and Harris, *RAND's Role in the Evolution*, p. 53.
44. Ibid., p. 54; Perry, *Origins of the USAF Space Program*, pp. 36–37.
45. Davies and Harris, *RAND's Role in the Evolution*, p. 57.
46. Perry, *Origins of the USAF Space Program*, p. 41; Davies and Harris, *RAND's Role in the Evolution*, p. 57.
47. Perry, *Origins of the USAF Space Program*, p. 42; Steve J. Heims, *John Von Neumann and Norbert Wiener: From Mathematics to the Technologies of Life and Death* (Cambridge, MA: MIT Press, 1980), pp. 271, 490 n. 97.
48. Perry, *Origins of the USAF Space Program*, pp. 42–43.
49. Davies and Harris, *RAND's Role in the Evolution*, p. 56; Michael R. Beschloss, *MAYDAY: Eisenhower, Khrushchev and the U-2 Affair* (New York: Harper & Row, 1986), p. 73.
50. Davies and Harris, *RAND's Role in the Evolution*, p. 61.
51. Ibid.; Interview.
52. Perry, *Origins of the USAF Space Program*, pp. 42–43.
53. Ibid., pp. 42–43.
54. Ibid., p. 43.
55. Ibid., pp. 55–56.
56. Ibid., p. 56.
57. Ibid.
58. Davies and Harris, *RAND's Role in the Evolution*, p. 55.
59. Telephone Interview with Amrom Katz, June 6, 1989.
60. Ibid.
61. Ibid.
62. Ibid.
63. Davies and Harris, *RAND's Role in the Evolution*, p. 55; Davies interview.

64. Amrom H. Katz, "Memorandum to J. L. Hult: Recommendations, Formal and Informal, on Reconnaissance Satellite Matters," May 11, 1959, p. 3.
65. Davies and Harris, *RAND's Role in the Evolution*, p. 69.
66. Ibid.
67. Ibid., p. 70.
68. Ibid.
69. Ibid.
70. William E. Burrows, *Deep Black: Space Espionage and National Security* (New York: Random House, 1986), p. 86.
71. Davies and Harris, *RAND's Role in the Evolution*, p. 70.
72. Klass, *Secret Sentries in Space*, p. 85.
73. Burrows, *Deep Black*, p. 87.
74. Merton Davies and Amrom Katz, Memorandum to Ed Barlow, October 12, 1956, p. 1, cited in Davies and Harris, *RAND's Role in the Evolution*, p. 75.
75. Davies and Harris, *RAND's Role in the Evolution*, p. 76.
76. Ibid., p. 78.
77. Ibid.
78. Ibid.
79. Ibid., pp. 78–79.
80. Ibid., p. 79.
81. Ibid., pp. 79, 81.
82. Ibid., p. 85, citing M. E. Davies, Memorandum to A. H. Katz, "Progress of Recoverable Satellite Study," September 10, 1957.
83. Anthony Kenden, "U.S. Reconnaissance Satellite Programmes," *Spaceflight*, 20, 7, 1978, pp. 243ff.
84. Fred Kaplan, *The Wizards of Armageddon* (New York: Simon & Schuster, 1983), p. 135.
85. Ibid.
86. Stephen E. Ambrose, *Eisenhower, Volume Two: The President* (New York: Simon & Schuster, 1984), p. 428, n. 39.
87. "USAF Pushes Pied Piper Space Vehicle," *Aviation Week*, October 14, 1957, p. 26.
88. *Chronology of Early Air Force Man-in-Space Activity, 1955–1960* (Washington DC: Air Force Systems Command, 1965), p. 5.
89. Davies and Harris, *RAND's Role in the Evolution*, pp. 86, 90.
90. Ibid., pp. 87–88.
91. Burrows, *Deep Black*, pp. 87–88; Amrom H. Katz and Merton Davies, "An Earlier Reconnaissance Satellite System," (Santa Monica, CA: RAND, 1957), Appendix p. 17; M. E. Davies and A. H. Katz, *A Family of Recoverable Satellites* (Santa Monica, CA: RAND, November 12, 1957), p. 63.

92. Katz and Davies, "An Earlier Reconnaissance Satellite System," Appendix, p. 2.
93. Ibid., Appendix, p. 18.
94. Ibid., Appendix, p. 18.
95. Interview with Andrew J. Goodpaster, November 9, 1988.
96. *Briefing on Army Satellite Program,* November 9, 1957, *Declassified Documents Reference System,* 1977-101B, p. 4.
97. Ibid.
98. Ibid., p. 3.
99. Ibid., p. 5.
100. Ibid., p. 6.
101. Ibid.
102. Ibid., p. 9.
103. National Security Council, NSC 5814, "U.S. Policy on Outer Space," June 20, 1958, p. 21
104. Interview.
105. Goodpaster interview.
106. Private information; John Prados, *The Soviet Estimate: U.S. Intelligence and Russian Military Strength* (New York: Dial, 1982), pp. 105–106.
107. Interview.
108. Ibid.
109. Peter Wyden, *Bay of Pigs: The Untold Story* (New York: Simon & Schuster, 1979), pp. 13, 18; U.S. Congress, Senate Select Committee to Study Governmental Operations with Respect to Intelligence Activities, *Final Report Book IV: Supplementary Detailed Staff Reports on Foreign and Military Intelligence* (Washington DC: U.S. Government Printing Office, 1976), pp. 58–59.
110. Wyden, *Bay of Pigs,* pp. 12–13.
111. Interview with Richard Bissell, Farmington, CT, January 6, 1984; Interview.
112. Davies interview.
113. Thomas Powers, *The Man Who Kept the Secrets,* (New York: Knopf, 1979), p. 97; Burrows, *Deep Black,* pp. 89–90.
114. Burrows, *Deep Black,* pp. 89–90; Interview.
115. NSC 5814, "U.S. Policy on Outer Space," pp. 8, 21.
116. Ibid., pp. 9, 21.
117. Kenden, "U.S. Satellite Reconnaissance Programmes"; "Work on Pied Piper Accelerated, Satellite Has Clam-Shell Nose Cone," *Aviation Week,* June 23, 1958, pp. 18–19.
118. Office of the Director of Defense Research and Engineering, *Military Space Projects Report No. 10,* (Washington DC: DOD, 1960), p. 1, *DDRS* 1980–36C.
119. Leonard Mosley, *Dulles: A Biography of Eleanor, Allen and John Foster and their Family Network* (New York: Dial, 1978), p. 432.

Notes For Chapter Two: A New Era

1. Letter from Neil McElroy to President Dwight D. Eisenhower, January 29, 1959, *DDRS* 1982–001538; "Discoverer Aborted," *Aviation Week*, March 2, 1959, p. 27.
2. "A Satellite Rocket is Fired on the West Coast," *New York Times*, March 1, 1959, pp. 1, 32.
3. "A Satellite Rocket is Fired"; J. E. D. Davies, "The Discoverer Programme," *Spaceflight*, November 1969, pp. 405–407; "Air Force Reports It is Receiving Signal from Discoverer Satellite," *New York Times*, March 2, 1959, pp. 1, 10.
4. John A. Osmunden, "Rivalry is Cited," *New York Times*, March 2, 1959, p. 10.
5. Curtis Peebles, *Guardians : Strategic Reconnaissance Satellites* (Novato, CA: Presidio, 1987), p. 47; Harry Schwartz, " 'Spying' in Space by U.S. Charged," *New York Times*, March 6, 1959, p. 10.
6. Advanced Research Projects Agency, *Military Space Projects, Quarter Ended 30 June 1959* (Washington DC: DOD, 1959), p. 4; John W. Finney, "Discoverer Shot Into Polar Orbit; Recovery is Aim," *New York Times*, April 14, 1959, pp. 1, 18.
7. Office of the Director of Defense Research and Engineering, *Military Space Projects Report No. 10 (March–April–May 1960)*, p. 6 in *DDRS* 1980–36C; "Agena B to Put Samos, Midas in Orbit," *Aviation Week*, February 8, 1960, pp. 73, 75.
8. Advanced Research Projects Agency, *Military Space Projects, Quarter Ended 30 June 1959*, p. 7.
9. Ibid., p. 4; Finney, "Discoverer Shot Into Polar Orbit."
10. Advanced Research Projects Agency, *Military Space Projects, Quarter Ended 30 June 1959*, p. 4; Finney, "Discoverer Shot into Polar Orbit"; Russell Hawkes, "USAF's Satellite Test Center Grows," *Aviation Week and Space Technology*, May 30, 1960, pp. 57–59.
11. Advanced Research Projects Agency, *Military Space Projects, Quarter Ended 30 June 1959*, p. 4; Richard Witkin, "Washington to Hail Retrieved Capsule in Ceremony Today," *New York Times*, August 13, 1960, pp. 1, 7; Anthony Kenden, "U.S. Reconnaissance Satellite Programmes," *Spaceflight* 20, 7, 1978, pp. 243ff.; "Capsule Hunt Halted," *New York Times*, April 23, 1959, p. 59; Telephone interview with Charles Mathison, December 8, 1988; Telephone interview with Richard Bissell, June 5, 1988; Interview.
12. John W. Finney, "U.S. Cancels Plan to Catch Capsule," *New York Times*, April 15, 1959, pp. 1, 17.
13. Advanced Research Projects Agency, *Military Space Projects, Quarter Ended 30 June 1959*, pp. 5, 6; Finney, "Discoverer Shot into Polar Orbit."

14. Philip J. Klass, *Secret Sentries in Space* (New York: Random House, 1971), p. 94; Peebles, *Guardians*, p. 50.

15. 'Satellite Fired into Polar Orbit," *New York Times*, November 8, 1959, p. 43.

16. "Discoverer Failure Caused by Inverter," *Aviation Week*, November 16, 1959, p. 33; Peebles, *Guardians*, p. 50.

17. "Satellite Fired into Polar Orbit."

18. "Catching Capsule is Complex Task," *New York Times*, November 29, 1959, p. 34; Klass, *Secret Sentries in Space* p. 95.

19. Peebles, *Guardians*, p. 50.

20. George B. Kistiakowsky, *A Scientist at the White House: The Private Diary of President Eisenhower's Special Assistant for Science and Technology* (Cambridge, MA: Harvard University Press, 1976), p. 196.

21. "Letter from James H. Douglas to President Dwight David Eisenhower," April 11, 1960, *DDRS* 1977–288A.

22. "Discoverer Fails in Rain of Debris," *New York Times*, February 20, 1960, p. 6.

23. Office of the Director of Defense Research and Engineering, *Military Space Projects No. 10*, p. 2; "Letter from James H. Douglas to Dwight D. Eisenhower."

24. Office of the Director of Defense Research and Engineering, *Military Space Projects Report No. 10*, p. 3.

25. Ibid.; Peebles, *Guardians*, p. 51.

26. Office of the Director of Defense Research and Engineering, *Military Space Projects Report No. 10*, p. 3.

27. Office of the Director of Defense Research and Engineering, *Military Space Projects Report No. 11*, (Washington DC: DOD, 1960), pp. 1–2 in *DDRS* 1980–36C; Larry Booda, "First Capsule Recovered from Satellite," *Aviation Week*, August 22, 1960, pp. 33–35.

28. Letter from Thomas S. Gates to President Dwight David Eisenhower, in Office of the Director of Defense Research and Engineering, *Military Space Projects Report No. 10*.

29. Leonard Mosley, *Dulles: A Biography of Eleanor, Allen, and John Foster and their Family Network* (New York: Dial, 1978) p. 432.

30. Booda, "First Capsule Recovered From Satellite"; Witkin, "Washington to Hail"; Interview.

31. Office of the Director of Defense Research and Engineering, *Military Space Projects No. 11*, p. 2 in DDRS 1980–36D; John W. Finney, "Copter Recovers Capsule Ejected by U.S. Satellite," *New York Times*, August 12, 1960, pp. 1, 3; Witkin, "Washington to Hail."

32. Booda, "First Capsule"; Witkin, "Washington to Hail"; Office of the Director of Defense Research and Engineering, *Military Space Projects No. 11*, p. 3.

33. "Frogman First on Scene," *New York Times*, August 12, 1960, p. 3; Witkin, "Washington to Hail."
34. "Frogman First on Scene;" Witkin, "Washington to Hail."
35. "Frogman First on Scene"; Witkin, "Washington to Hail"; Mathison interview.
36. Carl Berger, *The Air Force in Space Fiscal Year 1961*, (Washington, DC: USAF Historical Division Liaison Office, 1966), p. 49; Felix Belair Jr., "Eisenhower is Given Flag that Orbited the Earth," *New York Times*, August 16, 1960, pp. 1, 4.
37. Interview.
38. Donald Welzenbach, "Observation Balloons and Reconnaissance Satellites," *Studies in Intelligence*, Spring 1986, pp. 21–28.
39. W.W. Rostow, *Open Skies: Eisenhower's Proposal of July 21, 1955* (Austin, TX: University of Texas Press, 1982), pp. 192–94; Office of the Director of Defense Research Engineering, *Military Space Projects No. 11*, p. 4; "New Discoverer Shot into Orbit," *New York Times*, August 19, 1960, pp. 1, 13.
40. Interview; Office of the Director of Defense Research and Engineering, *Military Space Projects No. 11*, p. 4.
41. John Nammack, "C-119's Third Pass Snares Discoverer," *Aviation Week*, August 29, 1960, pp. 30–31; "Space Capsule Caught in Mid-Air by U.S. Plane on Re-entry from Orbit," *New York Times*, August 20, 1960, pp. 1, 7.
42. Nammack, "C-119's Third Pass Snares Discoverer."
43. Ibid.
44. Ibid.
45. "Space Capsule is Caught in Mid-Air by U.S. Plane on Re-entry from Orbit"; Peebles, *Guardians*, p. 56.
46. Interview; Peebles, *Guardians*, p. 56.
47. Interview.
48. Ibid.
49. Interview with Andrew Goodpaster, Washington DC, November 9, 1988.
50. Kistiakowsky, *A Scientist at the White House*, p. 245.
51. Ibid., p. 336.
52. Interview.
53. Ibid.
54. A. J. Goodpaster, Memorandum for the Honorable Gordon Gray, June 10, 1960.
55. Berger, *The Air Force in Space Fiscal Year 1961*, p. 34.
56. Kistiakowsky, *A Scientist at the White House*, p. 382.
57. Berger, *The Air Force in Space Fiscal Year 1961*, p. 35.

58. Ibid., p. 387; Goodpaster interview; Interview.
59. Kistiakowsky, *A Scientist at the White House*, p. 387; Goodpaster interview.
60. Berger, *The Air Force in Space Fiscal Year 1961*, p. 35.
61. Kistiakowsky, *A Scientist at the White House*, pp. 382–83, 394–95; Goodpaster interview; Berger, *The Air Force in Space Fiscal Year 1961*, pp. 34–35.
62. Interview with Richard Bissell, Farmington, CT, January 6, 1984; William E. Burrows, *Deep Black: Space Espionage and National Security* (New York: Random House, 1986), p. 206.
63. Secretary of the Air Force Order No. 115.1, Subj: Organization and Functions of the Office of Missile and Satellite Systems, August 31, 1960; Secretary of the Air Force Order No. 116.1, Subj: The Director of the SAMOS Project, August 31, 1960.
64. Berger, *The Air Force in Space Fiscal Year 1961*, p. 42.
65. Ibid., p. 43; A. J. Goodpaster, Memorandum for Record, October 1960, White House Office Staff Secretary, Subject Series, Alphabetical Subseries, Box 15, Intelligence Matters, Fldr 20, October 1960–Jan 1961, Dwight D. Eisenhower Library.
66. "Satellite Capsule Sighted, Then Lost," *New York Times*, September 16, 1960, p. 13.
67. "Discoverer Rocket Fails in Launching," *New York Times*, October 27, 1960, p. 17; Klass, *Secret Sentries in Space* p. 103.
68. "Capsule from Rocket Snagged from Sky," *Los Angeles Times*, November 15, 1960, pp. 1, 13; Klass, *Secret Sentries in Space*, pp. 103, 104; "Discoverer XVII Shot into Orbit, Recovery Attempt is Scheduled," *New York Times*, November 13, 1960, pp. 1, 27; "2nd Space Capsule Caught in Mid-Air," *New York Times*, November 15, 1960, pp. 1, 20.
69. "Discoverer May Give Sunspot Activity Data," *Los Angeles Times*, November 15, 1960, pp. 1, 13; "Drop of Capsule Put Off for Day," *New York Times*, November 14, 1960, p. 16.
70. "2nd Space Capsule Caught in Mid-Air."
71. G. Zhukov, "Space Espionage Plans and International Law," *International Affairs (Moscow)*, October 1960, pp. 53–57.
72. Ibid.
73. Kistiakowsky, *A Scientist at the White House*, p. 394.
74. Ibid., p. 394.
75. Ibid., p. 397.
76. U.S. Congress, House Committee on Science and Astronautics, *Science, Astronautics, and Defense* (Washington DC: U.S. Government Printing Office, 1961), p. 63.

77. Murray Snyder, Memorandum for Assistant Secretary of Defense (International Security Affairs), Public Affairs Plan for SAMOS Research and Development Test Firing in Early October, October 4, 1960, *DDRS* 1984-000819.
78. Kenden, "U.S. Satellite Reconnaissance Programs."
79. "SAMOS II Fact Sheet," (Washington DC: Department of Defense, 1961).
80. Arthur Sylvester, Memorandum for the President, The White House, Subject: SAMOS Launch, January 26, 1961 in *DDRS* 1981-364B.
81. Ibid.
82. Gerald M. Steinberg, *Satellite Reconnaissance: The Role of Informal Bargaining* (New York: Praeger, 1983), p. 41.
83. Ibid., pp. 27, 40, 42.
84. Ibid., pp. 41, 43.
85. Interview.
86. Steinberg, *Satellite Reconnaissance*, p. 43.
87. Ibid., p. 44.
88. Burrows, *Deep Black* p. 142.
89. Ibid., pp. 142, 143.
90. Ibid., p. 144.
91. Ibid.
92. Paul B. Stares, *The Militarization of Space, U.S. Policy 1945–1984* (Ithaca, NY: Cornell University Press, 1985), pp. 114–115.
93. Ibid., p. 115.
94. "Discoverer XX Fired into Orbit," *New York Times*, February 18, 1961, pp. 1, 5; "Discoverer Capsule Fails to Go in Orbit," *New York Times*, March 31, 1961, p. 10; "Discoverer XXIII Fired into Orbit," *New York Times*, April 9, 1961, p. 31; "Attempt to Orbit Discoverer Fails," *New York Times*, June 9, 1961, p. 14.
95. "Capsule of Discoverer Is Recovered in Pacific," *New York Times*, June 19, 1961, p. 24.
96. "Discoverer Cone is Caught in Air," *New York Times*, July 10, 1961, pp. 1, 5.
97. Klass, *Secret Sentries in Space*, p. 106; "Discoverer Satellite Silent," *New York Times*, September 21, 1961, p. 15.
98. Lawrence Freedman, *U.S. Intelligence and the Soviet Strategic Threat* 2nd ed. (Princeton, NJ: Princeton University Press, 1986), p. 73.
99. "U.S. Orbits Discoverer," *New York Times*, Oct 14, 1961, p. 3; "Industry Observer," *Aviation Week*, December 5, 1960, p. 23; "Discoverer Capsule Caught Over Pacific," *New York Times*, October 15, 1961, p. 57.
100. Lawrence C. McQuade, Memorandum for Mr. Nitze, Subj: But Where Did the Missile Gap Go?, (Washington DC: Assistant Secretary of Defense, International Security Affairs, May 31, 1963), p. 7.
101. Ibid.

102. Ibid., pp. 7–8.
103. Ibid., pp. 9–10.
104. Ibid., pp. 10–11.
105. Fred Kaplan, *The Wizards of Armageddon* (New York: Simon & Schuster, 1983), p. 287.
106. Ibid., p. 287.
107. Interview with Robert McNamara, Washington DC, January 19, 1989; John Newhouse, *War and Peace in the Nuclear Age* (New York: Knopf, 1989), p. 148.
108. McQuade, "Memorandum for Mr. Nitze," p. 14.
109. Ibid., p. 15.
110. Joseph A. Loftus, "Gilpatric Warns U.S. Can Destroy Atom Aggressor," *New York Times*, October 22, 1961, pp. 1, 6; McGeorge Bundy, *Danger and Survival: Choices About the Bomb in the First Fifty Years* (New York: Random house, 1988), pp. 361, 378–82; Interview with Roswell Gilpatric, New York, May 31, 1989; Interview with Daniel Ellsberg, Washington DC, June 22, 1989.
111. Interview.
112. Interview.
113. Kistiakowsky, *A Scientist at the White House*, pp. 394–95.
114. *The Reminiscences of Arthur C. Lundahl*, Oral History Research Office, Columbia University, 1982, pp. 11, 38, 42.
115. Ibid., pp. 51, 56, 57; Dino A. Brugioni and Robert F. McCourt, "Personality: Arthur C. Lundahl," *Photogrammetric Engineering & Remote Sensing*, 1988, pp. 271–72.
116. Telephone interview with Arthur Lundahl, July 28, 1989.
117. Ibid.; *The Reminiscences of Arthur C. Lundahl*, pp. 182, 187, 197; Jack Anderson, "Getting the Big Picture for the CIA," *Washington Post*, November 28, 1982, p. C7.
118. *The Reminiscences of Arthur C. Lundahl*, p. 221; Interview.
119. *The Reminiscences of Arthur C. Lundahl*, pp. 197, 201; John Prados, *The Soviet Estimate: U.S. Intelligence Analysis and Russian Military Strength* (New York: Dial, 1982), p. 110.
120. *The Reminiscences of Arthur C. Lundahl*, pp. 197–201, 229.
121. Prados, *The Soviet Estimate*, pp. 122–23.
122. *The Joint Study Group Report on Foreign Intelligence Activities of the United States Government*, December 15, 1960, pp. 1, 2.
123. Ibid., pp. 133, 137, 138, 140.
124. Ibid., pp. 53, 61.
125. *The Reminiscences of Arthur C. Lundahl*, Ibid., pp. 299–300.
126. Ibid., pp. 300–301.
127. Ibid., pp. 301–302; Telephone interview with Arthur Lundahl.

Notes For Chapter Three: Problems In Space, Confrontations On Earth

1. Gerald H. Cantwell, *The Air Force in Space 1964* (Washington DC: USAF Historical Division Liaison Office, 1967), pp. 92–93.
2. Gerald M. Steinberg, *Satellite Reconnaissance: The Role of Informal Bargaining* (New York: Praeger, 1983), p. 43; "Space Secrecy Muddle," *Aviation Week and Space Technology*, April 23, 1962, p. 21.
3. Interview.
4. Curtis Peebles, *Guardians: Strategic Reconnaissance Satellites* (Novato, CA: Presidio Press, 1987), pp. 76–77; Robert C. Berman and John C. Baker, *Soviet Strategic Forces: Requirements and Responses* (Washington DC: The Brookings Institution, 1982), pp. 104–105.
5. John L. Lewis and Xue Litai, *China Builds the Bomb* (Stanford, CA: Stanford University Press, 1988), pp. 75, 95, 97, 111, 175–80.
6. Central Intelligence Agency, National Intelligence Estimate 13-2-60, "The Chinese Communist Atomic Energy Program," December 13, 1960, p. 2.
7. Ibid.
8. Lewis and Litai, *China Builds the Bomb*, pp. 97, 98, 111–12.
9. Samuel Glasstone and Philip J. Dolan, *The Effects of Nuclear Weapons* (Washington DC: DOD/DOE, 1977 3rd.), pp. 45–47; Glenn T. Seaborg, *Kennedy, Khrushchev and the Test Ban* (Berkeley, CA: University of California Press, 1983), p. 158.
10. Interview with Herbert Scoville, McLean, Virginia, 1983.
11. Howard Simons, "Our Fantastic Eye in the Sky," *Washington Post*, December 8, 1963, pp. E1, E5; Oral History Interview with Robert Amory Jr., February 9, 1966, JFK Library, p. 115; Interview with Richard M. Bissell Jr., Farmington, Connecticut, January 6, 1984.
12. Interview.
13. Steinberg, *Satellite Reconnaissance*, pp. 53–54.
14. Ibid., p. 53.
15. Interview with U. Alexis Johnson, Washington DC, November 22, 1988; Raymond L. Garthoff, "Banning the Bomb in Outer Space," *International Security* 5, 3 1980/81, pp. 25–40.
16. Johnson interview; Garthoff, "Banning the Bomb in Outer Space."
17. Paul B. Stares, *The Militarization of Space : U.S. Policy 1945–1984* (Ithaca, NY: Cornell University Press, 1985), pp. 66–67.
18. Ibid., p. 67.
19. Johnson interview.
20. Garthoff, "Banning the Bomb in Outer Space."
21. Oral History Interview with Abraham Chayes, John F. Kennedy Library, p. 215.

22. Telephone interview with Richard S. Leghorn, September 12, 1989; Richard S. Leghorn, "Political Action and Satellite Reconnaissance," Itek Corporation, April 24, 1959.
23. Oral History Interview with Abraham Chayes, pp. 215–16.
24. U. Alexis Johnson with Jef Olivarius McAlister, *The Right Hand of Power* (Englewood Cliffs, NJ: Prentice Hall, 1984), p. 370; Stares, *The Militarization of Space*, p. 69.
25. Johnson interview; Garthoff., "Banning the Bomb in Outer Space;" Stares, *The Militarization of Space*, p. 69.
26. Johnson interview; Johnson with McAlister, *The Right Hand of Power*, p. 370.
27. Garthoff interview.
28. Ibid.
29. Johnson with McAlister, The Right Hand of Power, p. 370.
30. William C. Foster, Memorandum for the President: Arms Control Aspects of Proposed Satellite Reconnaissance Policy, July 6, 1962, p. 1.
31. Ibid., p. 2
32. Ibid.
33. Ibid.
34. Garthoff, "Banning the Bomb in Outer Space."
35. Foster, Memorandum for the President, pp. 3–4.
36. Ibid., p. 4.
37. Garthoff, "Banning the Bomb in Outer Space."
38. Philip Klass, *Secret Sentries in Space* (New York: Random House, 1971), p. 124.
39. Ibid., pp. 124–25.
40. "Statement by the Soviet Representative (Morozov) to the First Committee of the General Assembly: Peaceful Uses of Outer Space, December 3, 1962," in *Documents on Disarmament*, 1962, Volume I (Washington DC: U.S. Government Printing Office, 1963), pp. 1130–31.
41. "Statement by the United States Representative (Gore) to the First Committee of the General Assembly: Peaceful Uses of Outer Space [Extracts], December 3, 1962," in *Documents on Disarmament, 1962*, Volume I, p. 1121.
42. "Statement by the Soviet Representative," p. 1131.
43. *The Reminiscences of Arthur C. Lundahl*, Columbia University Oral History Research Office, 1982, p. 302.
44. Ibid., p. 303.
45. Ibid., p. 305.
46. Interview.
47. Ibid.
48. Ibid.
49. Berman and Baker, *Soviet Strategic Forces*, pp. 104–105.

50. Interview.
51. C.L. Sulzberger, "Those Who Spy Out the Land," *New York Times*, July 15, 1963, p. 28; Klass, *Secret Sentries in Space*, pp. 125, 127; U.S. Congress, Senate Committee on Aeronautical Sciences, *Manned Orbiting Laboratory* (Washington DC: U.S. Government Printing Office, 1966), pp. 75–76.
52. Mark E. Miller, *Soviet Strategic Power and Doctrine: The Quest for Superiority* (Washington DC: Advanced International Studies Institute, 1982), p. 84.
53. William Burrows, *Deep Black: Space Espionage and National Security* (New York: Random House, 1986), p. 206.
54. Ibid.; Scoville interview.
55. Interview; U.S. Congress, Senate Select Committee to Study Governmental Operations with Respect to Intelligence Activities, *Final Report, Book IV: Supplementary Detailed Staff Reports on Foreign and Military Intelligence* (Washington DC: U.S. Government Printing Office, 1976), p. 77.
56. Interview; Telephone interview with Richard Bissell, September 5, 1989.
57. Interview; John Prados, *The Soviet Estimate: U.S. Intelligence and Russian Military Strength* (New York: Dial Press, 1982), pp. 200–201.
58. Interview; U.S. Congress Senate Select Committee to Study Governmental Operations with Respect to Intelligence Activities, *Final Report, Book IV*, p. 77; John Ranelagh, *The Agency: The Rise and Decline of the CIA* (New York: Simon & Schuster, 1986), p. 491.
59. Interview.
60. Burrows, *Deep Black*, p. 207.
61. Ibid.
62. U.S. Congress, Senate Select Committee to Study Governmental Operations with Respect to Intelligence Activities, *Final Report, Book IV*, p. 75.
63. Ibid.
64. Klass, *Secret Sentries in Space*, p. 147; "President Approves DOD Development of Manned Orbiting Laboratory," OASD (PA), August 25, 1965; William J. Normyle, "Air Force Given Manned Space Role," *Aviation Week and Space Technology*, August 30, 1965, p. 23.
65. Manned Orbiting Laboratory Background Briefing, Attribute to Defense Officials, August 25, 1965.
66. Howard Simons and Chalmers M. Roberts, "Role in Arms Control Clinched MOL Victory," *Washington Post*, September 5, 1965, pp. A1, A5.
67. Aeronautical Systems Division, Air Force Systems Command, "Review and Summary of X-20 Military Application Studies," (Wright-Patterson AFB: ASD, December 14, 1963), p. 19.
68. Ibid.
69. Peebles, *Guardians*, p. 238.

70. Ibid., p. 239.
71. Ibid.
72. Ibid.
73. Ibid.
74. Ibid., p. 241.
75. Ibid., p. 242; Interview.
76. Simons and Roberts, "Role in Arms Control Clinched MOL Victory."
77. Klass, *Secret Sentries in Space*, pp. 147–48; Donald E. Fink, "CIA Control Bid Slowed Decision on MOL," *Aviation Week and Space Technology*, September 20, 1965, pp. 26–27.
78. Fink,"CIA Control Bid Slowed Decision on MOL"; Peebles, *Guardians*, p. 244.
79. "For $1.5 Billion . . . A New Air Force Eye in the Sky," *Newsweek*, September 6, 1965, pp. 46–47.

Notes For Chapter Four: New Programs, New Wars

1. Interview.
2. Curtis Peebles, *Guardians: Strategic Reconnaissance Satellites* (Novato, CA: Presidio Press, 1987), p. 91; Interview.
3. Philip Klass, "Military Satellites Gain Vital Data," *Aviation Week and Space Technology*, September 15, 1969, pp. 55–61.
4. U.S. Congress, Senate Committee on Aeronautical and Space Sciences, *Manned Orbiting Laboratory*, (Washington DC: U.S. Government Printing Office, 1966), p. 11; Albert Sehlstedt Jr., "Florida Ired as California is Winner of Space Project," *Baltimore Sun*, February 11, 1966, p. 7.
5. Cyrus Vance, Memorandum for the President, February 12, 1966, *DDRS*, 1980-37A.
6. U.S. Congress, Senate Committee on Aeronautical and Space Sciences, *Manned Orbiting Laboratory*, pp. 35, 36.
7. D. R. Andelin, *Evaluation of the Usefulness of the MOL to Accomplish Early NASA Mission Objectives, Volume 8: Support Systems, Gemini B, and Titan III-M*, Douglas Aircraft Company, May 1967, p. 39; Peebles, *Guardians* p. 245.
8. "Titan 3C Passes 6th Test, Furnishes MOL Support," *Aviation Week and Space Technology*, November 14, 1966, p. 30.
9. D.E. Charhut, *Evaluation of the Usefulness of the MOL to Accomplish Early NASA Mission Objectives, Volume I: Summary*, Douglas Aircraft Company, October 1967, p. 6.
10. Peebles, *Guardians*, p. 251.
11. Ibid.
12. William E. Burrows, *Deep Black: Space Espionage and National Security* (New York: Random House, 1986), p. 236.

13. Peebles, *Guardians,* p. 251. Defense Marketing Service, "MOL," *DMS Market Intelligence Report* (Greenwich, CT: DMS, 1967).

14. Defense Marking Service, "MOL".

15. Ibid.

16. Ibid.; "Industry Observer," *Aviation Week and Space Technology,* October 18, 1965, p. 13.

17. Peebles, *Guardians,* p. 252.

18. Andelin, *Evaluation of the Usefulness of the MOL,* p. 90.

19. U.S. Congress, Senate Committee on Appropriations, *Department of Defense Appropriations for Fiscal Year 1969, Part 1* (Washington DC: U.S. Government Printing Office, 1968), p. 512.

20. Mary L. Whittington, James Eastman, W. E. Alford, and Charles W. Dickens, *MAC Support of Project GEMINI 1963-1966* (Scott AFB: Ill., Military Airlift Command, April 1967), pp. 7–9.

21. Burrows, *Deep Black,* p. vii; Evert Clark, "Satellite Spying Cited by Johnson," *New York Times,* March 17, 1967, p. 13.

22. Robert P. Berman and John Baker, *Soviet Strategic Forces: Requirements and Responses* (Washington DC: The Brookings Institution, 1982), pp. 104–105; Lawrence Freedman, *U.S. Intelligence and the Soviet Strategic Threat,* 2nd ed. (Princeton, NJ: Princeton University Press, 1986), pp. 131–34.

23. Ibid., p. 88; Peebles, *Guardians,* p. 95.

24. Freedman, *U.S. Intelligence and the Soviet Strategic Threat,* p. 88.

25. Donald Neff, *Warriors for Jerusalem: The Six Days that Changed the Middle East* (New York: Simon & Schuster, 1984), pp. 201–202; Ze'ev Schiff, *A History of the Israeli Army: 1874 to the Present* (New York: Macmillan, 1985), pp. 127, 130.

26. Neff, *Warriors for Jerusalem,* p. 203.

27. Ibid.

28. Ibid., p. 204; Schiff, *A History of the Israeli Army,* p. 140.

29. Neff, *Warriors for Jerusalem,* p. 207.

30. Interview; Interview with Robert McNamara, January 20, 1989, Washington DC.

31. John Ranelagh, *The Agency: The Rise and Decline of the CIA* (New York: Simon & Schuster, 1986), p. 328; Interview.

32. DCID 1/13, "Committee on Imagery Requirements and Exploitation (COMIREX)," July 1, 1967, *DDRS,* 1980-132B.

33. Ibid.

34. Jiri Valenta, *Soviet Intervention in Czechoslovakia, 1968: Anatomy of a Decision* (Baltimore, MD: Johns Hopkins, 1979), pp. 11–12.

35. Ibid., pp. 15, 21, 23.

36. Interview.

37. Interview.

38. Interview.
39. *CIA: The Pike Report* (Nottingham: Spokesman Books, 1977), pp. 139–140.
40. Jonathan Steele, *Soviet Power: The Kremlin's Foreign Policy-Brezhnev to Chernenko* (New York: Touchstone/Simon & Schuster, 1984), p. 138.
41. Steele, *Soviet Power,* p. 138.
42. Ibid., p. 139.
43. Ibid.
44. Ibid.
45. Ibid., p. 141.
46. Ibid., p. 142; Seymour Hersh, *The Price of Power: Kissinger in the Nixon White House* (New York: Summit, 1983), p. 357.
47. Interview.
48. Ibid.
49. Peebles, *Guardians,* p. 252.
50. Ibid.
51. Ibid., p. 253.
52. "MOL Delayed by Funding Cut," *Aviation Week and Space Technology,* April 21, 1969, p. 17.
53. Cecil Brownlow, "Budget Cuts Threaten MOL Project," *Aviation Week and Space Technology,* May 5, 1969, pp. 22–23.
54. "MOL Delayed by Funding Cut."
55. "President By-Passed DOD & Joint Chiefs to Cancel MOL," *Space Daily,* December 16, 1969, p. 199; Peebles, p. 253.
56. Victor Marchetti and John D. Marks, *The CIA and the Cult of Intelligence* (New York: Knopf, 1974), p. 97; Interview.
57. Andelin, *Evaluation of the Usefulness of the MOL,* p. 57.
58. Peebles, *Guardians,* pp. 245, 248.
59. Freedman, *U.S. Intelligence and the Soviet Strategic Threat,* pp. 156, 214.
60. Ibid., pp. 156–57; John Prados, *The Soviet Estimate: U.S. Intelligence Analysis and Russian Military Strength* (New York: Doubleday, 1982), p. 169.
61. Peebles, *Guardians,* p. 100.

Notes For Chapter Five: The Big Bird Arrives

1. Curtis Peebles, *Guardians: Strategic Reconnaissance Satellites* (Novato, CA: Presidio Press, 1987), p. 108; Interview.
2. Interview.
3. Interview.
4. Peebles, *Guardians,* pp. 107–108.
5. Interview.

6. Interview.
7. Anthony Kenden, "U.S. Reconnaissance Satellite Programmes," *Spaceflight,* 20, 7, 1978, p. 243ff.
8. Interview.
9. Ibid; Allan S. Krass, *Verification: How Much is Enough?* (Philadelphia: Taylor & Francis, 1985), p. 30; Richard D. Hudson, Jr. and Jacqueline W. Hudson, "The Military Applications of Remote Sensing by Infrared," *Proceedings of the IEEE* 63,1 1975, pp. 104–128.
10. Interview.
11. Private information.
12. Interview; "Space Reconnaissance Dwindles," *Aviation Week and Space Technology,* October 6, 1980, pp. 18–20.
13. Interview.
14. Curtis Peebles, "The Guardians," *Spaceflight,* November 1978, p. 381ff.
15. William E. Burrows, *Deep Black: Space Espionage and National Security* (New York: Random House, 1986). p. 239.
16. Lawrence Freedman, *U.S. Intelligence and the Soviet Strategic Threat,* 2nd ed. (Princeton, NJ: Princeton University Press, 1986), p. 164.
17. Ibid., pp. 164-65; John Newhouse, *War and Peace in the Nuclear Age* (New York: Knopf, 1989), p. 226.
18. Stewart Alsop, "What's Going in the Holes?", *Newsweek,* May 10, 1971, p. 124.
19. Freedman, *U.S. Intelligence and the Soviet Strategic Threat* pp. 165-66; Michael Getler, "New Soviet Silo Building Seen as Protection for Two Missiles," *Washington Post,* May 27, 1971, p. A14.
20. Getler, "New Soviet Silo Building Seen as Protection for Two Missiles."
21. John Newhouse, *Cold Dawn: The Story of SALT* (New York: Holt, Reinhart & Winston, 1973), p. 15.
22. Private information.
23. Ibid.
24. Arms Control and Disarmament Agency, *Arms Control and Disarmament Agreements: Texts and Histories of Negotiations* (Washington DC: ACDA, 1980), pp. 148–49.
25. Anthony Kenden, "U.S. Reconnaissance Satellite Programmes," *Spaceflight,* 20, 7 1978, p. 243ff.
26. Fred Kaplan, *The Wizards of Armageddon* (New York: Simon & Schuster, 1983), p. 357.
27. James Schlesinger, *A Review of the Intelligence Community* (Washington DC: Office of Management and Budget, March 10, 1971), pp. 6, 20.
28. Interview.
29. Peebles, *Guardians,* p. 112.
30. Freedman, *U.S. Intelligence and the Soviet Strategic Threat,* pp. 174–75.

31. Ibid.
32. Stewart Steven, *The Spymasters of Israel* (New York: Macmillan, 1980), p. 297; Jeffrey Richelson, *Foreign Intelligence Organizations* (Cambridge, MA: Ballinger, 1988), pp. 215–17.
33. Ze'ev Schiff, *A History of the Israeli Army: 1874 to the Present* (New York: Macmillan, 1985), p. 212; Richard Deacon, *The Israeli Secret Service* (New York: Taplinger, 1977), pp. 262, 264; Steven, *The Spymasters of Israel,* p. 298.
34. Jacques Derogy and Henri Carmel, *The Untold Story of Israel* (New York: Grove Press, 1979), p. 281; Steven, *The Spymasters of Israel,* pp. 303–4.
35. Schiff, *A History of the Israeli Army: 1874 to the Present* (New York: Macmillan, 1985), pp. 207–208, 215.
36. Ibid., p. 214; Steve Weissman and Herbert Krosney, *The Islamic Bomb* (New York: Times Books, 1981), p. 107.
37. Ibid., p. 221.
38. Ibid., p. 208.
39. Kurt Gottfried and Bruce G. Blair (ed.), *Crisis Stability and Nuclear War* (New York: Oxford University Press, 1988), pp. 200–203; Newhouse, *War and Peace in the Nuclear Age,* p. 239.
40. U.S. Congress, Senate Select Committee to Study Governmental Operations with Respect to Intelligence Activities, *Final Report, Book I: Foreign and Military Intelligence* (Washington DC: U.S. Government Printing Office, 1976), p. 85.
41. Interview.
42. *CIA: The Pike Report* (Nottingham: Spokesman Books, 1977), p. 145.
43. Clarence A. Robinson, Jr., "Soviets Hiding Submarine Work," *Aviation Week and Space Technology,* November 11, 1974, pp. 14–16.
44. Ralph Kinney Bennett, "U.S. Eyes Over Russia: How Much Can We See?", *Reader's Digest,* October 1985, pp. 142–47.
45. Burrows, *Deep Black,* p. 241; Gloria Duffy (ed.), *Compliance and the Future of Arms Control* (Cambridge, MA: Ballinger, 1988), pp. 105–107.
46. Robert P. Berman and John C. Baker, *Soviet Strategic Forces: Requirements and Responses* (Washington DC: The Brookings Institution, 1982), pp. 104–105.
47. Ibid.
48. Peebles, *Guardians,* p. 116.
49. Norman Polmar (ed.), *Soviet Naval Developments* (Annapolis, MD: Nautical and Aviation Publishing Co. of America, 1979), p. 21.
50. Jan M. Lodal, "Verifying SALT," *Foreign Policy,* 24, Fall 1976, pp. 40–64.
51. Ibid.
52. Ibid.

Notes For Chapter Six: Quantum Leap

1. William E. Burrows, *Deep Black: Space Espionage and National Security* (New York: Random House, 1986), p. 242.
2. Robert P. Berman and John Baker, *Soviet Strategic Forces: Requirements and Responses* (Washington DC: The Brookings Institution, 1982), pp. 104–107.
3. Anthony Kenden, "U.S. Reconnaissance Satellite Programmes," *Spaceflight*, 20,7 1978, pp. 243ff.
4. Burrows, *Deep Black*, p. 227.
5. Interview.
6. Ibid.
7. United States of America vs. Samuel Loring Morison, United States District Court of Maryland, Case No. Y-84-00455, October 15, 1985, Direct Testimony of Roland Inlow, pp. 5–6.
8. Interview.
9. Ibid.
10. Ibid.
11. Ibid.
12. Ibid.
13. Ibid.
14. Ibid.
15. Ibid.
16. Ibid.
17. Ibid.
18. United States of America vs. Samuel Loring Morison, October 16, 1985, Direct Testimony of Richard J. Kerr, pp. 43–47.
19. Michael Getler, "New Spy Satellites Planned for Clearer, Instant Pictures," *Washington Post*, February 8, 1972, pp. A1, A9; "Industry Observer," *Aviation Week and Space Technology*, February 7, 1972, p. 9 and May 8, 1972, p. 9.
20. Interview.
21. John Noble Wilford, "Spy Satellite Reportedly Aided in Shuttle Flight," *New York Times*, October 20, 1981, p. C4.
22. James R. Janesick and Morley Blouke, "Sky on a Chip: The Fabulous CCD," *Sky & Telescope*, September 1987, pp. 238–42; Burrows, p. 244.
23. James R. Janesick and Morley M. Blouke, "Introduction to Charged Couple Device Imaging Sensors," in Kosta Tsipis (ed.), *Arms Control Verification: The Technologies that Make It Possible* (New York: Pergamon-Brassey's, 1985), p. 104; Curtis Peebles, *Guardians: Strategic Reconnaissance Satellites* (Novato, CA: Presidio Press, 1987), pp. 118–119.
24. Burrows, *Deep Black*, p. 244.
25. Ibid., pp. 244–45.

26. Ibid., p. 245.

27. Ibid., p. 247; Interview.

28. Interview.

29. Jeffrey Richelson, "The Satellite Data System," *Journal of the British Interplanetary Society*, 37,5 (1984), pp. 226–28.

30. John Pike, "Reagan Prepares for War in Outer Space," *CounterSpy*, 7, 1 September-November 1982, pp. 17–22; James Bamford, "America's Supersecret Eyes in Space," *The New York Times Magazine*, January 13, 1985, pp. 39ff.

31. *Organization and Functions Chart Book, Air Force Satellite Control Facility*, (Onizuka AFS, California: AFSCF, 1987), pp. 19–20.

32. Interview.

33. Telephone interview with Clarence A. Robinson, Jr., May 11, 1989; Interviews.

34. Burrows, *Deep Black*, p. 226.

35. Ibid.

36. Ibid.

37. Ibid.

38. Ibid.

39. Ibid., pp. 228–29.

40. Interview.

41. Interview.

42. Burrows, *Deep Black*, p. 218; George Wilson, " 'N-Pic'—CIA Technicians Ferret Out Secrets Behind Cemented Windows," *Los Angeles Times*, January 12, 1975, pp. 14–17.

43. NSCID No. 8, "Photographic Interpretation," February 17, 1972, *DDRS* 1976–253 G.

44. "The Art and Science of Photointerpretation," *IEEE Spectrum*, July 1986, pp 52–53; Translation of "Satellite Reconnaissance and Photographic Interpretation," *Aviation Knowledge*, July 1983, pp. 22–23.

45. Private information.

46. "The Art and Science of Photointerpretation."

47. Interview with Kevin Hussey, Pasadena, California, November 30, 1988.

48. Allan S. Krass, *Verification: How Much is Enough?* (Philadelphia: Taylor and Francis, 1986), p. 54.

49. "The Art and Science of Photointerpretation."

50. Ibid.; John F. Ebersole and James C. Wyant, "Real-time Optical Subtraction of Photographic Imagery for Difference Detection," *Applied Optics* 15,4 (1976), pp. 871–76.

51. "The Art and Science of Photointerpretation."

52. Ibid.

53. Ibid.

54. Ibid.

55. Leonard S. Spector, *Nuclear Proliferation Today: The Spread of Nuclear Weapons 1984*, (Cambridge, MA: Ballinger, 1984), p. 282.
56. Ibid., pp. 282–83.
57. CIA, "Prospects for Further Proliferation of Nuclear Weapons," DCI/NIO 1945/74, September 4, 1974, p. 3.
58. Spector, *Nuclear Proliferation Today*, pp. 291–92.
59. Ibid., p. 292; Interview.
60. Spector, *Nuclear Proliferation Today*, p. 292.
61. Dale Van Atta, "Death of the State Secret," draft, 1984.
62. "Remarks by President Carter at the Kennedy Space Center, Florida [Extract], October 1, 1978," *Documents on Disarmament 1978* (Washington DC: U.S. Government Printing Office, 1979), p. 586.
63. Robert C. Toth, "U.S. Facing Uproar in Spy Satellite Photos," *Los Angeles Times*, November 12, 1978, pp. 1, 27.
64. Interview with Robert McNamara, Washington DC, January 19, 1989; Interview.
65. Interview.
66. Interview with Paul Warnke, Washington DC, September 27, 1988; Telephone interview with Roger Molander.
67. Van Atta, "Death of the State Secret."
68. Ibid.
69. Ibid.
70. Ibid.
71. Warnke interview.
72. Telephone interview with Bobby R. Inman, June 27, 1989; Hans Mark, *The Space Station: A Personal Journey* (Durham, NC: Duke University Press, 1987), p. 81.
73. Toth, "U.S. Facing Uproar in Spy Satellite Photos."
74. Ibid.
75. Inman interview.
76. Ibid.

Notes For Chapter Seven: Betrayal

1. Andrew Tully, *Inside the FBI* (New York: Dell, 1987), p. 48; George Lardner, Jr., "Spy Rings of One," *Washington Post Magazine*, December 4, 1983, pp. 60–65.
2. Lardner, "Spy Rings of One"; Tully, *Inside the FBI*, p. 48.
3. Henry Hurt, "CIA in Crisis: The Kampiles Case," *Reader's Digest*, June 1979, pp. 65–72.
4. Lardner, "Spy Rings of One"; Tully, *Inside the FBI*, p. 45; Griffin Bell, *Taking Care of the Law* (New York: William Morrow, 1982), p. 119; Hurt, "CIA in Crisis."

5. United States of America vs. William Peter Kampiles, United States District Court, Northern District of Indiana, November 6, 1978, Direct Testimony of Donald E. Stukey.
6. Hurt, "The CIA in Crisis: The Kampiles Case."
7. United States of America vs. William Peter Kampiles, Direct Testimony of Ray Hart, p. 635.
8. Ibid., p. 636.
9. Tully, *Inside the FBI*, p. 45; Lardner, "Spy Rings of One"; Interview.
10. Stansfield Turner, *Secrecy and Democracy: The CIA in Transition* (Boston, MA: Houghton-Mifflin, 1985), p. 65; Interview.
11. Turner, *Secrecy and Democracy*, p. 69.
12. Interview; United States of America vs. William Peter Kampiles, Direct testimony of Donald E. Stukey, p. 804.
13. Ibid., p. 805; Peebles, *Guardians: Strategic Reconnaissance Satellites* (Novato, CA: Presidio, 1987), p. 120; Lardner, "Spy Rings of One."
14. Peebles, *Guardians*, p. 120; Lardner, "Spy Rings of One."
15. United States of America vs. William Peter Kampiles, Direct Testimony of Donald E. Stukey, p. 806.
16. Ibid., p. 807.
17. Ibid., p. 808; Lardner, "Spy Rings of One."
18. United States of America vs. William Peter Kampiles, Direct Testimony of Donald E. Stukey, p. 809.
19. Lardner, "Spy Rings of One"; United States of America vs. William Peter Kampiles, Direct Testimony of Vivian Psachos, p. 259.
20. Tully, *Inside the FBI*, p. 42.
21. Ibid., p. 42.
22. Ibid.
23. Tully, *Inside the FBI*, pp. 42–43; Bell, *Taking Care of the Law*, pp. 120–121.
24. Tully, *Inside the FBI*, p. 43.
25. Ibid.
26. Hurt, "The CIA in Crisis: The Kampiles Case."
27. Tully, *Inside the FBI*, p. 44.
28. Thomas O'Toole and Charles Babcock, "CIA 'Big Bird' Satellite Manual Was Allegedly Sold to the Soviets," *Washington Post*, August 23, 1978, pp. A1, A16; Michael Ledeen, "A Mole in Our Midst," *New York*, October 2, 1978, pp. 55–57.
29. James Ott, "Espionage Trial Highlights CIA Problems," *Aviation Week and Space Technology*, November 27, 1978, pp. 21–23.
30. Tully, *Inside the FBI*, p. 52.
31. Ibid.; United States of America vs. William Kampiles, Direct Testimony of James Murphy, p. 352; George Lardner Jr., "Former CIA Man Convicted as Spy in Sale of Secrets," *Washington Post*, November 18, 1978, pp. A1, A10.

32. Hurt, "The CIA in Crisis: The Kampiles Case"; Tully, *Inside the FBI*, p. 52; United States of America vs. William Peter Kampiles, Direct Testimony of James Murphy, pp. 349–50.
33. Tully, *Inside the FBI*, p. 53; United States of America vs. William Peter Kampiles, Direct Testimony of James Murphy, pp. 382–89.
34. Bell, *Taking Care of the Law*, p. 121.
35. Ibid., pp. 121–22.
36. Ibid., p. 122.
37. Ibid.
38. Ibid.
39. Ibid.
40. Ibid., pp. 122–23.
41. Ibid., p. 123; Tully, *Inside the FBI*, p. 51.
42. George Lardner Jr., "13 Copies of Classified Data Missing," *Washington Post*, November 7, 1978, pp. A1, A8.
43. Ibid.
44. Ibid.
45. Ibid.
46. William R. Corson, Susan B. Trento and Joseph J. Trento, *Widows* (New York: Crown, 1989), pp. 37–150.
47. Hurt, "The CIA in Crisis: The Kampiles Case."
48. Lardner, "13 Copies of Classified Data Missing."
49. Ott, "Espionage Trial Highlights CIA Problems."
50. Ibid.
51. Tully, *Inside the FBI*, p. 53.
52. Ibid.
53. Ott, "Espionage Trial Highlights CIA Problems."
54. United States of America vs. William Peter Kampiles, Direct Testimony of Leslie Dirks, pp. 6, 8.
55. Ibid., pp. 10, 12.
56. Ibid., p. 13.
57. Ott, "Espionage Trial Highlights CIA Problems."
58. Tully, *Inside the FBI*, p. 54.
59. Ott, "Espionage Trial Highlights CIA Problems."
60. Tully, *Inside the FBI*, p. 55.
61. Ibid., p. 56.
62. Bell, *Taking Care of the Law*, pp. 123–24
63. Robert C. Toth, "CIA 'Mighty Wurlitzer' is Now Silent," *Los Angeles Times*, December 30, 1980, pp. 1, 12.
64. Interview with Stansfield Turner, McLean, Virginia, May 30, 1984.
65. O'Toole and Babcock, "CIA 'Big Bird' Satellite Manual Was Allegedly Sold to Soviets."
66. Don Oberdorfer, "The 'Brigada': An Unwelcome Sighting in Cuba," *Washington Post*, September 9, 1979, pp. A1, A18.

67. Paul Crickmore, *Lockheed SR-71 Blackbird* (London: Osprey, 1986), p. 164.
68. Ibid.
69. Oberdorfer, "The 'Brigada'. "
70. Ibid.
71. Gloria Duffy, "Crisis Mangling and the Cuban Brigade," *International Security*, 8, 1 1983, pp. 67–87.
72. Ibid; Tad Szulc, "Russians in Cuba Now Put at 12,500," *New York Times*, June 20, 1963, p. 2.
73. Duffy, "Crisis Mangling and the Cuban Brigade."
74. Ibid.; David Binder, "Soviet Brigade: How the U.S. Traced It," *New York Times*, September 13, 1979, p. A16.
75. Oberdorfer, "The 'Brigada'."
76. Duffy, "Crisis Mangling and the Cuban Brigade."
77. Ibid.; Oberdorfer, "The 'Brigada'."
78. Oberdorfer, "The 'Brigada'."
79. David D. Newsom, *The Soviet Brigade in Cuba: A Study in Political Diplomacy* (Bloomington, IN: Indiana University Press, 1977), p. 21.
80. Duffy, "Crisis Mangling and the Cuban Brigade."
81. Newsom, *The Soviet Brigade in Cuba*, pp. 12–19, 22; Duffy, "Crisis Mangling and the Cuban Brigade"; John Newhouse, *War and Peace in the Nuclear Age* (New York: Knopf, 1989), p. 330; Martin Schram, "Response: Avoiding a Crisis Tone," *Washington Post*, September 9, 1979, p. A18.
82. Newhouse, *War and Peace in the Nuclear Age*, p. 330.
83. Arms Control and Disarmament Agency, *Arms Control and Disarmament Agreements: Texts and Histories of Negotiations* (Washington DC: ACDA, 1980), p. 210; Strobe Talbot, *Endgame: The Inside Story of SALT II* (New York: Harper & Row, 1979), p. 111.
84. Talbot, *Endgame*, pp. 111–12.
85. Ibid., p. 112.
86. Ibid., pp. 112–13.
87. Ibid., p. 113.
88. Ibid., p. 114.
89. ACDA, *Arms Control and Disarmament Agreements*, p. 212.
90. Gary Sick, *All Fall Down: America's Tragic Encounter with Iran* (New York: Penguin, 1986), p. 228.
91. U.S. Congress, House Permanent Select Committee on Intelligence, *Iran: Evaluation of U.S. Intelligence Performance Prior to November 1978* (Washington DC: U.S. Government Printing Office, 1979), pp. 6, 7; Central Intelligence Agency, *Iran in the 1980s* (Washington DC: CIA, August 1977), p. iii.
92. Interview.

93. Zbigniew Brzezinski, *Power and Principle: Memoirs of the National Security Adviser 1977–1981* (New York: Farrar, Straus, Giroux, 1983), pp. 479, 487.
94. Steve Emerson, *Secret Warriors: Inside the Covert Military Operations of the Reagan Era* (New York: G.P. Putnam's, 1988), p. 20.
95. Ibid.
96. Ibid., p. 21.
97. Interview.
98. Charlie A. Beckwith and Donald Knox, *Delta Force* (New York: Harcourt, Brace, Jovanovich, 1983), p. 220.
99. Ibid.
100. Ibid., pp. 254–55.
101. Ibid., p. 253.
102. Ibid.
103. Ibid.
104. Ibid., pp. 253–54.
105. Ibid., p. 254.
106. Sick, *All Fall Down*, p. 349.
107. David C. Martin and John Walcott, *Best Laid Plans: The Inside Story of America's War Against Terrorism* (New York: Harper & Row, 1988), p. 24.
108. Private information.

Notes For Chapter Eight: Watching The Evil Empire

1. Department of Defense, *Soviet Military Power: An Assessment of the Threat 1988* (Washington DC: U.S. Government Printing Office, 1988), pp. 50–51; David M. North and John D. Morocco, "Blackjack Shares Aspects of U.S. B-1B and XB-70," *Aviation Week and Space Technology*, August 15, 1988, pp. 16–18.
2. "Soviet Strategic Bomber Photographed at Ramenskoye," *Aviation Week and Space Technology*, December 14, 1981, p. 17.
3. Max White, "U.S. Satellite Reconnaissance During the Falklands War," Earth Satellite Research Unit, Department of Mathematics, University of Aston, p. 5.
4. "Possible Soviet Chemical Weapons Launcher is Tested," *Weekly Intelligence Summary*, July 23, 1982, pp. 4–5; "USSR Builds Aircraft Takeoff Ramp," *Weekly Intelligence Summary*, July 23, 1982, pp. 2–3.
5. "Tracked Multiple Rocket Launchers Noted in Beijing Military Region," *Weekly Intelligence Summary*, July 23, 1982, pp. 11–12; "Strategic Value of Simonstown Naval Base May Decrease," *Weekly Intelligence Summary*, July 23, 1982, pp. 23–25.

6. John Walcott, "U.S. Analysts Find Soviet Radars, Possibly Complicating Arms Pact Effort," *Wall Street Journal*, August 15, 1986, p. 2.

7. Charles Mohr, "U.S. and Soviets Discuss Whether Moscow Violated Terms of Arms Pacts," *New York Times*, October 5, 1983, p. A8.

8. Walcott, "U.S. Analysts Find Soviet Radars."

9. Philip Klass, "U.S. Scrutinizing New Soviet Radar," *Aviation Week and Space Technology*, August 22, 1983, pp. 19–20; Mohr, "U.S. and Soviets Discuss."

10. Gloria Duffy (ed.), *Compliance and the Future of Arms Control* (Cambridge, MA: Ballinger, 1988), p. 107.

11. Ibid., pp. 148–49; U.S. Arms Control and Disarmament Agency, *Soviet Noncompliance*, (Washington DC: ACDA, February 1, 1986), pp. 1–2.

12. Michael R. Gordon, "Soviet Finishing Large Radar Center in Siberia," *New York Times*, November 23, 1986, p. 6.

13. Mark Daly, "Krasnoyarsk: first picture suggests Treaty violation," *Jane's Defence Weekly*, April 11, 1987, pp. 620–21.

14. Philip Klass, "Soviets Test Defense Missile Reload," *Aviation Week and Space Technology*, August 29, 1983, p. 19.

15. Jeffrey T. Richelson, *The U.S. Intelligence Community* (Cambridge, MA: Ballinger, 1988, 2nd ed.), p. 435.

16. Steve Emerson, *Secret Warriors: Inside the Covert Military Operations of the Reagan Era,* (New York: G.P. Putnam's, 1988), p. 149.

17. *The President's Unclassified Report on Soviet Noncompliance with Arms Control Agreements*, December 2, 1989, p. 7; Bill Gertz, "Soviets Still Stalling on Removal of Disputed Radar, Cheney Says," *Washington Times*, June 9, 1989, p. A5.

18. Charles L. Smith, "Soviet Maskirovka," *Air Power Journal*, Summer 1988, pp. 28–39.

19. Ibid.; Roger Beaumont, *Maskirovka: Soviet Camouflage, Concealment and Deception* (College Station, TX: The Texas Engineering Experiment Station, Texas A&M, November, 1982), pp. 2–6.

20. George Soloveytchik, *Potemkin: Soldier, Statesman, Lover and Consort of Catherine of Russia* (New York: W.W. Norton & Co., 1947), pp. 78, 176, 186, 276–77.

21. David M. Glantz, "The Red Mask: The Nature and Legacy of Soviet Military Deception in the Second World War," *Intelligence and National Security*, 2, 3 July 1987, pp. 175–259.

22. Oleg Penkovskiy, *The Penkovskiy Papers* (New York: Doubleday, 1965), pp. 216–17, 342; Lawrence Freedman, *U.S. Intelligence and the Soviet Strategic Threat, 2nd ed.* (Princeton, NJ: Princeton University Press, 1986), p. 101; John Prados, *The Soviet Estimate: U.S. Intelligence Analysis and Russian Military Strength* (New York: Doubleday, 1982), p. 243;

Curtis Peebles, *Guardians: Strategic Reconnaissance Satellites* (Novato, CA: Presidio Press, 1987), p. 98.

23. Yossef Bodansky, "Ogarkov Maps Soviets' Strategy for Winnable War," *Washington Times*, July 23, 1985, pp. 1A, 10A.

24. Viktor Suvorov, "GUSM: The Soviet Service of Strategic Deception," *International Defense Review*, 8, 1985, pp. 1235–40.

25. Viktor Suvorov, *Inside the Soviet Army* (New York: Macmillan, 1982), p. 106.

26. Accounts of the NSC reports can be found in "How Russia Hides Its Missiles," *Foreign Report*, March 5, 1981, pp. 1–3 and Richard Burt, "U.S. Report Says Soviet Attempted Deception on Its Nuclear Strength," *New York Times*, September 26, 1979, p. A4. Additional material about submarine covers and dummy submarines can be found in Clarence A. Robinson, Jr., "Soviets Hiding Submarine Work," *Aviation Week and Space Technology*, November 11, 1974, pp. 14–16, and Andrew Cockburn, *The Threat: Inside the Soviet Military Machine* (New York: Random House, 1983), p. 277.

27. Burt, "U.S. Report Says Soviet Attempted Deception on its Nuclear Strength."

28. "Washington Roundup," *Aviation Week and Space Technology*, September 29, 1980, p. 17.

29. Frank Greve, "Soviets Trying to Hide New Missile System in Area of Jet Incident," *San Francisco Chronicle & Examiner*, September 11, 1983, p. A5.

30. "Washington Roundup," *Aviation Week and Space Technology*, October 31, 1983, p. 17.

31. Rick Atkinson, "Soviet Tunnels Could Hide Submarines," *Washington Post*, April 7, 1984, p. A4; Ted Agres, "Soviet Navy Completing Four Tunnels to Hide Subs," *Washington Times*, March 27, 1984, pp. 1A, 12A.

32. Bill Gertz, "Soviets Fill Craters, Dig New Ones to Fool U.S. on Missile Accuracy," *Washington Times*, August 7, 1985, pp. 1A, 10A.

33. "Arms Verification Issue at Heart of U.S. Debate," *New York Times*, November 24, 1985, p. 18.

34. Michael Gordon, "U.S. Says Soviet Complies on Some Arms Issues," *New York Times*, November 24, 1985, p. 18.

35. Chong-Pin Lin, *China's Nuclear Weapons Strategy: Tradition Within Evolution* (Lexington, MA: Lexington Books, 1988), p. 55.

36. Wolf Blitzer, *Territory of Lies* (New York: Harper & Row, 1989), p. 4.

37. Bruce D. Berkowitz and Allan E. Goodman, *Strategic Intelligence for American National Security* (Princeton, NJ: Princeton University Press, 1989), p. 18.

38. Central Intelligence Agency, *KGB and GRU* (Washington DC: CIA, 1984), p. 43.

39. United States of America vs. Samuel Loring Morison, District Court for Maryland, October 15, 1985, Direct Testimony of Richard J. Kerr, pp. 61–62.
40. William J. Broad, "U.S. Designs Spy Satellites To Be More Secret Than Ever," *New York Times,* November 3, 1987, pp. C1, C3; Dino A. Brugioni, "Hiding the Aircraft Factories," *Air Force Magazine,* March, 1983, pp. 112–15.
41. Stephen Engelberg, "Recent Setbacks Aside, C.I.A. Still Wants More 'Human' Spies," *New York Times,* November 14, 1985, p. A12.
42. Central Intelligence Agency, *KGB and GRU,* p. 43 n. 31.
43. Telephone interview with Bobby R. Inman, June 27, 1989.
44. Drew Middleton, "Soviet's Moves in Poland Are Studied," *New York Times,* January 9, 1981, p. 4.
45. "Space Reconnaissance Dwindles," *Aviation Week and Space Technology,* October 6, 1980, pp. 18–20; Interview.
46. "Sunny Debut for Snowstorm," *Time,* November 28, 1988, p. 80.
47. Nicholas Johnson, *The Soviet Year in Space 1988* (Colorado Springs, CO: Teledyne-Brown Engineering, 1989), p. 111.
48. Ibid., p. 108; Private information.
49. "Letters," *Aviation Week and Space Technology,* January 16, 1989, p. 76.
50. Thomas Y. Canby, "Are the Soviets Ahead in Space?" *National Geographic,* October 1986, pp. 420-458; "Soviets Ready New Boosters at Tyuratam," *Aviation Week and Space Technology,* August 27, 1984, pp. 18–19.
51. White, "U.S. Satellite Reconnaissance During the Falklands Conflict."
52. "Satellite Pictures Show Soviet CVN Towering Above Nikolaev Shipyard," *Jane's Defence Weekly,* August 11, 1984, pp. 171–73.
53. Thomas B. Allen and Norman Polmar, *Merchants of Treason: America's Secrets for Sale* (New York: Delacorte, 1988), p. 185; "The Dark Side of Moonlighting," *Security Awareness Bulletin,* June 1987, pp. 1–12.
54. Allen and Polmar, *Merchants of Treason,* p. 187.
55. Richard Halloran, "U.S. Says Blast Hit Soviet Arms Base," *New York Times,* June 23, 1984, p. 3; Rick Atkinson, "Soviet Arms Disaster Reported," *Washington Post,* June 22, 1984, pp. A1, A12.
56. Atkinson, "Soviet Arms Disaster Reported."
57. Ibid.; Halloran, "U.S. Says Blast Hit Soviet Arms Base."
58. "Soviet Naval Blast Called Crippling," *New York Times,* July 11, 1984, p. 6.
59. *Documents from the Espionage Den (52): USSR: The Aggressive East, Section 3–2* (Teheran: Muslim Students Following the Line of the Imam, n.d.), pp. 40–41.

Notes For Chapter Nine: Disaster and Recovery

1. Personal correspondence with Anthony Kenden, May 23, 1985.
2. Ibid.
3. *USAF Mishap Report 86-4-18-701* (Los Angeles, CA: Space Division, June 8, 1986).
4. "Titan 34D Booster Failed Following Premature Shutdown of Aerojet Engine," *Aviation Week and Space Technology*, November 18, 1985, p. 26; *USAF Mishap Report 85-8-28-701* (Los Angeles, CA: Space Division, October 25, 1985), p. A-1.
5. "Titan 34D Booster Failed . . ."; *USAF Mishap Report 85-8-28-701*, p. A-1.
6. *USAF Mishap Report 85-8-28-701*, p. A-1.
7. Joseph J. Trento, *Prescription for Disaster* (New York: Crown, 1987), pp. 289–91.
8. Ibid., pp. 98–101.
9. Ibid., p. 101.
10. "Washington Roundup," *Aviation Week and Space Technology*, June 4, 1979, p. 11; Private information.
11. Trento, *Prescription for Disaster*, p. 110; Thomas O'Toole, "AF Study Questions Economy of Space Shuttle," *Washington Post*, May 27, 1971, p. 10.
12. Trento, *Prescription for Disaster*, pp. 155–56.
13. Ibid., p. 167.
14. Paul B. Stares, *The Militarization of Space: U.S. Policy 1945-84* (Ithaca, NY: Cornell University Press, 1985), p. 185; Hans Mark, *The Space Station: A Personal Journey* (Durham, NC: Duke University Press, 1987), p. 79.
15. "Space Reconnaissance Dwindles," *Aviation Week and Space Technology*, October 6, 1980, pp. 18–20.
16. William E. Burrows, *Deep Black: Space Espionage and National Security* (New York: Random House, 1987), p. 304; Jack Cushman, "Space Shuttle Explosion Throws Military Programs Into Disarray," *Defense Week*, February 3, 1986, p. 2–5.
17. Burrows, *Deep Black*, pp. 304–305; *USAF Mishap Report 86-4-18-701*, pp. A-1, E-1; "Explosion Sequence Photos Depict Titan 34D Launch Failure," *Aviation Week and Space Technology*, November 3, 1986, p. 43.
18. U.S. Congress, Senate Armed Services Committee, *Department of Defense Authorization of Appropriations for Fiscal Year 1989, Part 6* (Washington DC: U.S. Government Printing Office, 1988), p. 308.
19. "Rocket Explodes; 58 Sent to Hospital," *Mansfield News-Journal*, April 19, 1986, p. 1–A.

20. Jeff Snyder, "Vandenberg Rocket Blows Up," *L.A. Daily News*, April 19, 1986, pp. 1, 11.
21. Theresa M. Foley, "Titan Analysis Unable to Show Cause of Insulation Separation from Casing," *Aviation Week and Space Technology*, November 3, 1986, p. 43.
22. Interview.
23. Nigel Hawkes, Geoffrey Lean, David Leigh, Robin McKie, Peter Pringle and Andrew Wilson, *Chernobyl: The End of the Nuclear Dream* (New York: Vintage, 1986), pp. 99–103.
24. Stephen Engelberg, "U.S. Says Intelligence Units Did Not Detect the Accident," *New York Times*, May 2, 1986, p. A9.
25. Hawkes, Lean, Leigh, McKie, Pringle and Wilson, *Chernobyl*, p. 122.
26. "Meltdown," *Newsweek*, May 12, 1986, pp. 20–35; Boyce Rensberg, "Explosion, Graphite Fire Suspected," *Washington Post*, April 30, 1986, pp. A1, A17; Carl M. Cannon and Mark Thompson, "Threat to Soviets grows, U.S. spy photos indicate," *Miami Herald*, April 30, 1986, pp. 1A, 14A.
27. "Meltdown."
28. Ibid.
29. Robert C. Toth, "Satellites Keep Eye on Reactor," *Los Angeles Times*, May 2, 1986, p. 22.
30. "Meltdown"; Bernard Gwertzman, "Fire in Reactor May Be Out, New U.S. Pictures Indicate; Soviet Says Fallout is Cut," *New York Times*, May 2, 1986, pp. A1, A8.
31. Philip M. Boffey, "U.S. Panel Calls the Disaster in the Ukraine the Worst Ever," *New York Times*, May 4, 1986, pp. 1, 20.
32. Serge Schemann, "Soviet Mobilizes a Vast Operation to Overcome the Disaster," *New York Times*, May 19, 1986, p. A8.
33. Michael Wines, "Soviet Secrecy Hides Extent of Disaster," *Los Angeles Times*, May 2, 1986, pp. 1,22.
34. Bill Gertz, "New Soviet Deployments May Breach ABM Treaty," *Washington Times*, July 31, 1987, p. A10.
35. Ibid.; "Pawn Shop and Flat Twin Radars: Are They Obsolete?" *Jane's Defence Weekly*, January 9, 1988, p. 28.
36. Department of Defense, *Soviet Military Power 1987* (Washington DC: U.S. Government Printing Office, 1987), p. 30.
37. William B. Scott, "Titan Mission Success Based on Tighter Heavy Booster Standards," *Aviation Week and Space Technology*, November 2, 1987, pp. 25–26; Craig Covault, "U.S. Air Force Titan Launch Restarts Heavy Booster Flights," *Aviation Week and Space Technology*, November 2, 1987, pp. 24–26.
38. Covault, "U.S. Air Force Titan Launch . . ."; "Air Force Launches a Titan Rocket Probably Carrying a Spy Satellite," *Washington Times*, October 27,

1987, p. 6A; William J. Broad, "2 Years of Failure End as U.S. Lofts Big Titan Rocket," *New York Times*, October 27, 1987, pp A1, C4; Kathy Sawyer, "Air Force Orbits Satellite," *Washington Post*, October 27, 1987, pp. A1, A7; U.N. Committee on the Peaceful Uses of Outer Space, Information Furnished in Conformity with the Convention in Registration of Objects Launched into Outer Space ST/SG/SER.E/192, 1 December 1988, p. 3.

39. Scott, "Titan Mission Success . . ."

40. Covault, "U.S. Air Force Titan Launch. . . ."

41. Department of Defense, *Soviet Military Power: An Assessment of the Threat 1988* (Washington DC: U.S. Government Printing Office, 1988), pp. 47, 50; Bill Gertz, "U.S. Satellites Detect Marked Increase in Mobile Soviet ICBMs," *Washington Times*, October 14, 1988, p. A6.

42. "New Soviet Aircraft Carrier in Detail," *Jane's Defence Weekly*, November 5, 1988, p. 1147.

43. Rowland Evans and Robert Novak, "The Radars of Perestroika," *Washington Post*, December 21, 1988, p. A19; Bill Gertz, "CIA Warns of Verification Woes in Future Treaty," *Washington Times*, December 21, 1988, p. A3.

44. NBC News Report, At-The-Hour, 4 P.M., December 9, 1987.

45. David B. Ottaway, "Behind the New Battle with Libya," *Washington Post*, January 8, 1989, pp. C1, C4.

46. Ibid.

47. Ibid.; "Showdown with Libya," *Newsweek*, January 16, 1989, pp. 16–17.

48. William Tuohy, "U.S. Shows Photos of Libya Plant; Europeans Have Doubts," *Los Angeles Times*, January 3, 1989, p. 12; "Showdown with Libya."

49. "Showdown with Libya."

50. Private information.

51. "The Flight of Atlantis," *Newsweek*, December 12, 1988, p. 68; Bill Gertz, "Plan to Delay Spy Satellite Will Be Costly, Sources Say," *Washington Times*, April 17, 1989, p. A4; Interview.

52. Bhupendra Jasani and Christer Larsson, "Remote Sensing, Arms Control and Crisis Observation," *International Journal of Imaging, Remote Sensing and Integrated Geographical Systems*, 1, 1 1987, pp. 31–41; Charles Elachi, "Radar Images of the Earth from Space," *Scientific American*, December 1982, pp. 54–61.

53. William J. Broad, "New Satellite Is the First in a Class of All-Weather Spies, Experts Say," *New York Times*, December 4, 1988, pp. 1, 15; Elachi, "Radar Images of the Earth from Space."

54. Burrows, *Deep Black*, p. 313.

55. Ibid.

56. Ibid., p. 314.

57. Bob Woodward and Walter Pincus, "At CIA, a Rebuilder 'Goes With the Flow,'" *Washington Post*, August 10, 1988, pp. A1, A8.

58. Interview.
59. Bob Woodward, *Veil: The Secret Wars of the CIA 1981–1987* (New York: Simon & Schuster, 1987), pp. 221–23.
60. Ibid., pp. 223–24.
61. Ibid., p. 224.
62. Private information.
63. Bill Gertz, "Senate Panel Asks for Radar Funds," *Washington Times*, April 5, 1988, p. A43; Rowland Evans and Robert Novak, "The Indigo-Lacrosse Satellite Gets the Nod," *Washington Post*, April 6, 1988, p. A25; "INF: The Politics of Ratification," *Newsweek*, May 16, 1988, p. 22; Rowland Evans and Robert Novak, " 'Eyes' vs. Arms," *Washington Post*, May 2, 1988, p. A21.
64. Gertz, "Senate Panel Asks for Radar Funds"; Floyd C. Painter, "The Tracking and Data Relay Satellite System," *Defense Electronics*, June 1989, pp. 115–20.
65. Painter, "The Tracking and Data Relay Satellite."
66. "Mission 27 Launch Set for December 1," *Aviation Week and Space Technology*, November 21, 1988, p. 23.
67. Robert C. Toth, "Anaheim Firm May Have Sought Spy Satellite Data," October 10, 1982, pp. 1, 32.
68. Burrows, *Deep Black*, p. 309.
69. Senate Armed Services Committee, *Air Force Space Launch Policy and Plans* (Washington DC: U.S. Government Printing Office, 1988), p. 8.
70. Burrows, *Deep Black*, p. 309.
71. John Noble Wilford, "Shuttle Flight Readied Behind a Curtain of Secrecy," *New York Times*, November 28, 1988, p. A21; "The Flight of Atlantis."
72. John M. Broder, "Shuttle Fired Into Orbit on Spy Mission," *Los Angeles Times*, December 3, 1988, pp. 1, 22–23.
73. "Sketches of 5 on the Shuttle Mission," *New York Times*, December 3, 1989, p. 10.
74. Kathy Sawyer, "Weather Delays Space Shuttle Launch," *Washington Post*, December 2, 1988, p. A4.
75. "Grapevine," *Time*, November 28, 1988, p. 24.
76. Kathy Sawyer, "Shuttle Atlantis Lifts Off After Racing the Clock," *Washington Post*, December 3, 1988, pp. A1, A11; John Noble Wilford, "Shuttle With Spy Craft Lifts Off, As Winds Ease with Minute Left," *New York Times*, December 3, 1988, pp. 1, 10.
77. Sawyer, "Shuttle Atlantis Lifts Off After Racing the Clock"; Broder, "Shuttle Fired Into Orbit on Spy Mission"; Wilford, "Shuttle with Spy Craft Lifts Off"; Craig Covault, "Atlantis' Radar Satellite Payload Opens New Reconnaissance Era," *Aviation Week and Space Technology*, December 12, 1988, pp. 26–28.

78. Sawyer, "Shuttle Atlantis Lifts Off After Racing the Clock."
79. "Satellite with 150-ft. Span Set for Launch on Mission 27," *Aviation Week and Space Technology*, November 7, 1988, p. 25; Covault, "Atlantis' Radar Satellite Payload Opens New Reconnaissance Era."
80. "Satellite with 150-ft."; Covault, "Atlantis' Radar Satellite"
81. Covault, "Atlantis' Radar Satellite; "Secret Photographs," *Aviation Week and Space Technology*, January 23, 1989, p. 11.
82. Letter to the Author from J.C. Runyon, Naval Intelligence Command Freedom of Information Act Coordinator, February 8, 1989; "The Flight of Atlantis."
83. "Amateurs Keep Eye on 'Secret' Satellite," *Washington Times*, December 16, 1988, p. A9.
84. Covault, "Atlantis' Radar Satellite"; "Space Watchers Had a Key," *New York Times*, December 10, 1988, p. 50.
85. Sandra Blackeslee, "Shuttle Returns from Secret Work," *New York Times*, December 7, 1988, p. A17; Thomas H. Maugh II, "Atlantis Completes Secret Spy Satellite Mission," *Los Angeles Times*, December 7, 1988, pp. 1, 36; William B. Scott, "Atlantis Returns from Secret Mission With Substantial Thermal Tile Damage," *Aviation Week and Space Technology*, December 12, 1988, p. 29.
86. Broder, "Shuttle Fired into Orbit on Spy Mission."
87. "Lacrosse Orbit," *Aviation Week and Space Technology*, January 16 1989, p. 11; Bill Gertz, "New Spy Satellite Needed to Monitor Treaty," *Washington Times*, October 20, 1987, p. A5; Kathy Sawyer, "Shuttle Crew is Believed to Deploy Spy Satellite," *Washington Post*, December 4, 1988, p. A3; "Satellite With 150-ft."; Bill Gertz, "Atlantis Shuttle to Carry Aloft Superspy All-Weather Satellite," *Washington Times*, November 7, 1988, p. A7; "Washington Roundup," *Aviation Week and Space Technology*, June 4, 1979, p. 11; Toth, "Anaheim Firm May Have Sought Spy Satellite Data."
88. Interview; Private information.

Notes For Chapter Ten: New Systems, New Targets

1. Bill Gertz, "Bush Plan to Slight Satellites and Boren"; *Washington Times*, March 30, 1989, p. A3; Matthew Bunn, "Spy Satellite Controversy Resolved," *Arms Control Today*, May 1989, p. 23.
2. Gertz, "Bush Plan to Slight Satellites and Boren"; Bunn, "Spy Satellite Controversy Resolved."
3. Bill Gertz, "Plan to Delay Spy Satellite Will Be Costly, Sources Say," *Washington Times*, April 17, 1989, p. A4.
4. Susan F. Rasky, "Bush is Accused of Backing Away from Promise on 1988 Arms Pact," *New York Times*, April 7, 1989, pp. A1, A9.

5. Rowland Evans and Robert Novak, "Putting Up New Spy Satellites," *Washington Post*, April 12, 1989, p. A23.
6. "U.S. to Modernize Spy Satellites," *Washington Times*, April 18, 1989, p. A2; Bunn, "Spy Satellite Controversy Resolved."
7. Edward C. Kolcum, "Orbiting of Advanced Imaging Satellite Bolsters U.S. Intelligence Capabilities," *Aviation Week and Space Technology*, August 14, 1989, pp. 30–31.
8. Interview.
9. "KH-11 Overruns Said to Slow Development of Follow-On Spacecraft," *Aerospace Daily*, January 23, 1984, pp. 16–17; Richard D. Hudson and Jacqueline W. Hudson, "The Military Applications of Remote Sensing by Infrared," *Proceedings of the IEEE* 63, 1 (1975), pp. 104–28.
10. Interview.
11. "New Payload Could Boost Shuttle Cost," *Aviation Week and Space Technology*, August 14, 1978, pp. 16–17.
12. William E. Burrows, *Deep Black: Space Espionage and National Security* (New York: Random House, 1986), pp. 307–308.
13. Ibid., p. 308; Private information; "C-5As Converted for Secret Cargo," *Armed Forces*, May 1987, p. 200; "Massive Satellite," *Aviation Week and Space Technology*, September 5, 1988, p. 23.
14. Walter Pincus, "Hill Conferees Propose Test of Space Arms," *Washington Post*, July 11, 1984, pp. A1, A13.
15. Kolcum, "Orbiting of Advanced Imaging Satellite Bolsters U.S. Intelligence Capabilities."
16. Breck W. Henderson, "Lockheed Develops Threat Warning System for U.S. Military Satellites," *Aviation Week & Space Technology*, July 3, 1989, pp. 61–62; Burrows, *Deep Black*, p. 260; Robert B. Giffen, *U.S. Space System Survivability: Strategic Alternatives for the 1990s* (Washington DC: National Defense University Press, 1982), p. ix; Paul Stares, *Space and National Security*, Washington DC: The Brookings Institution, 1987), p. 79.
17. Nicholas Johnson, *The Soviet Year in Space 1987* (Colorado Springs, CO: Teledyne-Brown Engineering, 1988), pp. 78–81; Nicholas Johnson, *Soviet Military Strategy in Space* (London: Jane's, 1987), p. 154; Craig Covault, "Soviet Strategic Laser Sites Imaged by French Spot Satellite," *Aviation Week and Space Technology*, October 26, 1987, pp. 26–27; Richard Halloran, "General Describes Soviet Laser Threat," *New York Times*, October 24, 1987, p. 62; "White House Assesses Reports of Soviet Asat Laser Facilities," *Aviation Week and Space Technology*, September 15, 1986, p. 21; "Soviets Display Laser Facility at Sary Shagan," *Aviation Week and Space Technology*, July 17, 1989, p. 27; Michael J. Ybarra, "Soviet 'Star Wars' Laser Facility Still Poses Threat, Pentagon Says," *Los Angeles Times*, July 13, 1989, p. 11.

18. William J. Broad, "U.S. Designs Spy Satellites To Be More Secret Than Ever," *New York Times,* November 3, 1987, pp. C1, C3.
19. Ibid.
20. Burrows, *Deep Black*, p. 308.
21. Ibid.,p. 298.
22. Ibid.
23. Ibid., p. 299.
24. Ibid., pp. 299–300.
25. James W. Canan, "Coming Back in Space," *Air Force Magazine,* February 1987, pp. 45–52; Interview with Robert NcNamara, Washington DC, January 19, 1989.
26. U.S. Congress, House Armed Services Committee, *Hearings on H.R. 5167: Department of Defense Authorization of Appropriations for Fiscal Year 1985, Part 2* (Washington DC: U.S. Government Printing Office, 1984), pp. 8, 12.
27. Burrows, *Deep Black,* p. 304; James W. Canan, "Recovery in Space," *Air Force Magazine,* August 1988, pp. 68–73.
28. Kathy Sawyer, "New Titan IV Rocket Orbits Secret Satellite," *Washington Post,* June 15, 1989, p. A12.
29. A.C. Morrissey, *Space Launch Systems* (Denver, CO: Martin-Marietta, October 19, 1988), unpaginated; Sawyer, "New Titan IV Rocket Orbits Secret Satellite."
30. William J. Broad, "Biggest U.S. Unmanned Rocket, Rival of Shuttle, Soars into Space," *New York Times,* June 15, 1989, pp. A1, B11.
31. Earl Lane, "Shuttle Columbia Returns to Space," *N.Y. Newsday,* August 9, 1989, pp. 7, 24.
32. William J. Broad, "Pentagon Leaves the Shuttle Program," *New York Times,* August 9, 1989, p. A13; Michael Cassutt, "The Manned Spaceflight Engineer Programme," *Spaceflight,* January 1989, pp. 26–33.
33. Canan "Recovery in Space"; U.S. Congress, Senate Committee on Armed Services, *Department of Defense Authorization of Appropriations for Fiscal Year 1989, Part 6* (Washington DC: U.S. Government Printing Office, 1988), pp. 306–307.
34. Morrissey, *Space Launch Systems*; Edward H. Kolcum, "Air Force, Contractors Predict Long-Life for Heavy Lift," *Aviation Week and Space Technology,* July 17, 1989, pp. 32–34.
35. Burrows, *Deep Black,* p. 323; James B. Schultz, "TRW to Deliver MILSTAR Payload and House Votes to Kill Satellite Program," *Armed Forces Journal International,* September, 1989, pp. 75–78.
36. Dan Charles, "Spy Satellites: Entering a New Era," *Science,* March 24, 1989, pp. 1541–43.
37. Interview.
38. "USAF $24.9 b Request for Satellite Centre," *Jane's Defence Weekly,* March 19, 1988.

39. Ibid.
40. U.S. Congress, House Committee on Appropriations, *Military Construction Appropriations for 1983, Part 4* (Washington DC: U.S. Government Printing Office, 1982), pp. 454–55.
41. Department of Defense, *Soviet Military Power: Prospects for Change 1989* (Washington DC: U.S. Government Printing Office, 1989), p. 45
42. "New Soviet Missile Sites Reported," *New York Times*, April 23, 1985, p. A4; Drew Middleton, "Soviet Said to Deploy a New Missile," *New York Times*, October 22, 1984, p. A3.
43. William Drozdiak, "NATO Backs U.S. on Arms Charges," *Washington Post*, October 30, 1985, p. A3.
44. Department of Defense, *Soviet Military Power: An Assessment of the Threat 1988*, (Washington DC: U.S. Government Printing Office, 1988). p. 52.
45. *Treaty Between the United States of America and the Union of Soviet Socialist Republics on the Elimination of their Intermediate-Range and Shorter-Range Missiles* (Washington DC: Department of State, December 1987), pp. 1–7.
46. Ibid; "Further details released on SS-20 Saber missile," *Jane's Defence Weekly*, January 30, 1988, pp. 182–83.
47. "Industry Observer," *Aviation Week and Space Technology*, December 21, 1987, p. 15; Nicholas Johnson, *The Soviet Year in Space 1988* (Colorado Springs, CO: Teledyne-Brown Engineering, 1989), p. 16.
48. William M. Arkin and Richard Fieldhouse, *Nuclear Battlefields: Global Links in the Arms Race* (Cambridge, MA: Ballinger, 1985), pp. 260, 263.
49. Ibid; Central Intelligence Agency, *The Soviet Weapons Industry: An Overview* (Washington DC: CIA, 1986), p. 3.
50. "Soviets Build Directed-Energy Weapon," *Aviation Week and Space Technology*, July 28, 1980, pp. 47–50.
51. David C. Morrison, "Radar Diplomacy," *National Journal*, January 3, 1987, pp. 17–21.
52. U.S. Congress, House Committee on Appropriations, *Department of Defense Appropriations for 1984, Part 2* (Washington DC: U.S. Government Printing Office, 1983), p. 608.
53. "Strong and Silent," *Newsweek*, September 12, 1988, pp. 27–28.
54. Private information.
55. Ibid; Francis X. Clines, "Uzbek Violence Continues As Gunfire Wounds Scores," *New York Times*, June 10, 1989, p. 3; "Thousands Fleeing Uzbek Riots Stay in Dismal Camp," *Washington Times*, June 12, 1989, p. A7.
56. "Background for Speech to Rotary Club," December 9, 1988, p. 2.
57. Arkin and Fieldhouse, *Nuclear Battlefields*, pp. 290–91.
58. Thomas L. Friedman, "200,000 Troops Near Beijing, U.S. Says," *New York Times*, June 8, 1989, p. A13; "Reign of Terror," *Newsweek*, June 19, 1989, pp. 14–22.

59. Interview.
60. See W. Seth Carus, *The Military Balance: The Threat to Israel's Air Bases* (Washington DC: AIPAC 1985) for a description of the targets.
61. Leonard S. Spector, *The Undeclared Bomb: The Spread of Nuclear Weapons 1987-1988* (Cambridge, MA: Ballinger, 1988), p. 238; Stephen Engelberg, "U.S. Sees Pakistan Seeking an A-Bomb," *New York Times,* June 11, 1989, p. 5; Don Oberdorfer, "North Koreans Pursue Nuclear Weapons," *Washington Post,* July 29, 1989, p. A9; John J. Fialka, "North Korea May Be Developing Ability to Build Nuclear Weapons," *Wall Street Journal,* July 19, 1989, p. A16.
62. Spector, *The Undeclared Bomb,* p. 238.
63. Stephen Engelberg, "C.I.A.'s Chief Campaigns Against Missile-Making by Third World," *New York Times,* March 31, 1989, p. A6.
64. Glenn Frankel, "Iraq Said Developing A-Weapons," *Washington Post,* March 31, 1989, pp. A1, A32.
65. Spector, *The Undeclared Bomb,* pp. 32, 64 n.11; Bill Gertz, "S. Africa on the Brink of Ballistic Missile Test," *Washington Times,* June 20, 1989, pp. A1, A9.
66. Richard M. Weinraub, "India Tests Mid-Range 'Agni' Missile," *Washington Post,* May 23, 1989, pp. A1, A21; Spector, *The Undeclared Bomb,* p. 32.
67. Private information.
68. David B. Ottaway, "Saudis Hid Acquisition of Missiles," *Washington Post,* March 29, 1988, pp. A1, A3.
69. Ibid.
70. Jack Anderson and Dale van Atta, "Israel May Hit Syrian Nerve Gas Plant," *Washington Post,* February 24, 1988, p. D14.
71. Jonathan C. Randal, "Iran Pours Reinforcements Into Bridgeheads in Iraq," *Washington Post,* February 14, 1986, pp. A23, A29.
72. Xinhau General Overseas News Service, May 17, 1985.
73. Private information.
74. Center for National Security Studies, et.al., Plaintiffs vs. Central Intelligence Agency, et. al., Defendants, Civil Action, No. 80–1235, United States District Court for the District of Columbia DGC 80–08250, p. 11n.
75. Interview with Richard Bissell, January 6, 1984; Private information.
76. Testimony of Dino Brugioni, in U.S. Congress, House Committee on Science and Technology, *The Role of Information Technology in Emergency Management* (Washington DC: U.S. Government Printing Office, 1984), pp. 54–55.
77. Ibid.; p. 55; Allan Sloan, "Big Brother Strikes Again," *Forbes,* May 12, 1980, pp. 50–51; Interview.
78. Roland S. Inlow, "An Appraisal of the Morison Espionage Trial," *First Principles,* 11, 4 (May 1986), pp. 1, 2–5.

79. United States of America vs. Samuel Loring Morison, United States District Court for Maryland, October 15, 1985, Direct Testimony of Roland Inlow, p. 6.
80. Ibid., pp. 6–7.
81. Interview; James Bamford, "America's Supersect Eyes in Space," *New York Times Magazine,* January 13, 1985, pp. 38ff.
82. Charles, "Spy Satellites: Entering a New Era."
83. CINCPAC Instruction S3822.1E, "Pacom Imagery Reconnaissance Procedures and Responsibilities," July 5, 1983, p. 1; Department of the Air Force, *Department of the Air Force Organization and Functions* (Washington DC: USAF, March 1, 1986), pp. 6–23; U.S. Congress, House Permanent Select Committee on Intelligence, *Annual Report* (Washington DC: U.S. Government Printing Office, 1978), p. 54; Office of the Assistant Chief of Staff for Intelligence [U.S. Army], *Annual Historical Review, 1 October 1983–30 September 1984,* pp. 2–14.
84. Interview.
85. United States of America vs. Samuel Loring Morison, Direct Testimony of Roland Inlow, p. 15.
86. Interview.
87. Bamford, "America's Supersecret Eyes in Space."

Notes For Chapter Eleven: Still Secret After All These Years

1. Interview with Congressman George E. Brown, Jr., October 3, 1988.
2. Ibid.; "Washington Roundup," *Aviation Week and Space Technology,* November 23, 1987, p. 21; George E. Brown, Jr., "Outer Space: Frontier for Exploration or Battleground for Conflict?" *Congressional Record,* February 26, 1987, pp. H849–H854; George E. Brown, Jr., "International Cooperation in Space: Enhancing the World's Security," n.d.
3. Brown interview.
4. Ibid.
5. "Washington Round Up"; "Rep. Brown Announces Withdrawal from Intelligence Committee," Press Release, November 18, 1987.
6. Brown interview.
7. See William E. Burrows, "A Study of Space Reconnaissance: Methodology for Researching a Classified System," presented at the Symposium on Space History, National Air and Space Museum, Smithsonian Institution, Washington DC, June 12, 1987.
8. Center for National Security Studies, et al., Plaintiffs vs. Central Intelligence Agency et al., Defendants, Civil Action No. 80–1235, United States District Court for the District of Columbia, DGC 80–08250, pp. 4–5.

9. Ibid., p. 13.
10. William Colby with Peter Forbath, *Honorable Men: My Life in the CIA* (New York: Simon and Schuster, 1978), p. 357.
11. Philip Taubman, "Showing Secrets: A U.S. Compromise," *New York Times*, March 24, 1983, p. A21; Richard Burt, "Arms Treaty: How to Verify Moscow's Compliance," *New York Times*, March 21, 1979, p. 8.
12. Telephone interview with Bobby R. Inman, June 27, 1989.
13. Philip Taubman, "Secrecy of U.S. Reconnaissance Office is Challenged," *New York Times*, March 1, 1981, p. 10.
14. Telephone interview with Bobby R. Inman, June 27, 1989.
15. William J. Broad, "Soviet Photos of U.S. Were for Spying," *New York Times*, January 30, 1989, p. A12.
16. William M. Arkin and Richard Fieldhouse, *Nuclear Battlefields: Global Links in the Arms Race* (Cambridge, MA: Ballinger, 1985); Department of Defense, *Soviet Military Power: An Assessment of the Threat 1988* (Washington DC: U.S. Government Printing Office, 1988), pp. 35, 40, 52, 60, 143.
17. United States of America vs. Samuel Loring Morison, United States District Court for Maryland, October 15, 1985, Testimony of General Parker Hazard, pp. 32–42; Testimony of Richard Evan Hineman, pp. 51–52.
18. Thomas B. Allen and Norman Polmar, *Merchants of Treason* (New York: Delacorte Press, 1988), p. 189.
19. Ibid.
20. Stuart A. Cohen, "The Evolution of Soviet Views on SALT Verification," in William C. Potter (ed.), *Verification and SALT: The Challenge of Strategic Deception* (Boulder, CO: Westview, 1980), p. 69 n.13.
21. Jeffrey Richelson, "Impact for Nations without Space-Based Intelligence Capabilities," in Michael Krepon, Peter Zimmerman, Leonard Spector and Mary Umberger (eds.), *Commercial Observation Satellites and International Security* (New York: St. Martin's, forthcoming).
22. Letter from Hans Mark to William E. Burrows, May 7, 1985, cited in William E. Burrows, *Deep Black: Space Espionage and National Security* (New York: Random House, 1986), p. xii.
23. Deborah M. Kyle, "SACEUR General Rogers Urges US to Release Threat Photos," *Armed Forces Journal International*, April 1984; Tom Philpott, "Gen. Rogers: Show Spy Photos in Europe Debate," *Air Force Times*, July 23, 1984, p. 6.
24. Inman interview.
25. Colby, *Honorable Men*, pp. 356–357; Interview with William Colby, Washington DC, June 20, 1989.
26. Stansfield Turner, *Secrecy and Democracy: The CIA in Transition* (Boston: Houghton-Mifflin, 1985), pp. 256–57.

27. Craig Covault, "Military Space Capabilities Expanding, But Excess Secrecy Limits Progress," *Aviation Week and Space Technology*, April 17, 1989, pp. 18–19.
28. Interview with Frederick J. Doyle, Reston, Virginia, November 15, 1988.
29. Ibid.
30. Interview.
31. U.S. Congress, House Committee on Science and Technology, *Information Technology in Emergency Management* (Washington DC: U.S. Government Printing Office, 1984), p. 227.
32. Ibid., p. 325.
33. Ibid., p. 354.
34. Interview; Brown interview.
35. Interview.
36. Brown Interview.
37. Ibid.

SOURCES

Interviews

Richard M. Bissell, Jr.	Farmington, CT. Telephone	January 6, 1984* June 5, 1989
Rep. George E. Brown, Jr.	Washington DC	October 3, 1988
McGeorge Bundy	New York, NY	October 5, 1988
William Colby	Washington DC	June 20, 1989
Merton Davies	Santa Monica, CA Telephone	December 1, 1988 February 22, 1989
Frederick J. Doyle	Reston, VA	November 15, 1988
Daniel Ellsberg	Washington DC	June 22, 1989
Raymond Garthoff	Washington DC	September 20, 1988
Roswell Gilpatric	New York, NY	May 31, 1989
Andrew J. Goodpaster	Washington DC	November 9, 1988
Kevin Hussey	Pasadena, CA	November 30, 1988
Bobby R. Inman	Telephone	June 27, 1989
James R. Janesick	Pasadena, CA	November 30, 1988
U. Alexis Johnson	Washington DC	November 22, 1988

Amrom Katz	Telephone	June 6, 1989
Spurgeon Keeny	Washington DC	May 9, 1989
Richard S. Leghorn	Telephone	September 12, 1989
Arthur C. Lundahl	Telephone	July 28, 1989 September 6, 1989
Robert McNamara	Washington DC	January 19, 1989
Charles C. Mathison	Telephone	December 8, 1988
Roger Molander	Telephone	November 1, 1988
Clarence A. Robinson, Jr.	Telephone	May 11, 1989
Walt W. Rostow	Telephone	June 27, 1989
Herbert Scoville, Jr.	McLean, VA	1983*
Stansfield Turner	McLean, VA	May 30, 1984*
Paul Warnke	Washington DC	September 27, 1988

* Interviewed for earlier projects.

Government and Contractor Documents

Advanced Research Projects Agency. *Military Space Projects, Quarter Ended 30 June 1959* (Washington DC: ARPA, 1959).

Aeronautical Systems Division, Air Force Systems Command. "Review and Summary of X-20 Military Application Studies" (Wright Patterson AFB, Ohio: ASD/AFSC, December 14, 1963).

Andelin, D.R. *Evaluation of the Usefulness of the MOL to Accomplish Early NASA Mission Objectives, Volume 8: Support Systems, Gemini B, and Titan III-M* (Douglas Aircraft Company, May 1967).

Arms Control and Disarmament Agency. *Documents on Disarmament 1962, Volume I Jan.–June 1962* (Washington DC: U.S. Government Printing Office, 1963).

Arms Control and Disarmament Agency. *Documents on Disarmament, 1978* (Washington DC: U.S. Government Printing Office, 1979).

Arms Control and Disarmament Agency. *Arms Control and Disarmament Agreements: Texts and Histories of Negotiations* (Washington DC: ACDA, 1980).

Arms Control and Disarmament Agency. *Soviet Noncompliance* (Washington DC: ACDA, February 1, 1986).

Berger, Carl. *The Air Force in Space Fiscal Year 1961* (Washington DC: USAF Historical Division Liaison Office, 1966).

Bowen, Lee. *The Threshold of Space: The Air Force in the National Space Program, 1945–1959* (Washington DC: USAF Historical Division Liaison Office, September 1960).

Briefing on Army Satellite Program. November 19, 1957. In *Declassified Documents Reference System*, 1977–101B.

Brown, Rep. George E., Jr. "Outer Space: Frontier for Exploration or Battleground for Conflict?" *Congressional Record*, February 26, 1987, pp. H849–H854.

Brown, Rep. George E., Jr. "International Cooperation in Space; Enhancing the World's Common Security." n.d.

Cantwell, Gerald H. *The Air Force in Space 1964* (Washington DC: USAF Historical Division Liaison Office, 1967).

Center for National Security Studies et al., Plaintiffs vs. Central Intelligence Agency et al., Defendants, Civil Action No. 80–1235, United States District Court for the District of Columbia, DGC 80–08250.

Central Intelligence Agency. National Intelligence Estimate 13–2–60, "The Chinese Communist Atomic Energy Program." December 13, 1960.

Central Intelligence Agency. "Prospects for Further Proliferation of Nuclear Weapons." DCI/NIO 1945/74, September 4, 1974.

Central Intelligence Agency. *Iran in the 1980s* (Washington DC: CIA, August 1977).

Central Intelligence Agency. *KGB and GRU* (Washington DC: CIA, 1984).

Central Intelligence Agency. *The Soviet Weapons Industry: An Overview* (Washington DC: CIA, 1986).

Charhut, D.E. *Evaluation of the Usefulness of the MOL to Accomplish Early NASA Mission Objectives, Volume I: Summary* (Douglas Aircraft Company, October 1967).

Chronology of Early Air Force Man-in-Space Activity 1955–1960 (Washington DC: Air Force Systems Command, 1965).

CINCPAC Instruction S3822.1E, "PACOM Imagery Reconnaissance Procedures and Responsibilities." July 5, 1983.

Collbohm, F.R. "An Earlier Reconnaissance Satellite System." (Santa Monica, CA: RAND Corporation, November 12, 1957).

Davies, Merton E. and William R. Harris. *RAND's Role in the Evolution of Balloon and Satellite Observation Systems and Related U.S. Space Technology* (Santa Monica, CA: RAND, 1988).

DCID 1/13, "Committee on Imagery Requirements and Exploitation (COMIREX)." July 1, 1967, in *Declassified Documents Reference System*, 1980–132B.

Department of Air Force. *Department of the Air Force Organization and Functions* (Washington DC: DAF, 1986).

Department of Defense. *Soviet Military Power 1987* (Washington DC: U.S. Government Printing Office, 1987).

Department of Defense. *Soviet Military Power: An Assessment of the Threat 1988* (Washington DC: U.S. Government Printing Office, 1988).

Department of Defense, *Soviet Military Power: Prospects for Change 1989* (Washington DC: U.S. Government Printing Office, 1989).

Directorate of Intelligence. DCS/O to Assistant for Evaluation, DCS/D subj: Research and Development on Proposed RAND Satellite Reconnaissance Vehicle. March 17, 1951, RG 341, Entry 214, File 2–19900 to 2–19999, Modern Military Branch, National Archives.

Douglas Aircraft Company. *Preliminary Design of an Experimental World-Circling Spaceship* (Santa Monica, CA: DAC, 1946).

Foster, William C. "Memorandum for the President: Arms Control Aspects of Proposed Satellite Reconnaissance Policy." July 6, 1962.

Giffen, Robert B. *U.S. Space System Survivability: Strategic Alternatives for the 1990s* (Washington DC: National Defense University Press, 1982).

Glickman, Rep. Dan. "Background for Speech to the Rotary Club." December 9, 1988.

Goodpaster, A.J. "Memorandum for the Honorable Gordon Gray." June 10, 1960.

Goodpaster, A.J. "Memorandum for Record, October 1960," White Office Staff Secretary, Subject Series, Alphabetical Subseries, Box 15, Intelligence Matters, Folder 20, October 1960–January 1961, Dwight D. Eisenhower Library.

Greenfield, S.M., and W.W. Kellogg. *Inquiry into the Feasibility of Weather Reconnaissance from a Satellite Vehicle* (Santa Monica, CA: RAND, 1951).

Jernigan, Roger A. *Air Force Satellite Control Facility, Historical Brief and Chronology* (Sunnyvale, CA: AFSCF History Office, n.d.).

The Joint Study Group Report on Foreign Intelligence Activities of the United States Government. December 15, 1960.

Katz, Amrom. "Reconnaissance Satellites and Comments on Your WM-2553" (Santa Monica, CA: RAND Corporation, January 3, 1958).

Katz, Amrom H. "Memorandum to J. L. Hult: Recommendations, Formal and Informal, on Reconnaissance Satellite Matters," (Santa Monica, CA: RAND Corporation, May 11, 1959).

Leghorn, Richard S. "Political Action and Satellite Reconnaissance" (Cambridge, MA: Itek Corporation, 1959).

"Letter from James H. Douglas to President Dwight D. Eisenhower." April 11, 1960 in *Declassified Documents Reference System*, 1977–288A.

"Letter from Neil McElroy to President Dwight D. Eisenhower." January 29, 1959 in *Declassified Documents Reference System*, 1982–001538.

Lipp, J.E., R.M. Salter, R.S. Wehner, R.R. Carhart, C.R. Culp, S.L. Gendler, W.J. Howard, and J.S. Thompson. *Utility of a Satellite Vehicle for Reconnaissance* (Santa Monica, CA: RAND, 1951).

Lipp, J.E., R.M. Salter, and R.S. Wehner (eds.). *Project FEEDBACK Summary Report, Volume II* (Santa Monica, CA: RAND, 1954).

"Manned Orbiting Laboratory Background Briefing." August 25, 1965.

McQuade, Lawrence C. "Memorandum for Mr. Nitze, Subj: But Where Did the Missile Gap Go?" (Washington DC: Assistant Secretary of Defense, International Security Affairs, May 31, 1963).

Morrissey, A.C. *Space Launch Systems* (Denver, CO: Martin-Marietta, October 19, 1988).

NSAM 156, "Negotiation on Disarmament and Peaceful Uses of Outer Space." May 26, 1962.

National Security Council, NSC 5814. "U.S. Policy on Outer Space." June 20, 1958.

NSCID No. 8, "Photographic Interpretation." February 17, 1972, in *Declassified Documents Reference System*, 1976–253G.

Office of the Assistant Chief of Staff for Intelligence [U.S. Army]. *Annual Historical Review 1 October 1983–30 September 1984* (Washington DC: OACSI, n.d).

Office of the Director of Defense Research and Engineering. *Military Space Projects Report No. 10, March–April–May 1960* (Washington, DC: DOD, 1960), in *Declassified Documents Reference System*, 1980–36C.

Office of the Director of Defense Research and Engineering. *Military Space Projects No. 11: Report of Progress for June–July–August 1960*; In *Declassified Documents Reference System*, 1980–36D.

Organizations and Functions Chart Book, Air Force Satellite Control Facility (Onizuka, AFS, CA, AFSCF, 1987).

Perry, Robert. *Origins of the USAF Space Program, 1945–1956* (Washington DC: Air Force Systems Command, 1962).

"Possible Soviet Chemical Weapons Launcher is Tested," *Weekly Intelligence Summary*, July 23, 1982, pp. 4–5.

"President Approves DOD Development of Manned Orbiting Laboratory." OASD (PA), August 25, 1965.

The President's Unclassified Report on Soviet Noncompliance with Arms Control Agreements (Washington DC: White House, December 2, 1987).

RAND 25th Anniversary Volume (Santa Monica, CA: RAND, 1973).

"Rep. Brown Announces Withdrawal from Intelligence Committee." Press Release, November 18, 1987.

"SAMOS II Fact Sheet" (Washington DC: Department of Defense, 1961).

Schlesinger, James. *A Review of the Intelligence Community* (Washington DC: Office of Management and Budget, March 10, 1971).

Sharp, Dudley C. Secretary of the Air Force Order No. 115.1, Subj: Organization and Functions of the Office of Missile and Satellite Systems. August 31, 1960.

Sharp, Dudley C. Secretary of the Air Force Order No. 116.1, Subj: The Director of the SAMOS Project. August 31, 1960.

Snyder, Murray. "Memorandum for Assistant Secretary of Defense (International Security Affairs), Public Affairs Plan for SAMOS Research & Development Test Firing in Early October." October 4, 1960, in *Declassified Documents Reference System*, 1984–000819.

"Strategic Value of Simonstown Naval Base May Decrease." *Weekly Intelligence Summary*, July 23, 1982, pp. 23–25.

Sylvester, Arthur. "Memorandum for the President, The White House, Subject: SAMOS Launch." January 26, 1961 in *Declassified Documents Reference System*.

"Tracked Multiple Rocket Launchers Noted in Beijing Military Region." *Weekly Intelligence Summary*, July 23, 1982, pp. 2–3.

Treaty Between the United States of America and the Union of Soviet Socialist Republics on the Elimination of their Intermediate-Range and Shorter-Range Missiles (Washington DC: Department of State, December 1987).

United States of America vs. William Peter Kampiles, United States District Court, Northern District of Indiana, November 6, 1978.

United States of America vs. Samuel Loring Morison, United States District Court for Maryland, October, 1985.

U.S. Congress, House Committee on Appropriations. *Department of Defense Appropriations for 1969, Part I* (Washington DC: U.S. Government Printing Office, 1968).

U.S. Congress, House Committee on Appropriations. *Military Construction Appropriations for 1983* (Washington DC: U.S. Government Printing Office, 1982).

U.S. Congress, House Committee on Appropriations. *Department of Defense Appropriations for 1984, Part 2* (Washington DC: U.S. Government Printing Office, 1983).

U.S. Congress, House Committee on Armed Services. *Hearings on H.R. 5167: Department of Defense Authorization of Appropriations for Fiscal Year 1985, Part 2* (Washington DC: U.S. Government Printing Office, 1984).

U.S. Congress, House Committee on Science and Technology. *The Role of Information Technology in Emergency Management* (Washington DC: U.S. Government Printing Office, 1984).

U.S. Congress, House Permanent Select Committee on Intelligence. *Annual Report* (Washington DC: U.S. Government Printing Office, 1978).

U.S. Congress, House Permanent Select Committee of Intelligence. *Iran: Evaluation of U.S. Intelligence Performance Prior to November 1978* (Washington DC: U.S. Government Printing Office, 1979).

U.S. Congress, House Committee on Science and Astronautics. *Science, Astronautics, and Defense* (Washington DC: U.S. Government Printing Office, 1961).

U.S. Congress, Senate Committee on Aeronautical and Space Sciences. *Manned Orbiting Laboratory* (Washington DC: U.S. Government Printing Office, 1966).

U.S. Congress, Senate Committee on Armed Services. *Air Force Space Launch Policy and Plans* (Washington DC: U.S. Government Printing Office, 1988).

U.S. Congress, Senate Committee on Armed Services. *Department of Defense Authorization of Appropriations for Fiscal Year 1989, Part 6* (Washington DC: U.S. Government Printing Office, 1988).

U.S. Congress, Senate Select Committee to Study Governmental Operations with Respect to Intelligence Activities. *Final Report, Book I: Foreign and Military Intelligence* (Washington DC: U.S. Government Printing Office, 1976).

U.S. Congress, Senate Select Committee to Study Governmental Operations with Respect to Intelligence Activities. *Final Report, Book IV: Supplementary Detailed Staff Reports on Foreign and Military Intelligence* (Washington DC: U.S. Government Printing Office, 1976).

USAF Mishap Report 85-8-28-701 (Los Angeles, CA: USAF Space Division, October 25, 1985).

USAF Mishap Report 86-4-18-701 (Los Angeles, CA: USAF Space Division, June 8, 1986.)

"USSR Builds Takeoff Ramp." *Weekly Intelligence Summary*, July 23, 1982, pp. 2–3.

Vance, Cyrus. "Memorandum for the President." February 12, 1966, in *Declassified Documents Reference System*, 1980–37A.

White, Thomas. "Memorandum for Deputy Chief of Staff, Development, Subject: [deleted] Satellite Vehicles." December 18, 1952, File 2–36300 through 2–36399, RG–341, Entry 214, Modern Military Branch, National Archives.

Whittington, Mary L., James Eastman, W.E. Alford, and Charles W. Dickens. *MAC Support of Project GEMINI, 1963–1966* (Scott, AFB, IL: Military Airlift Command, 1967).

Books and Reports

Allen, Thomas B., and Norman Polmar. *Merchants of Treason: America's Secrets for Sale* (New York: Delacorte, 1988).

Ambrose, Stephen. *Eisenhower, Volume Two: The President* (New York: Simon and Schuster, 1984).

Arkin, William M., and Richard Fieldhouse. *Nuclear Battlefields: Global Links in the Arms Race* (Cambridge, MA: Ballinger, 1985).

Beaumont, Roger. *Maskirovka: Soviet Camouflage, Concealment and Deception* (College Station, TX: Texas Engineering Experiment Station, Texas A & M, 1982).

Bell, Griffin. *Taking Care of the Law* (New York: William Morrow, 1982).

Beschloss, Michael R. *MAYDAY: Eisenhower, Khrushchev and the U-2 Affair* (New York: Harper & Row, 1986).

Beckwith, Charlie, and Donald Knox. *Delta Force* (New York: Harcourt, Brace and Jovanovich, 1983).

Berkowitz, Bruce D. and Allen E. Goodman. *Strategic Intelligence for American National Security* (Princeton, NJ: Princeton University Press, 1989).

Berman, Robert, and John C. Baker. *Soviet Strategic Forces: Requirements and Responses* (Washington DC: The Brookings Institution, 1982).

Blitzer, Wolf. *Territory of Lies* (New York: Harper & Row, 1989).

Brzezinski, Zbigniew. *Power and Principle: Memoirs of the National Security Advisor 1977–1981* (New York: Farrar, Straus, Giroux, 1983).

Bundy, McGeorge. *Danger and Survival: Choices About the Bomb in the First Fifty Years* (New York: Random House, 1988).

Burrows, William E. *Deep Black: Space Espionage and National Security* (New York: Random House, 1986).

Carus, W. Seth. *The Military Balance: The Threat to Israel's Air Bases* (Washington DC: American-Israel Public Affairs Committee, 1985).

CIA: The Pike Report (Nottingham: Spokesman Books, 1977).

Cleator, P.E. *Rockets through Space: The Dawn of Interplanetary Travel* (New York: Simon & Schuster, 1936).

Cochran, Thomas, William M. Arkin, Robert S. Norris and Jeffrey I. Sands. *Nuclear Weapons Databook, Volume IV: Soviet Nuclear Weapons* (New York: Harper & Row, 1989).

Cockburn, Andrew. *The Threat: Inside the Soviet Military Machine* (New York: Random House, 1983).

Colby, William, with Peter Forbath. *Honorable Men: My Life in the CIA* (New York: Simon and Schuster, 1978).

Corson, William R., Susan B. Trento, and Joseph J. Trento. *Widows* (New York: Crown, 1989).

Crickmore, Paul. *Lockheed SR-71 Blackbird* (London: Osprey, 1986).

Deacon, Richard. *The Israeli Secret Service* (New York: Taplinger, 1977).

Derogy, Jacques and Henri Carmel, *The Untold Story of Israel* (New York: Grove Press, 1979).

Documents from the Espionage Den (52): USSR: The Aggressive East, Section 3-2 (Teheran: Muslim Students Following the Line of the Imam, n.d.).

Duffy, Gloria, ed. *Compliance and the Future of Arms Control* (Cambridge, MA: Ballinger, 1988).

Emerson, Steve. *Secret Warriors: Inside the Covert Military Operations of the Reagan Era* (New York: G.P. Putnam's, 1988).

Freedman, Lawrence. *U.S. Intelligence and the Soviet Strategic Threat* 2nd ed. (Princeton, NJ: Princeton University Press, 1986).

Glasstone, Samuel, and Philip J. Dolan. *The Effects of Nuclear Weapons* (Washington DC: DOD/DOE, 3rd ed. 1977).

Gottfried, Kurt, and Bruce G. Blair. *Crisis Stability and Nuclear War* (New York: Oxford University Press, 1988).

Hawkes, Nigel, Geoffrey Lean, David Leigh, Robin McKie, Peter Pringle and Andrew Wilson. *Chernobyl: The End of the Nuclear Dream* (New York: Vintage, 1986).

Heims, Steve J. *John Von Neumann and Norbert Weiner: From Mathematics to the Technologies of Life and Death* (Cambridge, MA: MIT Press, 1980).

Hersh, Seymour. *The Price of Power: Kissinger in the Nixon White House* (New York: Summit, 1983).

Johnson, Nicholas. *The Soviet Year in Space 1987* (Colorado Springs, CO: Teledyne-Brown Engineering, 1988).

Johnson, Nicholas. *Soviet Military Strategy in Space* (London: Jane's, 1987).

Johnson, Nicholas. *The Soviet Year in Space 1988* (Colorado Springs, CO: Teledyne-Brown Engineering, 1989).

Johnson, U. Alexis, with Jef Olivarius McAlister. *The Right Hand of Power* (Englewood Cliffs, NJ: Prentice-Hall, 1984).

Kaplan, Fred. *The Wizards of Armageddon* (New York: Simon & Schuster, 1983).

Kistiakowsky, George B. *A Scientist at the White House: The Private Diary of President Eisenhower's Special Assistant for Science and Technology* (Cambridge, MA: Harvard University Press, 1976).

Klass, Philip J. *Secret Sentries in Space* (New York: Random House, 1971).

Krass, Allan S. *Verification: How Much is Enough?* (Philadelphia, PA: Taylor and Francis, 1985).

Lewis, John L., and Xue Litai. *China Builds the Bomb* (Stanford, CA: Stanford University Press, 1988).

Lin, Chong-Pin. *China's Nuclear Weapons Strategy: Tradition Within Evolution* (Lexington, MA: Lexington Books, 1988).

Marchetti, Victor and John D. Marks. *The CIA and the Cult of Intelligence* (New York: Knopf, 1974).

Mark, Hans. *The Space Station: A Personal Journey* (Durham, NC: Duke University Press, 1987).

Martin, David C. and John Walcott. *Best Laid Plans: The Inside Story of America's War Against Terrorism* (New York: Harper & Row, 1988).

Miller, Mark E. *Soviet Strategic Power and Doctrine: The Quest of Superiority* (Washington DC: Advanced International Studies Institute, 1982).

Mosley, Leonard. *Dulles: A Biography of Eleanor, Allen and John Foster and Their Family Network* (New York: Dial, 1978).

Neff, Donald. *Warriors for Jerusalem: The Six Days that Changed the Middle East* (New York: Simon and Schuster, 1984).

Newhouse, John. *Cold Dawn: The Story of SALT* (New York: Holt, Rinehart & Winston, 1973).

Newhouse, John. *War and Peace in the Nuclear Age* (New York: Knopf, 1989).

Newsom, David. *The Soviet Brigade in Cuba: A Study in Political Diplomacy* (Bloomington, IN: Indiana University Press, 1977).

Peebles, Curtis. *Guardians: Strategic Reconnaissance Satellites* (Novato, CA: Presidio Press, 1987).

Penkovskiy, Oleg. *The Penkovskiy Papers* (New York: Doubleday, 1965).

Polmar, Norman, ed. *Soviet Naval Developments* (Annapolis, MD: Nautical and Aviation Publishing Co. of America, 1979).

Powers, Thomas. *The Man Who Kept the Secrets: Richard Helms and the CIA* (New York: Knopf, 1979).

Prados, John. *The Soviet Estimate: U.S. Intelligence Analysis and Russian Military Strength* (New York: Dial, 1982).

Ranelagh, John. *The Agency: The Rise and Decline of the CIA* (New York: Simon and Schuster, 1986).

Richelson, Jeffrey. *American Espionage and the Soviet Target* (New York: Quill, 1988).

Richelson, Jeffrey. *Foreign Intelligence Organizations* (Cambridge, MA: Ballinger, 1988).

Richelson, Jeffrey. *The U.S. Intelligence Community* (Cambridge, MA: Ballinger, 1988, 2nd ed.).

Rostow, W.W. *Open Skies: Eisenhower's Proposal of July 21, 1955* (Austin, TX: University of Texas Press, 1982).

Schiff, Ze'ev. *A History of the Israeli Army: 1874 to the Present* (New York: Macmillan, 1985).

Seaborg, Glenn T. *Kennedy, Khrushchev and the Test Ban* (Berkeley, CA: University of California Press, 1983).

Sick, Gary. *All Fall Down: America's Tragic Encounter with Iran* (New York: Penguin, 1986).

Soloveytchik, George. *Potemkin: Soldier, Statesman, Lover and Consort of Catherine of Russia* (New York: W.W. Norton & Co., 1947).

Spector, Leonard S. *Nuclear Proliferation Today: The Spread of Nuclear Weapons 1984* (Cambridge, MA: Ballinger, 1984).

Spector, Leonard S. *The Undeclared Bomb: The Spread of Nuclear Weapons 1987–1988* (Cambridge, MA: Ballinger, 1988).

Stares, Paul B. *The Militarization of Space, U.S. Policy 1945–1984* (Ithaca, NY: Cornell University Press, 1985).

Stares, Paul B. *Space and National Security* (Washington DC: Brookings Institute, 1987).

Steele, Jonathan. *Soviet Power: The Kremlin's Foreign Policy—Brezhnev to Chernenko* (New York: Touchstone/Simon and Schuster, 1984).

Steinberg, Gerald M. *Satellite Reconnaissance: The Role of Informal Bargaining* (New York: Praeger, 1983).

Steven, Stewart. *The Spymasters of Israel* (New York: Macmillan, 1980).

Suvorov, Viktor. *Inside the Soviet Army* (New York: Macmillan, 1982).

Talbot, Strobe. *Endgame: The Inside Story of SALT II* (New York: Harper & Row, 1979).

Trento, Joseph J. *Prescription for Disaster* (New York: Crown, 1987).

Tully, Andrew. *Inside the FBI* (New York: Dell, 1987).

Turner, Stansfield. *Secrecy and Democracy: The CIA in Transition* (Boston, MA: Houghton-Mifflin, 1985).

Valenta, Jiri, *Soviet Intervention in Czechoslovakia, 1968: Anatomy of a Decision* (Baltimore, MD: Johns Hopkins, 1979).

Weissman, Steve, and Herbert Krosney. *The Islamic Bomb* (New York: Times Books, 1981).

Woodward, Bob. *Veil: The Secret Wars of the CIA 1981–1987* (New York: Simon and Schuster, 1987).

Wyden, Peter. *Bay of Pigs: The Untold Story* (New York: Simon and Schuster, 1979).

Articles and Papers

"A Satellite Rocket is Fired on the West Coast," *New York Times*, March 1, 1959, pp. 1, 32.

"Agena B to Put Samos, Midas in Orbit," *Aviation Week*, February 8, 1960, pp. 73, 75.

Agres, Ted. "Soviet Navy Completing Four Tunnels to Hide Subs," *Washington Times*, March 27, 1984, pp. 1A, 12A.

"Air Force Launches Titan Rocket Probably Carrying a Spy Satellite," *Washington Times*, October 27, 1987, p. 6A.

"Air Force Reports It Is Receiving Signal from Discoverer Satellite," *New York Times*, March 2, 1959, pp. 1, 10.

Alsop, Stewart. "What's Going in the Holes?" *Newsweek*, May 10, 1971, p. 124.

"Amateurs Keep Eye on 'Secret' Satellite," *Washington Times*, December 16, 1988, p. A9.

Anderson, Jack. "Getting the Big Picture for the CIA," *Washington Post*, November 28, 1982, p. C7.

Anderson, Jack and Dale Van Atta. "Israel May Hit Syrian Nerve Gas Plant," *Washington Post*, February 24, 1988, p. D14.

"Arms Verification Issue at Heart of U.S. Debate," *New York Times*, November 24, 1985, p. 18.

"The Art and Science of Photointerpretation," *IEEE Spectrum*, July 1986, pp. 52–53.

Atkinson, Rick. "Soviet Tunnels Could Hide Submarines," *Washington Post*, April 7, 1984, p. A4.

Atkinson, Rick. "Soviet Arms Disaster Reported," *Washington Post*, June 22, 1984, pp. A1, A12.

"Attempt to Orbit Discoverer Fails," *New York Times*, June 9, 1961, p. 14.

Bamford, James. "America's Supersecret Eyes in Space," *The New York Times Magazine*, January 13, 1985, pp. 39ff.

Belair, Felix Jr. "Eisenhower is Given Flag that Orbited the Earth," *New York Times*, August 16, 1960, pp. 1, 4.

Bennett, Ralph Kinney. "U.S. Eyes Over Russia: How Much Can We See?" *Reader's Digest*, October 1985, pp. 142–47.

Binder, David. "Soviet Brigade: How the U.S. Traced It," *New York Times*, September 13, 1979, p. A16.

Blackeslee, Sandra. "Shuttle Returns from Secret Work," *New York Times*, December 7, 1988, p. A17.

Bodansky, Yossef. "Ogarkov Maps Soviets' Strategy for Winnable War," *Washington Times*, July 23, 1985, pp. 1A, 10A.

Boffey, Philip M. "U.S. Panel Calls the Disaster in the Ukraine the Worst Ever," *New York Times*, May 4, 1986, pp. 1, 20.

Booda, Larry. "First Capsule Recovered from Satellite," *Aviation Week*, August 22, 1960, pp. 33–35.

Broad, William J. "New Clues on a Soviet Laser Complex," *New York Times*, October 2, 1987.

Broad, William J. "2 Years of Failure End as U.S. Lofts Big Titan Rocket," *New York Times*, October 27, 1987, pp. A1, C4.

Broad, William J. "U.S. Designs Spy Satellites to Be More Secret Than Ever," *New York Times*, November 3, 1987, pp. C1, C3.

Broad, William J. "New Satellite is the First in a Class of All-Weather Spies, Experts Say," *New York Times*, December 4, 1988, pp. 1, 15.

Broad, William J. "Soviet Photos of U.S. Were for Spying," *New York Times*, January 30, 1989, p. A12.

Broad, William J. "Biggest U.S. Unmanned Rocket, Rival of Shuttle, Soars into Space," *New York Times*, June 15, 1989, pp. A1, B11.

Broad, William J. "Pentagon Leaves the Shuttle Program," *New York Times*, August 9, 1989, p. A13.

Broder, John M. "Shuttle Fired Into Orbit on Spy Mission," *Los Angeles Times*, December 3, 1988, pp. 1, 22–23.

Brownlow, Cecil. "Budget Cuts Threaten MOL Project," *Aviation Week and Space Technology*, May 5, 1969, pp. 22–23.

Brugioni, Dino A. "Hiding the Aircraft Factories," *Air Force Magazine*, March 1983, pp. 112–15.

Brugioni, Dino A. "Why Didn't the Feds Block the West's Floods," *Washington Post*, July 3, 1983.

Brugioni, Dino A. and Robert F. McCourt. "Personality: Arthur C. Lundahl," *Photogrammetric Engineering and Remote Sensing* 1988, pp. 271–72.

Burrows, William E. "A Study of Space Reconnaissance: Methodology for Researching a Classified System," Paper Presented at the Symposium on Space History, National Air and Space Museum, Smithsonian Institution, Washington DC, June 12, 1987.

Burt, Richard. "Arms Treaty: How to Verify Moscow's Compliance," *New York Times*, March 21, 1979, p. 8.

Burt, Richard. "U.S. Report Says Soviet Attempted Deception on Its Nuclear Strength," *New York Times*, September 26, 1979, p. A4.

Bunn, Matthew. "Spy Satellite Controversy Resolved," *Arms Control Today*, May 1989, p. 23.

"C-5As Converted for Secret Cargo," *Armed Forces*, May 1987, p. 200.

Canan, James W. "Coming Back in Space," *Air Force Magazine*, February 1987, pp. 45–52.

Canan, James W. "Recovery in Space," *Air Force Magazine*, August 1988, pp. 68–73.

Cannon, Carl M. and Mark Thompson. "Threat to Soviets Grows, U.S. Spy Photos Indicate," *Miami Herald*, April 30, 1986, pp. 1A, 4A.

"Capsule from Rocket Snagged from Sky," *Los Angeles Times*, November 15, 1960, pp. 1, 13.

"Capsule Hunt Halted," *New York Times*, April 23, 1959, p. 59.

"Capsule of Discoverer is Recovered in Pacific," *New York Times*, June 19, 1961, p. 24.

Canby, Thomas Y. "Are the Soviets Ahead in Space?," *National Geographic*, October 1986, pp. 420–458.

Cassutt, Michael. "The Manned Space Flight Engineer Programme," *Spaceflight*, January 1989, pp. 26–33.

"Catching Capsule is Complex Task," *New York Times*, November 29, 1959, p. 34.

Charles, Dan. "Spy Satellites: Entering a New Era," *Science*, March 24, 1989, pp. 1541–43.

Clark, Evert. "Satellite Spying Cited by Johnson," *New York Times*, March 17, 1967, p. 13.

Clines, Francis X. "Uzbek Violence Continues as Gunfire Wounds Scores," *New York Times*, June 10, 1989, p. 3.

Cohen, Stuart. "The Evolution of Soviet Views on SALT Verification," in William C. Potter (ed.), *Verification and SALT: The Challenge of Strategic Deception* (Boulder, CO: Westview, 1980).

Covault, Craig. "USAF, NASA Discuss Shuttle Use for Satellite Maintenance," *Aviation Week and Space Technology*, December 17, 1984, pp. 14–16.

Covault, Craig. "Soviet Strategic Laser Sites Imaged by French Spot Satellite," *Aviation Week and Space Technology*, October 26, 1987, pp. 26–27.

Covault, Craig. "U.S. Air Force Titan Launch Restarts Heavy Booster Flights," *Aviation Week and Space Technology*, November 2, 1987, pp. 24–26.

Covault, Craig. "Atlantis Radar Satellite Payload Opens New Reconnaissance Era," *Aviation Week and Space Technology*, December 12, 1988, pp. 26–28.

Covault, Craig. "Military Space Capabilities Expanding, but Excess Secrecy Limits Progress," *Aviation Week and Space Technology*, April 17, 1989, pp. 18–19.

Cushman, Jack. "Space Shuttle Explosion Throws Military Programs into Disarray," *Defense Week*, February 3, 1986, pp. 2–5.

Daly, Mark. "Krasnoyarsk: First Picture Suggests Treaty Violation," *Jane's Defence Weekly*, April 11, 1987, pp. 620–21.

"The Dark Side of Moonlighting," *Security Awareness Bulletin*, June 1987, pp. 1–12.

Davies, J.E.D. "The Discoverer Programme," *Spaceflight*, November 1969, pp. 405–07.

"Discoverer Aborted," *Aviation Week*, March 2, 1959, p. 27.

"Discoverer Capsule Fails to Go in Orbit," *New York Times*, March 31, 1961, p. 10.

"Discoverer Cone is Caught in Air," *New York Times*, July 10, 1961, pp. 1, 5.

"Discoverer Fails in Rain of Debris," *New York Times*, February 20, 1960, p. 6.

"Discoverer Failure Caused by Inverter," *Aviation Week*, November 16, 1959, p. 33.

"Discoverer May Give Sunspot Activity Data," *Los Angeles Times*, November 15, 1960, pp. 1, 13.

"Discoverer Rocket Fails in Launching," *New York Times*, October 27, 1960, p. 17.

"Discoverer Satellite Silent," *New York Times*, September 21, 1961, p. 15.

"Discoverer XVII Shot into Orbit, Recovery Attempt is Scheduled," *New York Times*, November 13, 1960, pp. 1, 27.

"Discoverer XX Fired into Orbit," *New York Times*, February 18, 1961, pp. 1, 5.

"Discoverer XXIII Fired into Orbit," *New York Times*, April 9, 1961, p. 31.

"Drop of Capsule Put Off for Day," *New York Times*, November 14, 1960, p. 16.

Drozdiak, William. "NATO Backs U.S. on Arms Charges," *Washington Post*, October 30, 1985, p. A3.

Duffy, Gloria. "Crisis Mangling and the Cuban Brigade," *International Security* 8, 1, 1983, pp. 67–87.

Ebersole, John F. and James C. Wyant. "Real-time Optical Subtraction of Photographic Imagery for Difference Detection," *Applied Optics*, 15, 4 (1976), pp. 871–76.

Elachi, Charles, "Radar Images of the Earth from Space," *Scientific American*, December 1982, pp. 54–61.

Engelberg, Stephen. "Recent Setbacks Aside, C.I.A. Still Wants More 'Human' Spies," *New York Times*, November 14, 1985, p. A12.

Engelberg, Stephen. "U.S. Says Intelligence Units Did Not Detect the Accident," *New York Times*, May 2, 1986, p. A9.

Engelberg, Stephen. "C.I.A.'s Chief Campaigns Against Missile-Making by Third World," *New York Times*, March 31, 1989, p. A6.

Engelberg, Stephen. "U.S. Sees Pakistan Seeking an A-Bomb," *New York Times*, June 11, 1989, p. 5.

Evans, Rowland and Robert Novak. "The Indigo-Lacrosse Satellite Gets the Nod," *Washington Post*, April 6, 1988, p. A25.

Evans, Rowland and Robert Novak. "'Eyes' vs. Arms," *Washington Post*, May 2, 1988, p. A21.

Evans, Rowland and Robert Novak. "The Radars of Perestroika," *Washington Post*, December 21, 1988, p. A19.

Evans, Rowland and Robert Novak. "Putting Up New Spy Satellites," *Washington Post*, April 12, 1989, p. A23.

"Explosion Sequence Photos Depict Titan 34D Launch Failure," *Aviation Week and Space Technology*, November 3, 1986, p. 43.

Fialka, John J. "North Korea May Be Developing Ability to Build Nuclear Weapons," *Wall Street Journal*, July 17, 1989, p. A16.

Fink, Donald E. "Defense Department Expands Capability of MOL," *Aviation Week and Space Technology*, February 15, 1965, pp. 16–17.

Fink, Donald E. "First Manned MOL Mission Slips to 1968," *Aviation Week and Space Technology*, April 5, 1965, pp. 26–27.

Fink, Donald E. "CIA Control Bid Slowed Decision on MOL," *Aviation Week and Space Technology*, September 20, 1965, pp. 26–27.

Finney, John W. "Discoverer Shot Into Polar Orbit, Recovery is Aim," *New York Times*, April 14, 1959, pp. 1, 18.

Finney, John W. "U.S. Cancels Plan to Catch Capsule," *New York Times*, April 15, 1959, pp. 1, 17.

Finney, John W. "Copter Recovers Capsule Ejected by U.S. Satellite," *New York Times*, August 12, 1960, pp. 1, 3.

Fishman, Charles. "Latest Failure Shocks Space Program," *Washington Post*, May 5, 1986, pp. A1, A10.

"The Flight of Atlantis," *Newsweek*, December 12, 1988, p. 68.

Foley, Theresa M. "Titan Analysis Unable to Show Cause of Insulation Separation from Casing," *Aviation Week and Space Technology*, July 7, 1986, pp. 22–23.

"For 1.5 Billion . . . A New Air Force Eye in the Sky," *Newsweek*, September 6, 1965, pp. 46–47.

Frankel, Glenn. "Iraq Said Developing A-Weapons," *Washington Post*, March 31, 1989, pp. A1, A32.

Friedman, Thomas L. "200,000 Troops Near Beijing, U.S. Says," *New York Times*, June 8, 1989, p. A13.

"Frogmen First on Scene," *New York Times*, August 12, 1960, p. 3.

"Fund Cuts Force 2-Year Stretch on MOL," *Aviation Week and Space Technology*, November 27, 1967, p. 22.

"Further Details Released on SS–20 Saber Missile," *Jane's Defence Weekly*, January 30, 1988, pp. 182–83.

Garthoff, Raymond L. "Banning the Bomb in Outer Space," *International Security*, 5, 3 1980/81, pp. 25–40.

Gertz, Bill. "Soviets Fill Craters, Dig New Ones to Fool U.S. on Missile Accuracy," *Washington Times*, August 7, 1985, pp. 1A, 10A.

Gertz, Bill. "New Soviet Deployments May Breach ABM Treaty," *Washington Times*, July 31, 1987, p. A10.

Gertz, Bill. "New Spy Satellite Needed to Monitor Treaty, Sits on Ground," *Washington Times*, October 20, 1987, p. A5.

Gertz, Bill. "Senate Panel Asks for Radar Funds," *Washington Times*, April 15, 1988, p. A4.

Gertz, Bill. "U.S. Satellites Detect Marked Increase in Mobile Soviet ICBMs," *Washington Times*, October 14, 1988, p. A6.

Gertz, Bill. "Atlantis Shuttle to Carry Aloft Superspy All-Weather Satellite," *Washington Times*, November 7, 1988, p. A7.

Gertz, Bill. "CIA Warns of Verification Woes in Future Treaty," *Washington Times*, December 21, 1988, p. A3.

Gertz, Bill. "Bush Plan to Slight Satellites and Boren," *Washington Times*, March 30, 1989, p. A3.

Gertz, Bill. "Plan to Delay Spy Satellite Will Be Costly, Sources Say," *Washington Times*, April 17, 1989, p. A4.

Gertz, Bill. "Soviets Still Stalling on Removal of Disputed Radar, Cheney Says," *Washington Times*, June 9, 1989, p. A5.

Gertz, Bill. "S. Africa on the Brink of Ballistic Missile Test," *Washington Times*, June 20, 1989, pp. A1, A9.

Getler, Michael. "New Soviet Silo Building Seen as Protection for Two Missiles," *Washington Post*, May 27, 1971, p. A14.

Getler, Michael. "New Spy Satellites Planned for Clearer, Instant Pictures," *Washington Post*, February 8, 1972, pp. A1, A9.

Glantz, David M. "The Red Mask: the Nature and Legacy of Soviet Military Deception in the Second World War," *Intelligence and National Security*, 2, 3 1987, pp. 175–259.

Gordon, Michael R. "U.S. Says Soviet Complies on Some Arms Issues," *New York Times*, November 24, 1985, p. 18.

Gordon, Michael R. "Soviets Finishing Large Radar Center in Siberia," *New York Times*, November 23, 1986, p. 6.

"Grapevine," *Time*, November 28, 1988, p. 24.

Greve, Frank. "Soviets Trying to Hide New Missile System in Area of Jet Incident," *San Francisco Chronicle & Examiner*, September 11, 1983, p. A5.

Gwertzman, Bernard. "Fire in Reactor May Be Out, New U.S. Pictures Indicate; Soviet Says Fallout is Cut," *New York Times*, May 2, 1986, pp. A1, A8.

Halloran, Richard. "U.S. Says Blast Hit Soviet Arms Base," *New York Times*, June 23, 1984, p. 3.

Halloran, Richard. "General Describes Soviet Laser Threat," *New York Times*, October 24, 1987, p. 62.

Hawkes, Russell. "USAF's Satellite Test Center Grows," *Aviation Week and Space Technology*, May 30, 1960, pp. 57–59.

Healy, Melissa. "Shuttle Launched with Secret Military Payload," *Los Angeles Times*, August 9, 1989, p. 20.

Henderson, Breck W. "Lockheed Develops Threat Warning System for U.S. Military Satellites," *Aviation Week and Space Technology*, July 3, 1989, pp. 61–62.

"How Russia Hides Its Missiles," *Foreign Report*, March 5, 1981, pp. 1–3.

Hudson, Richard D. Jr., and Jacqueline W. Hudson, "The Military Applications of Remote Sensing by Infrared," *Proceedings of the IEEE* 63, 1 1975, pp. 104–28.

Hurt, Henry. "CIA in Crisis: The Kampiles Case," *Reader's Digest*, June 1979, pp. 65–72.

"Industry Observer," *Aviation Week*, December 5, 1960, p. 23.

"Industry Observer," *Aviation Week and Space Technology*, October 18, 1965, p. 13.

"Industry Observer," *Aviation Week and Space Technology*, February 7, 1972, p. 9.

"Industry Observer," *Aviation Week and Space Technology*, May 8, 1972, p. 9.

"Industry Observer," *Aviation Week and Space Technology*, December 21, 1987, p. 15.

"INF: The Politics of Ratification," *Newsweek*, May 16, 1988, p. 22.

Inlow, Roland S. "An Appraisal of the Morison Espionage Trial," *First Principles* 11, 4 (May 1986), pp. 1, 2–5.

Janesick, James R., and Morley M. Blouke. "Introduction to Charge Coupled Device Image Sensors," in Kosta Tsipis (ed.), *Arms Control Verification: the Technologies that Make it Possible* (New York: Pergamon-Brassey's, 1985), pp. 104–20.

Janesick, James R., and Morley M. Blouke. "Sky on a Chip: The Fabulous CCD." *Sky and Telescope*, September 1987, pp. 238–42.

Jasani, Bhupendra and Christer Larsson. "Remote Sensing, Arms Control and Crisis Observation," *International Journal of Imaging, Remote Sensing and Integrated Geographical Systems* 1, 1 1987, pp. 31–41.

Kenden, Anthony. "U.S. Reconnaissance Satellite Programmes," *Spaceflight* 20, 7 (1978), pp. 243ff.

"KH-11 Overruns Said to Slow Development of Follow-On Spacecraft," *Aerospace Daily*, January 23, 1984, pp. 16–17.

"KH-11 Recon Satellite, Navstar Launched from Vandenberg," *Aviation Week and Space Technology*, February 18, 1980, p. 23.

Klass, Philip J. "Military Satellites Gain Vital Data," *Aviation Week and Space Technology*, September 15, 1969, pp. 55–61.

Klass, Philip J. "U.S. Scrutinizing New Soviet Radar," *Aviation Week and Space Technology*, August 22, 1983, pp. 19–20.

Klass, Philip J. "Soviets Test Defense Missile Reload," *Aviation Week and Space Technology*, August 29,1983, p. 19.

Kolcum, Edward H. "Air Force, Contractors Predict Long Life for Heavy Lift," *Aviation Week and Space Technology*, July 17, 1989, pp. 32–34.

Kolcum, Edward H. "Orbiting of Advanced Imaging Satellite Bolsters U.S. Intelligence Capabilities," *Aviation Week and Space Technology*, August 14, 1989, pp. 30–31.

Kyle, Deborah. "SACEUR General Rogers Urges US to Release Threat Photos," *Armed Forces Journal International*, April 1984.

"Lacrosse Orbit," *Aviation Week and Space Technology*, January 16, 1989, p. 11.

Lane, Earl. "Shuttle Columbia Returns to Space," *N.Y. Newsday*, August 9, 1989, pp. 7,24.

Lane, Earl. "Spy Satellites Threaten Data Jam," *N.Y. Newsday*, August 10, 1989.

Lardner, George Jr. "13 Copies of Classified Data Missing," *Washington Post*, November 7, 1978, pp. A1, A8.

Lardner, George Jr. "Spy Rings of One," *Washington Post*, December 4, 1983, pp. 60–65.

Lardner, George Jr. "Satellite Unchanged from Manual Bought by Soviets, U.S. Officials Say," *Washington Post*, October 10, 1985, p. A20.

Ledeen, Michael. "A Mole in Our Midst," *New York*, October 2, 1978, pp. 55–57.

"Letters," *Aviation Week and Space Technology*, January 16, 1978, p. 76.

Lodal, Jan M. "Verifying SALT," *Foreign Policy*, 24, Fall 1976, pp. 40–64.

Loftus, Joseph A. "Gilpatric Warns U.S. Can Destroy Atom Aggressor," *New York Times*, October 22, 1961, pp. 1, 6.

"Massive Satellite," *Aviation Week and Space Technology*, September 5, 1988, p. 23.

Maugh, Thomas H. II. "Atlantis Completes Secret Spy Satellite Mission," *Los Angeles Times*, December 7, 1988, pp. 1, 36.

"Meltdown," *Newsweek*, May 12, 1986, pp. 20–35.

Middleton, Drew. "Soviet's Moves in Poland Are Studied," *New York Times*, January 9, 1981, p. 4.

Middleton, Drew. "Soviet Said to Deploy a New Missile," *New York Times*, October 22, 1984, p. A3.

"Mission 27 Launch Set for December 1," *Aviation Week and Space Technology*, November, 21, 1988, p. 23.

Mohr, Charles, "U.S. and Soviet Discuss Whether Moscow Violated Terms of Arms Pacts," *New York Times*, October 5, 1983, p. A8.

"MOL Delayed by Funding Cut," *Aviation Week and Space Technology*, April 21, 1969, p. 17.

Morrison, David C. "Radar Diplomacy," *National Journal*, January 3, 1987, pp. 17–21.

Nammack, John. "C-119's Third Pass Snares Discoverer," *Aviation Week and Space Technology*, August 29, 1960, pp. 30–31.

"NASA Dishes Up Menu, No News on Atlantis' Day." *Los Angeles Times*, December 4, 1988, p. 24.

"NASA Mum on Columbia, Other Than Its Doing Well," *Washington Times*, August 10, 1989, p. A7.

"New Discoverer Shot into Orbit," *New York Times*, August 19, 1960, pp. 1, 13.

"New Payload Could Boost Shuttle Cost," *Aviation Week and Space Technology*, August 14, 1978, pp. 16–17.

"New Soviet Aircraft Carrier in Detail," *Jane's Defence Weekly*, November 5, p. 1147.

"New Soviet Missile Sites Reported," *New York Times*, April 23, 1985, p. A4.

Normyle, William J. "Air Force Given Manned Space Role," *Aviation Week and Space Technology*, August 30, 1965, p. 23.

North, David M., and John D. Morocco. "Blackjack Shares Aspects of U.S. B-1B and XB-70," *Aviation Week and Space Technology*, August 15, 1988, pp. 16–18.

Oberdorfer, Don. "The 'Brigada:' An Unwelcome Sighting in Cuba," *Washington Post*, September 9, 1979, pp. A1, A18.

Oberdorfer, Don. "North Koreans Pursue Nuclear Arms," *Washington Post*, July 29, 1989, p. A9.

Osmunden, John A. "Rivalry is Cited," *New York Times*, March 2, 1959, p. 10.

O'Toole, Thomas. "AF Study Questions Economy of Space Shuttle," *Washington Post*, May 27, 1971, p. 10.

O'Toole, Thomas, and Charles Babcock. "CIA 'Big Bird' Satellite Manual Was Allegedly Sold to Soviets," *Washington Post*, August 23, 1978, p. A1, A16.

Ott, James. "Espionage Trial Highlights CIA Problems," *Aviation Week and Space Technology*, November 27, 1978, pp. 21–23.

Ottaway, David B. "Saudis Hid Acquisition of Missiles," *Washington Post*, March 29, 1988, pp. A1, A13.

Ottaway, David B. "Behind the New Battle with Libya," *Washington Post*, January 8, 1989, pp. C1, C4.

Painter, Floyd C. "The Tracking and Data Relay Satellite System," *Defense Electronics*, June 1989, pp. 115–20.

"Pawn Shop and Flat Twin Radars: Are They Obsolete?" *Jane's Defence Weekly*, January 9, 1988, p. 28.

Peebles, Curtis. "The Guardians," *Spaceflight*, November 1978, pp. 381ff.

Philpott, Tom. "Gen. Rogers: Show Spy Photos in Europe Debate," *Air Force Times*, July 23, 1984, p. 6.

Pike, John. "Reagan Prepares for War in Outer Space," *Counter Spy*, September–November 1982, pp. 17–22.

Pincus, Walter. "Hill Conferees Propose Test of Space Arms," *Washington Post*, July 11, 1984, pp. A1, A13.

"President By-Passed DOD & Joint Chiefs to Cancel MOL," *Space Daily*, December 16, 1969, p. 199.

Randal, Jonathan C. "Iran Pours Reinforcements into Bridgeheads in Iraq," *Washington Post*, February 14, 1986, pp. A23, A29.

Rasky, Susan F. "Bush is Accused of Backing Away from Promise on 1988 Arms Pact," *New York Times*, April 7, 1989, pp. A1, A9.

"Reign of Terror," *Newsweek*, June 19, 1989, pp. 14–22.

Rensberger, Boyce. "Explosion, Graphite Fire Suspected," *Washington Post*, April 30, 1986, pp. A1, A17.

Richelson, Jeffrey. "The Satellite Data System," *Journal of the British Interplanetary Society*, 37, 5 (1984), pp. 226–28.

Richelson, Jeffrey. "Implications for Nations without Space-Based Intelligence Capabilities," in Michael Krepon, Peter Zimmerman, Leonard Spector and Mary Umberger, *Commercial Observation Satellites and International Security* (New York: St. Martin's Press, forthcoming).

Robinson, Clarence A. Jr. "Soviets Hiding Submarine Work," *Aviation Week and Space Technology*, November 11, 1974, pp. 14–16.

Robinson, Clarence A. Jr. "USSR Cuban Force Clouds Debate on SALT," *Aviation Week and Space Technology*, September 10, 1979, pp. 16–19.

Robinson, Clarence A. Jr. "Soviets Accelerate Missile Defense," *Aviation Week and Space Technology*, January 16, 1984, pp. 14–16.

"Rocket Explodes; 58 Sent to Hospital," *Mansfield News-Journal*, April 19, 1986, p. 1-A.

"Satellite Capsule Sighted, Then Lost," *New York Times*, September 16, 1960, p. 13.

"Satellite Fired into Polar Orbit," *New York Times*, November 8, 1959, p. 43.

"Satellite Pictures Show Soviet CVN Towering Above Nikolaev Shipyard," *Jane's Defence Weekly*, August 11, 1984, pp. 171–73.

"Satellite Reconnaissance and Photographic Interpretation," *Aviation Knowledge*, July 1983, pp. 22–23.

"Satellite with 150-ft. Span Set for Launch on Mission 27," *Aviation Week and Space Technology*, November 7, 1988, p. 25.

Sawyer, Kathy. "Air Force Orbits Satellite," *Washington Post*, October 27, 1987, pp. A1, A7.

Sawyer, Kathy. "Weather Delays Space Shuttle Launch," *Washington Post*, December 2, 1988, p. A4.

Sawyer, Kathy. "Shuttle Atlantis Lifts Off After Racing the Clock," *Washington Post*, December 3, 1988, pp. A1, A11.

Sawyer, Kathy. "Shuttle Crew is Believed to Deploy Spy Satellite," *Washington Post*, December 4, 1988, p. A3.

Sawyer, Kathy. "New Titan IV Rocket Orbits Secret Satellite," *Washington Post*, June 15, 1989, p. A12.

Schemann, Serge. "Soviet Mobilizes a Vast Operation to Overcome the Disaster," *New York Times*, May 19, 1986, p. A8.

Schram, Martin. "Response: Avoiding a Crisis Tone," *Washington Post*, September 9, 1979, p. A18.

Schultz, James B. "TRW to Deliver MILSTAR Payload and House Votes to Kill Satellite Program," *Armed Forces Journal International*, September 1989, pp. 75–78.

Schwartz, Harry. "'Spying' in Space by U.S. Charged," *New York Times*, March 6, 1959, p. 10.

Scott, William B. "Titan Mission Success Based on Tighter Heavy Booster Standards," *Aviation Week and Space Technology*, November 2, 1987, pp. 25–26.

Scott, William B. "Atlantis Returns from Secret Mission with Substantial Thermal Tile Damage," *Aviation Week and Space Technology*, December 12, 1988, p. 29.

"2nd Space Capsule Caught in Mid-Air," *New York Times*, November 15, 1960, pp. 1, 20.

"Secret Photographs," *Aviation Week and Space Technology*, January 23, 1989, p. 11.

Sehlstedt, Albert Jr. "Florida Ired as California is Winner of Space Project," *Baltimore Sun*, February 11, 1966, p. 7.

"Showdown with Libya," *Newsweek*, January 16, 1989, pp. 16–17.

Simons, Howard. "Our Fantastic Eye in the Sky," *Washington Post*, December 8, 1963, pp. E1, E5.

Simons, Howard, and Chalmers M. Roberts. "Role in Arms Control Clinched MOL Victory," *Washington Post*, September 5, 1965, pp. A1, A5.

"Sketches of 5 on the Shuttle Mission," *New York Times*, December 3, 1989, p. 52.

Sloan, Allan. "Big Brother Strikes Again," *Forbes*, May 12, 1980, pp. 50–51.

Smith, Charles. "Soviet Maskirovka," *Air Power Journal*, Summer 1988, pp. 28–39.

"Soviet Naval Blast Called Crippling," *New York Times*, July 11, 1984, p. 6.

"Soviet Strategic Bomber Photographed at Ramenskoye," *Aviation Week and Space Technology*, December 14, 1981, p. 17.

"Soviets Build Directed-Energy Weapon," *Aviation Week and Space Technology*, July 28, 1980, pp. 47–50.

"Soviets Display Laser Facility at Sary Shagan," *Aviation Week and Space Technology*, July 17, 1989, p. 27.

"Soviets Ready New Boosters at Tyuratam," *Aviation Week and Space Technology*, August 27, 1984, pp. 18–21.

"Soviets Test Defense Missile Reload," *Aviation Week and Space Technology*, August 29, 1983, p. 19.

"Space Capsule Caught in Mid-Air by U.S. Plane on Reentry from Orbit," *New York Times*, August 20, 1960 pp. 1, 7.

"Space Reconnaissance Dwindles," *Aviation Week and Space Technology*, October 6, 1980, pp. 18–20.

"Space Secrecy Muddle," *Aviation Week and Space Technology*, April 23, 1962, p. 21.

"Space Watchers Had a Key," *New York Times*, December 10, 1988, p. 50.

"Strong and Silent," *Newsweek*, September 12, 1988, pp. 27–28.

Sulzberger, C. L. "Those Who Spy Out the Land," *New York Times*, July 15, 1963, p. 28.

"Sunny Debut for Snowstorm," *Time*, November 28, 1988, p. 80.

Suvorov, Viktor. "GUSM: The Soviet Service of Strategic Deception," *International Defense Review* 8, 1985, pp. 1235–40.

Szulc, Tad. "Russians in Cuba Now Put at 12,500," *New York Times*, June 20, 1963, p. 2.

Taubman, Philip. "Secrecy of U.S. Reconnaissance Office is Challenged," *New York Times*, March 1, 1981, p. 10.

Taubman, Philip. "Showing Secrets: A U.S. Compromise," *New York Times*, March 24, 1983, p. A21.

"3rd Space Capsule Caught in Mid-Air," *New York Times*, December 11, 1960, pp. 1, 64.

"Thousands Fleeing Uzbek Riots Stay in Dismal Camp," *Washington Times*, June 12, 1989, p. A7.

"Titan 3C Passes 6th Test, Furnishes MOL Support," *Aviation Week and Space Technology*, November 14, 1966, p. 30.

"Titan 34D Booster Failed Following Premature Shutdown of Aerojet Engine," *Aviation Week and Space Technology*, November 18, 1985, p. 26.

Toth, Robert C. "U.S. Facing Uproar in Spy Satellite Photos," *Los Angeles Times*, November 12, 1978, pp. 1, 27.

Toth, Robert C. "CIA 'Mighty Wurlitzer' is Now Silent," *Los Angeles Times*, December 30, 1980, pp. 1, 12.

Toth, Robert C. "Anaheim Firm May Have Sought Spy Satellite Data," *Los Angeles Times*, October 10, 1982, pp. 1, 32.

Toth, Robert C. "Satellites Keep Eye on Reactor," *Los Angeles Times*, May 2, 1986, p. 22.

Tuohy, William. "U.S. Shows Photos of Libya Plant; Europeans Have Doubts," *Los Angeles Times*, January 3, 1989, p. 12.

"U.S. Orbits Discoverer," *New York Times*, October 14, 1961, p. 3.

"U.S. to Modernize Spy Satellites," *Washington Times*, April 18, 1989, p. A2.

"USAF Push Pied Piper Space Vehicle," *Aviation Week*, October 14, 1957, p. 26.

"USAF $24.9b Request for Satellite Centre," *Jane's Defence Weekly*, March 19, 1988.

Walcott, John. "U.S. Analysts Find Soviet Radars, Possibly Complicating Arms Pact Effort," *Wall Street Journal*, August 15, 1986, pp. 2.

"Washington Roundup," *Aviation Week and Space Technology*, June 4, 1979, p. 11.

"Washington Roundup," *Aviation Week and Space Technology*, September 29, 1980, p. 17.

"Washington Roundup," *Aviation Week and Space Technology*, October 31, 1983, p. 17.

"Washington Roundup," *Aviation Week and Space Technology*, November 23, 1987, p. 21.

Weinraub, Richard M. "India Tests Mid-Range 'Agni' Missile," *Washington Post*, May 23, 1989, pp. A1, A21.

Welzenbach, Donald. "Observation Balloons and Reconnaissance Satellites," *Studies in Intelligence*, Spring 1986, pp. 21–28.

White, Max. "U.S. Satellite Reconnaissance During the Falklands War," Earth Satellite Research Unit, Department of Mathematics, University of Aston.

"White House Assesses Reports of Soviet Asat Laser Facilities," *Aviation Week and Space Technology*, September 15, 1986, p. 21.

Wilford, John Noble. "Spy Satellite Reportedly Aided in Shuttle Flight," *New York Times*, October 20, 1981, p. C4.

Wilford, John Noble. "Shuttle Flight Readied Behind a Curtain of Secrecy," *New York Times*, November 28, 1988, p. A21.

Wilford, John Noble. "Shuttle with Spy Craft Lifts Off, As Winds Ease with Minutes Left," *New York Times*, December 3, 1988, pp. 1, 52.

Wilson, George "'N-Pic'—CIA Technicians Ferret Out Secrets Behind Cemented Windows," *Los Angeles Times*, January 12, 1975, p. 14–17.

Wines, Michael. "Soviet Secrecy Hides Extent of Disaster," *Los Angeles Times*, May 2, 1986, pp. 1, 22.

Witkin, Richard. "Washington to Hail Returned Capsule in Ceremony Today," *New York Times*, August 13, 1960, pp. 1, 7.

Woodward, Bob and Walter Pincus. "At CIA, a Rebuilder 'Goes with the Flow'," *Washington Post*, August 10, 1988, pp. A1, A8.

"Work on Pied Piper Accelerated, Satellite Has Clam-Shelled Nose Cone," *Aviation Week*, June 23, 1958, pp. 18–19.

Ybarra, Michael J. "Soviet 'Star Wars' Laser Facility Still Poses Threat, Pentagon Says," *Los Angeles Times*, July 13, 1989, p. 11.

Zhukov, G. "Space Espionage Plans and International Law," *International Affairs (Moscow)*, October 1960, pp. 53–57.

APPENDIX A:

Chronology

May 2, 1946   RAND publishes *Preliminary Design for an Experimental World Circling Spaceship*.

February 1, 1947   RAND publishes *Reference Papers Relating to a Satellite Study*.

February 1948   The Air Force requests RAND to establish a satellite evaluation project.

November 1950   RAND recommends further research into satellite reconnaissance.

April 1951   RAND publishes *Inquiry into the Feasibility of Weather Reconnaissance from a Satellite Vehicle* and *Utility of a Satellite Vehicle for Reconnaissance*.

December 1953   Air Research and Development Command pulls together ongoing satellite work as Project 409–40 "Satellite Component Study" and unofficially assigns the designation WS-117L for ultimate system development.

March 1, 1954   RAND publishes *Project FEEDBACK Summary Report*.

November 27, 1954   Western Development Division issues System Requirement No. 5, "System Requirement for an Advanced Reconnaissance System."

February 14, 1955   President Eisenhower receives the report of the Technological Capabilities Panel.

March 16, 1955   The Air Force issues General Operational Requirement No. 80, establishing the requirement for a reconnaissance satellite.

June 26, 1956   RAND issues *Physical Recovery of a Satellite Payload: A Preliminary Investigation*.

October 4, 1957   The Soviet Union launches Sputnik I.

October 26, 1957   The Air Force awards a contract to the Lockheed Missile Systems Division to develop WS-117L, the Advanced Reconnaissance System.

November 3, 1957   The Soviet Union launches Sputnik II.

November 12, 1957   RAND formally recommends to the Air Staff development of a recoverable reconnaissance satellite; publishes *A Family of Recoverable Satellites*.

February 7, 1958   The CIA and Deputy Director for Plans Richard Bissell is given responsibility for developing a recoverable satellite. The project is codenamed CORONA. The recoverable satellite portion of WS-117L is cancelled.

November 1958   The Defense Department reveals that 117L consists of three elements—DISCOVERER (the cover for CORONA), MIDAS (an early-warning satellite), and SENTRY (the Air Force reconnaissance satellite project in which data would be radioed to ground stations).

February 28, 1959   Discoverer I is launched; East German radio denounces the U.S. for espionage.

April 13, 1959   Discoverer II is launched. Due to error the capsule lands on the island of Spitzbergen and is apparently recovered by the Soviet Union.

June 3, 1959   Discoverer III fails to attain orbit.

April 15, 1960   Discoverer XI is launched. The failure to recover its capsule represents over a year of failure in recovery attempts.

May 26, 1960   President Eisenhower orders reevaluation of the satellite reconnaissance program.

August 10, 1960   Discoverer XIII is launched.

August 11, 1960   Discoverer XIII's capsule is ejected and falls to earth. While aerial recovery fails it is recovered from the ocean, becoming the first capsule recovered.

August 15, 1960   The Discoverer XIII capsule is shown to President Eisenhower at the White House.

August 18, 1960   Discoverer XIV is launched, the first Discoverer to carry a camera.

August 19, 1960   The Discoverer XIV capsule is recovered in the air by a C-119 Flying Boxcar. It contains the first satellite photos of the Soviet Union.

August 25, 1960   The National Reconnaissance Office is established. Its existence is classified.

August 31, 1960   The Office of Missile and Satellite Systems is established within the Office of the Secretary of the Air Force to provide cover for NRO.

October 11, 1960   An attempted launch of the Air Force's reconnaissance satellite, now named SAMOS, fails.

January 18, 1961   NSCID (National Security Council Intelligence Directive) No. 8, "Photographic Interpretation," establishes the National Photographic Interpretation Center. Arthur Lundahl is named as director.

January 31, 1961   SAMOS 2 is successfully placed in orbit.

September 21, 1961   National Intelligence Estimate 11-8/1-61, based on the results of CORONA missions, reduces the estimate of Soviet ICBMs to 10–25. Previous versions had estimated hundreds.

December 22, 1961   The last publicly identified SAMOS launch, SAMOS 5, takes place.

February 27, 1962   The final publicly acknowledged DISCOVERER is launched. It is subsequently known as Program 162 until its termination with the launch of April 27, 1964.

March 23, 1962   A Defense Department directive directs that all military space activities be classified secret.

May 26, 1962   President Kennedy signs National Security Action Memorandum 156, "Negotiation on Disarmament and Peaceful Uses of Outer Space."

July 1, 1962   The NSAM 156 committee delivers its study, *Political and Informational Aspects of Satellite Reconnaissance Programs*.

May 18, 1963   The first launch of a new camera system for the CORONA program takes place. The camera is designated the KH-4A.

July 12, 1963   The first launch of the GAMBIT program, designed to provide close-look photos, takes place. The camera system is designated the KH-7.

August 25, 1963 President Johnson announces that the Air Force will build and deploy the Manned Orbiting Laboratory. Unstated is the MOLs main objective—reconnaissance employing cameras and people. The program to use the MOL for reconnaissance is code-named DORIAN. The camera system is designated the KH-10.

July 29, 1966 A Titan 3B-Agena D places a new camera system for the GAMBIT program, the KH-8, into orbit.

August 8, 1966 A Long Tank Thrust Augmented Thor places a new CORONA camera, the KH-4B, into space.

March 16, 1967 President Johnson tells a group of educators of the value of space reconnaissance.

March 30, 1967 The last satellite carrying a KH-4A reenters the atmosphere.

June 4, 1967 The final KH-7 satellite is launched.

June 5, 1967 Israel launches a surprise attack on Egypt and Syria.

July 1, 1967 Director of Central Intelligence Directive 1/13 establishes the Committee on Imagery Requirements and Exploitation (COMIREX).

August 21, 1968 Warsaw Pact troops invade Czechoslovakia, ending the Prague Spring. CORONA photos indicating an imminent invasion are not returned until after the invasion.

June 10, 1969 It is announced that the MOL program has been terminated.

June 15, 1971 The successor to the CORONA area surveillance program begins. The new program, designated HEXAGON, employs the new KH-9 camera system. The KH-9 can take pictures of wider areas than the KH-4B, with greater detail.

May 25, 1972 The final KH-4B is launched as Richard Nixon and Leonid Brezhnev sign the first SALT agreement.

October 6, 1973 Egypt and Syria initiate the Yom Kippur War. No satellite photographs are received during the war.

December 19, 1976 The first KH-11 satellite for the KENNAN program is launched. Rather than returning film, capsule imagery is produced by an electro-optical system. The imagery can be relayed back to a Washington area ground station almost instantaneously.

January 21, 1977   Acting DCI E. Henry Knoche shows President Carter the initial KH-11 photos, received the day before.

August 6, 1977   The Soviet Union, based on its satellite reconnaissance activities, informs the U.S. that South Africa is preparing to conduct a nuclear test. KEYHOLE photographs confirm the Soviet claim.

March 2, 1978   Former CIA employee William Kampiles sells Soviet agent Michael Zavali the KH-11 manual.

October 1, 1978   In a speech at Cape Canaveral, President Carter acknowledges that the U.S. operates photographic reconnaissance satellites.

December 22, 1978   William Kampiles is sentenced to 40 years in prison.

June 18, 1979   Jimmy Carter and Leonid Brezhnev sign the SALT II agreement.

August 1979   A KH-11 photographs the maneuvers of a Soviet brigade in Cuba. The brigade becomes the center of a major political controversy.

November 4, 1979   The U.S. embassy in Teheran is seized.

April 24, 1980   A mission to rescue the hostages fails. KH-11 photographs, essential to planning the mission, are left behind at Desert One.

December 14, 1981   The first leaked satellite photo, showing a Soviet BLACKJACK bomber, appears in *Aviation Week and Space Technology*.

June-August 1983   KEYHOLE photography uncovers a new phased-array radar being constructed at Kranoyarsk, in violation of the ABM Treaty.

April 17, 1984   The final KH-8 is launched.

June 25, 1984   The final KH-9 mission begins.

August 11, 1984   *Jane's Defense Weekly* publishes KH-11 photos of the first Soviet nuclear aircraft carrier under construction. Naval intelligence analyst Samuel Loring Morison receives a 2-year sentence for leaking the photos.

December 4, 1984   The sixth KH-11 is launched.

August 13, 1985   The fifth KH-11 is deorbited.

August 28, 1985   The replacement for the fifth KH-11 fails to obtain orbit and is destroyed.

January 26, 1986   The space shuttle orbiter Challenger explodes shortly after launch.

April 18, 1986   A Titan 34D carrying the last KH-9 explodes shortly after launch from Vandenberg AFB.

April 26, 1986   A nuclear accident occurs at the Chernobyl nuclear power plant.

April 29, 1986   A KH-11 obtains the first good photographs of the damaged reactor.

October 26, 1987   A new KH-11 is placed into orbit, the first KEY-HOLE satellite to reach orbit since the sixth KH-11 on December 4, 1984.

December 1987   A KH-11 detects construction at a site at Rabta, Libya. It is later determined to be a chemical weapons plant.

December 2, 1988   The space shuttle Atlantis lifts off carrying the first LACROSSE satellite. LACROSSE produces imagery using radar rather than visible light photography.

August 8, 1989   The space shuttle Columbia deploys the first Advanced KENNAN/Improved CRYSTAL satellite, with infrared imaging capability.

APPENDIX B:

Program Summaries, KH-4A through KH-11[*]

Program Codename:	CORONA
Numerical Designation:	N/A
Optical System Designation:	KH-4A
Optical System Equipment:	Camera
Booster:	Thor-Agena D
Resolution:	10 feet
Swath Width:	N/A
Data Return Method:	Capsules (2)
Attempted Launches:	51
Successful Launches:	46
Mean Perigee (Miles):	107
Mean Apogee (Miles):	242
Mean Inclination (Degrees):	79
Mean Lifetime (Days):	23.6
First Launch:	February 28, 1963
Last Launch:	March 30, 1967

KH-4A LAUNCH DATA

Date	Lifetime	Inclination	Perigee	Apogee
Feb. 28, 1963	0	Launch Failure		
Mar. 18, 1963	0	Launch Failure		
May 18, 1963	8	74.54	95	308
June 13, 1963	29	81.87	119	260
June 27, 1963	30	81.60	121	246
July 19, 1963	26	90.37	117	228
July 31, 1963	12	74.95	97	255
Aug. 25, 1963	19	75.01	100	198
Sept. 23, 1963	18	74.90	100	273
Oct. 29, 1963	84	89.90	171	191

* Excluding KH–5, KH–6

353

KH-4A LAUNCH DATA

Date	Lifetime	Inclination	Perigee	Apogee
Nov. 27, 1963	17	69.90	109	239
Dec. 21, 1963	18	64.94	109	220
Feb. 15, 1964	23	74.95	111	275
Mar. 24, 1964		Launch Failure		
Apr. 27, 1964	28	79.93	110	277
June 4, 1964	14	79.96	92	266
June 19, 1964	27	85.00	109	286
July 10, 1964	27	84.94	112	286
Aug. 5, 1964	26	79.96	113	270
Sept. 14, 1964	22	84.96	107	289
Oct. 5, 1964	21	79.97	113	273
Oct. 17, 1964	17	90.59	117	258
Nov. 2, 1964	25	79.95	112	278
Nov. 18, 1964	17	70.02	112	210
Dec. 19, 1964	26	74.97	113	254
Dec. 21, 1964	22	70.08	148	164
Jan. 15, 1965	25	74.95	112	260
Feb. 25, 1965	21	75.08	110	234
Mar. 25, 1965	10	96.08	115	164
Apr. 29, 1965	27	85.05	110	293
May 18, 1965	28	75.01	123	205
June 9, 1965	13	75.07	109	224
July 19, 1965	29	85.05	113	288
Aug. 17, 1965	54	70.04	112	252
Sept. 2, 1965		Launch Failure		
Sept. 22, 1965	18	80.01	118	226
Oct. 5, 1965	24	75.05	126	200
Oct. 28, 1965	20	74.97	109	267
Dec. 9, 1965	17	80.04	113	271
Dec. 24, 1965	27	80.01	110	277
Feb. 2, 1966	25	75.05	115	177
Mar. 9, 1966	20	75.03	110	268
Apr. 7, 1966	18	75.06	120	193
May 3 ,1966		Launch Failure		
May 24, 1966	16	66.04	111	168
June 21, 1966	22	80.10	120	228
Sept. 20, 1966	22	85.13	117	274
Nov. 8, 1966	21	100.09	107	197
Jan. 14, 1967	19	80.07	112	236
Feb. 22, 1967	17	80.03	112	236
Mar. 30, 1967	18	85.03	104	202

Program Codename: CORONA
Numerical Designation: N/A
Optical System Designation: KH-4B
Optical System Equipment: Camera
Booster: Long Tank Thrust Augmented Thor-Agena D
Resolution: 5 feet
Swath Width: 40 × 180 miles
Data Return Method: Capsules (2)
Attempted Launches: 34
Successful Launches: 33
Mean Perigee (Miles): 105.9
Mean Apogee (Miles): 197.9
Mean Inclination (Degrees): 83.50
Mean Lifetime (Days): 23.4
First Launch: August 9, 1966
Last Launch: May 25, 1972

KH-4B LAUNCH DATA

Date	Lifetime	Inclination	Perigee	Apogee
Aug. 9, 1966	32	100.12	120	178
Nov. 8, 1966	21	100.09	107	197
Jan. 14, 1967	19	80.07	112	236
Feb. 22, 1967	17	80.03	112	236
Mar. 30, 1967	18	85.03	104	202
May 9, 1967	64	85.10	124	482
June 16, 1967	33	80.02	112	228
Aug. 7, 1967	25	79.94	108	215
Sept. 15, 1967	19	80.07	93	241
Nov. 2, 1967	30	81.53	113	254
Dec. 9, 1967	15	81.65	98	147
Jan. 24, 1968	34	81.48	109	267
Mar. 14, 1968	26	83.01	110	242
May 1, 1968	14	83.05	102	151
June 20, 1968	25	85.10	120	202
Aug. 7, 1968	19	82.11	94	160
Sept. 18, 1968	19	83.02	104	244
Nov. 3, 1968	20	82.15	93	179

KH-4B LAUNCH DATA

Date	Lifetime	Inclination	Perigee	Apogee
Dec. 12, 1968	16	81.02	105	154
Feb. 5, 1969	19	81.54	110	148
Mar. 19, 1969	4	83.04	111	149
May 2, 1969	21	89.54	111	202
July 24, 1969	30	74.98	110	136
Sept. 22, 1969	20	85.03	110	157
Dec. 4, 1969	36	81.46	99	156
Mar. 4, 1970	22	88.02	104	160
May 20, 1970	28	83.00	100	153
July 23, 1970	27	60.00	98	247
Nov. 18, 1970	23	82.99	115	144
Feb. 17, 1971	0	Launch Failure		
Mar. 24, 1971	19	88.56	97	153
Sept. 10, 1971	25	74.95	97	151
Apr. 19, 1972	23	81.48	96	172
May 25, 1972	10	96.34	98	189

Program Codename:	GAMBIT
Numerical Designation:	N/A
Optical System Designation:	KH-7
Optical System Equipment:	Camera
Booster:	Atlas-Agena D
Resolution:	18 inches
Swath Width:	N/A
Data Return Method:	Capsules (2)
Attempted Launches:	38
Successful Launches:	36
Mean Perigee (Miles):	92.22
Mean Apogee (Miles):	186.86
Mean Inclination (Degrees):	97.21
Mean Lifetime (Days):	5.47
First Launch:	July 12, 1963
Last Launch:	June 4, 1967

KH-7 LAUNCH DATA

Date	Lifetime	Inclination	Perigee	Apogee
July 12, 1963	5	95.37	102	102
Sept. 6, 1963	7	94.37	104	163
Oct. 25, 1963	4	99.05	87	169
Dec. 18, 1963	1	88.48	76	165
Feb. 25, 1964	4	74.95	102	201
Mar. 11, 1964	4	88.20	101	126
Apr. 23, 1964	5	103.56	93	208
May 19, 1964	3	101.12	87	236
July 6, 1964	2	92.89	75	215
Aug. 14, 1964	9	95.52	92	190
Sept. 23, 1964	5	92.91	90	188
Oct. 8, 1964		Launch Failure		
Oct. 23, 1964	5	95.55	86	168
Dec. 4, 1964	1	97.02	98	221
Jan. 23, 1965	5	102.50	91	180
Mar. 12, 1965	5	107.69	96	153
Apr. 28, 1965	5	95.60	112	166
May 27, 1965	5	95.78	92	166
June 25, 1965	5	107.64	94	175
July 12, 1965		Launch Failure		
Aug. 3, 1965	4	107.47	92	190

KH-7 LAUNCH DATA

Date	Lifetime	Inclination	Perigee`	Apogee
Sept. 30, 1965	5	95.60	98	164
Nov. 8, 1965	3	93.88	90	172
Jan. 19, 1966	6	93.86	93	167
Feb. 15, 1966	7	96.54	92	182
Mar. 18, 1966	5	101.01	94	176
Apr. 19, 1966	6	116.95	90	247
May 14, 1966	6	110.55	82	222
June 3, 1966	6	87.01	89	179
July 12, 1966	7	95.52	85	146
Aug. 16, 1966	8	93.24	91	222
Sept. 16, 1966	6	93.98	92	206
Oct. 12, 1966	8	90.80	112	160
Nov. 2, 1966	7	90.96	99	189
Dec. 5, 1966	8	104.63	85	241
Feb. 2, 1967	9	102.96	84	221
May 22, 1967	8	91.49	84	182
June 4, 1967	8	104.87	90	269

Program Designation	GAMBIT
Numerical Designation:	1700
Optical System Designation:	KH-8
Optical System Equipment:	Camera
Booster:	Titan 3B-Agena D
Swath Width:	N/A
Method of Film Return	Capsules (2)
Attempted Launches:	51
Successful Launches:	50
Mean Perigee (Miles):	83.6
Mean Apogee (Miles):	254.3
Mean Inclination (Degrees):	104.2
Mean Lifetime (Days):	30.4
First Launch:	July 29, 1966
Last Launch:	April 17, 1984

KH-8 LAUNCH DATA

Date	Lifetime	Inclination	Perigee	Apogee
July 29, 1966	7	94.12	98	155
Sept. 28, 1966	9	93.98	94	184
Dec. 14, 1966	9	109.56	86	229
Feb. 24, 1967	10	106.98	84	257
Apr. 26, 1967	0	Failed to orbit		
June 20, 1967	10	111.40	79	202
Aug. 16, 1967	13	111.88	88	278
Sept. 19, 1967	10	106.10	76	249
Oct. 25, 1967	9	111.57	84	266
Dec. 5, 1967	11	109.55	85	267
Jan. 18, 1968	17	111.53	86	250
Mar. 13, 1968	11	99.87	79	252
Apr. 17, 1968	12	111.49	88	272
June 5, 1968	12	110.52	76	283
Aug. 6, 1968	9	110.00	88	245
Sept. 10, 1968	15	106.06	78	250
Nov. 6, 1968	14	106.00	81	242
Dec. 4, 1968	8	106.24	84	456
Jan. 22, 1969	12	106.12	88	675
Mar. 4, 1969	14	92.00	83	286

KH-8 LAUNCH DATA

Date	Lifetime	Inclination	Perigee	Apogee
Apr. 15, 1969	15	108.75	84	254
June 3, 1969	11	110.00	85	257
Aug. 22, 1969	16	107.99	84	115
Oct. 24, 1969	15	108.04	84	459
Jan. 14, 1970	18	109.60	83	237
Apr. 15, 1970	21	110.96	81	253
June 25, 1970	11	108.87	79	244
Aug. 18, 1970	16	110.95	94	275
Oct. 23, 1970	19	111.06	84	246
Jan. 21, 1971	19	110.86	86	259
Apr. 22, 1971	21	110.93	82	249
Aug. 12, 1971	22	111.00	85	263
Oct. 23, 1971	25	110.94	83	258
Mar. 17, 1972	25	110.98	81	254
Sept. 1, 1972	29	110.50	87	236
Dec. 1, 1972	33	110.44	82	247
May 16, 1973	28	110.51	86	247
Sept. 27, 1973	32	110.48	81	239
Feb. 13, 1974	32	110.44	83	244
June 6, 1974	47	110.49	84	244
Aug. 14, 1974	46	110.51	84	249
Apr. 18, 1975	48	110.54	83	249
Oct. 9, 1975	52	96.41	78	221
Mar. 22, 1976	57	96.40	78	215
Sept. 15, 1976	51	96.39	84	205
Mar. 13, 1977	74	96.40	77	216
Sept. 23, 1977	76	96.49	78	218
May 28, 1979	90	96.41	81	177
Feb. 28, 1981	112	96.38	86	208
Apr. 15, 1983	128	96.52	89	184
Apr. 17, 1984	118	96.40	79	193

Program Codename:	HEXAGON
Numerical Designation:	1900
Optical System Designation:	KH-9
Optical System Equipment:	2 60-inch cameras, 1 12-inch mapping camera on 5 missions
Boosters:	Titan 3D, Titan 34D (1984 and after)
Resolution:	2 feet
Swath Width:	80 × 360 miles
Data Return Method:	Capsules (4)
Attempted Launches:	20
Successful Launches:	19
Mean Perigee (Miles):	100.9
Mean Apogee (Miles):	159
Mean Inclination (Degrees):	96.4
Mean Lifetime (Days):	137.8
First Launch:	June 15, 1971
Last Launch:	Apr. 18, 1986

KH-9 LAUNCH DATA

Date	Lifetime	Inclination	Perigee	Apogee
June 15, 1971	52	96.41	114	186
Jan. 20, 1972	40	97.00	92	213
July 7, 1972	68	96.88	108	156
Oct. 10, 1972	90	96.46	102	166
Mar. 9, 1973	71	95.70	94	167
July 13, 1973	91	96.21	97	167
Nov. 10, 1973	123	96.93	99	171
Apr. 10, 1974	109	94.52	95	177
Oct. 29, 1974	141	96.69	100	168
June 8, 1975	150	96.38	95	167
Dec. 4, 1975	119	96.27	97	145
July 8, 1976	158	97.00	99	150
June 27, 1977	179	97.02	96	148
Mar. 16, 1978	179	96.43	99	149
Mar. 16, 1979	190	96.39	105	160
June 18, 1980	261	96.46	105	164
May 11, 1982	208	96.41	110	162
June 20, 1983	275	96.45	105	142
June 25, 1984	115	96.43	105	163
Apr. 18, 1986	0	Exploded after Take-off		

Program Codename:	KENNAN, CRYSTAL
Numerical Designation:	5500
Optical System Designation:	KH-11
Optical System:	Electro-Optical
Boosters:	Titan 3D, Titan 34D (1984 and after)
Resolution:	6 inches
Swath Width:	N/A
Data Return Method:	Digital, via Relay satellites
Attempted Launches:	8
Successful Launches:	7
Mean Perigee (Miles):	161
Mean Apogee (Miles):	377
Mean Inclination (Degree):	97.1
Mean Lifetime[*] (Days):	1053
First Launch:	December 19, 1976
Last Launch:	October 26, 1987

*As of June 30, 1989.

KH-11 LAUNCH DATA

Date	Lifetime	Inclination	Perigee	Apogee
Dec. 19, 1976	770	96.93	164	329
June 14, 1978	1166	96.82	171	316
Feb. 7, 1980	993	97.05	192	311
Sept. 3, 1981	1175	96.99	151	326
Nov. 17, 1982	987	96.97	174	322
Dec. 4, 1984	1671*	97.10	186	403
Aug. 25, 1985	0	Launch Failure		
Oct. 26, 1987	613*	97.80	95	653

* Still operating as of June 30, 1989.

Index